Fortschritte der Chemie organischer Naturstoffe

Progress in the Chemistry of Organic Natural Products

50

Founded by L. Zechmeister

Edited by W. Herz, H. Grisebach, G.W. Kirby, and Ch. Tamm

Authors:

H. Inouye, L. Jaenicke,
M. Lounasmaa, F.-J. Marner,
U. Séquin, P. Somersalo, S. Uesato,
R. M. Wenger

Springer-Verlag

Wien New York 1986

Dr. W. Herz, Professor of Chemistry, Department of Chemistry,
The Florida State University, Tallahassee, Florida, U.S.A.

Prof. Dr. H. Grisebach, Biologisches Institut II, Lehrstuhl für Biochemie der Pflanzen,
Albert-Ludwigs-Universität, Freiburg i. Br., Federal Republic of Germany

G.W. Kirby, Sc. D., Regius Professor of Chemistry, Chemistry Department,
The University, Glasgow, Scotland

Prof. Dr. Ch. Tamm, Institut für Organische Chemie der Universität Basel,
Basel, Switzerland

© 1986 by Springer-Verlag/Wien

Softcover reprint of the hardcover 1st edition 1986

Library of Congress Catalog Card Number AC 39-1015

ISSN 0071-7886

ISBN-13:978-3-7091-8890-3   e-ISBN-13:978-3-7091-8888-0
DOI: 10.1007/978-3-7091-8888-0

Contents

Contents

List of Contributors

INOUYE, Professor H., Faculty of Pharmaceutical Sciences, Kyoto University, Sakyo-ku, Kyoto 606, Japan.

JAENICKE, Professor Dr. L., Institut für Biochemie, Universität zu Köln, An der Bottmühle 2, D-5000 Köln 1, Federal Republic of Germany.

LOUNASMAA, Professor Dr. M., Laboratory for Organic and Bioorganic Chemistry, Department of Chemistry, Technical University of Helsinki, SF-02150 Espoo 15, Finland.

MARNER, Dr. rer. nat. F.J., Institut für Biochemie, Universität zu Köln, An der Bottmühle 2, D-5000 Köln 1, Federal Republic of Germany.

SÉQUIN, PD Dr. U., Institut für Organische Chemie der Universität, St. Johannsring 19, CH-4056 Basel, Switzerland.

SOMERSALO, Lic. Techn. P., Laboratory for Organic and Bioorganic Chemistry, Department of Chemistry, Technical University of Helsinki, SF-02150 Espoo 15, Finland.

UESATO, Dr. S., Faculty of Pharmaceutical Sciences, Kyoto University, Sakyo-ku, Kyoto 606, Japan.

WENGER, Dr. R.M., Preclinical Research, Sandoz Ltd., CH-4002 Basel, Switzerland.

The Irones and Their Precursors

By L. JAENICKE and F.-J. MARNER, Institut für Biochemie,
Universität zu Köln, Federal Republic of Germany

With 10 Figures

To Prof. Dr. H. Grisebach on the occasion of his 60th birthday

Contents

I. Introduction

The sword lilies or irises are monocotyledons, perennial plants of
great beauty like their close relatives in the order Liliflorae, the lilies
and the amaryllises. They are widely distributed all over the northern
hemisphere from swamp land to arid zones, from coastal to mountain
habitats, and propagate by rhizomes, tubers, or bulbs from which the
long sessile grass- or sword-shaped, parallel-ribbed leaves sprout which

give the family its name. The radial, showy flowers have the typical three-fold symmetry of Lilideae: brightly purple, blue, yellow or white coloured petals, each set often different and, hanging or upright, they stand in two circles of three. The flowers are visited by bumble bees in the beginning, later on mostly by hovering flies or bees; accordingly, the construction of the petals, stamina and stigmata is hercogamous so that self-pollination is excluded. They possess only one triad of stamina (in contrast to the lilies proper), a low standing ovary, forming on maturation a three-locular fruit capsule that opens to various degrees of dehidescence to shed brown, red or cream-coloured disc-like seeds. Most sword lilies bloom in May or June and the seeds mature and spread in high summer.

The Iridaceae are represented in Central Europe only by few species and cultivated forms. The latter are popular, unassuming and trouble-free garden plants.

Except for numerous flavones and flavonoid glycosides which have been isolated from various *Iris* species, very little is known about secondary metabolites. The rhizomes have a pungent and somewhat bitter taste. The calmus-like Yellow Flag Iris, *I. pseudacorus,* which grows in swampy habitats is poisonous to man and animal in the fresh or dried state, causing unpleasant irritations of the mucous membranes, vomiting and even bloody diarrheas which in farm animals has caused death. *Rhizoma Iridis* was a constituent of various pharmacopoeias and has been used formerly as a strong and efficient purgative of last resort, when all other means have proved inefficient (*1*).

Extracts of *Iris* (*pseudacorus*) rhizomes at D_2 to D_3 are prescribed in homoeopathic medicine for different forms of psychic ailments, depressive migraine (Sunday migraine), for neuralgia of the trigeminus, also in hyperemesis and hyperacidity of the stomach (*2*).

I. pallida and, less often, *I. germanica* are grown commercially in some areas of Italy, France and Morocco for production of an essential oil used in perfumery. After harvesting and sun-drying the root-stocks assume a pleasant violet-like scent which increases with time (*3*). The oil usually obtained by steam distillation yields the different blends of "essence d'iris" or "orris root oil" which are rather precious ingredients of scents, perfumes and other cosmetics (*4*).

The odoriferous principle was first isolated by TIEMANN in 1893 and named "irone". TIEMANN assumed that irone was the compound also responsible for the fragrance of the sweet violet. In a fascinating and – considering the unsophisticated state of the art at that time – brilliant investigation he concluded from various reactions and by comparison with α-ionone (**1a**), synthesized for the first time during this study, that the structure of irone was (**1b**) (*5*).

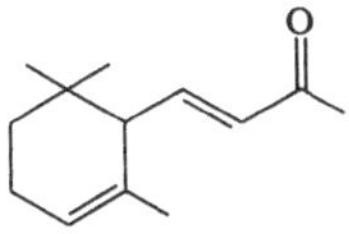

(1a) 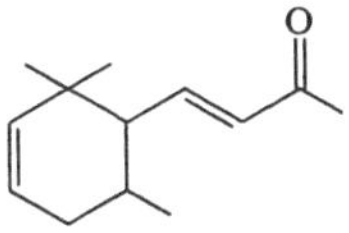 (1b)

It was not until several decades later that α-ionone (**1a**) $C_{13}H_{20}O$, was shown to be, in fact, a component of the scent of the garden violet flowers (*6*), but not of iris-oil. Unfortunately, TIEMANN's elemental analysis of irone was incorrect; 30 years later RUZICKA established the true composition as $C_{14}H_{22}O$ (*7*). He and, independently, NAVES found that at least three isomers of irone are present in natural iris-oil, their structure being determined as (**2a**), (**b**) and (**c**) (*8*).

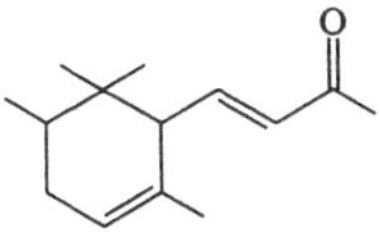

(2a) 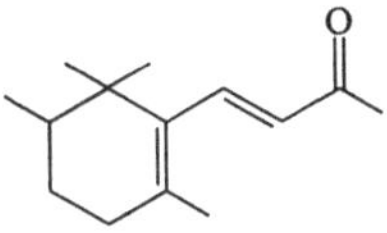 (2b) 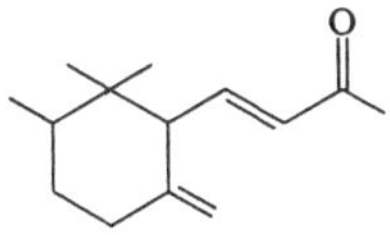 (2c)

In 1971 RAUTENSTRAUCH and OHLOFF finally completed the structure determination by elucidating the stereochemistry of irones which they had reisolated from the Italian iris oil (*9*) originally used by RUZICKA. The oil was shown to contain the four isomers (**3**)–(**6**). In a later study RAUTENSTRAUCH also detected traces of the *trans*-γ-isomer (**7**) together with some other isomers – probably with Z-geometry of the side chain (*10*). The stereochemistry of (**7**) is still unknown.

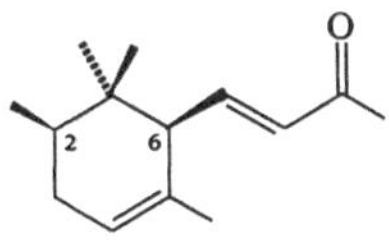

(3) 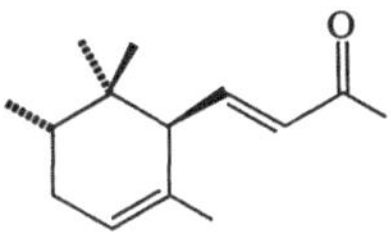 (4) 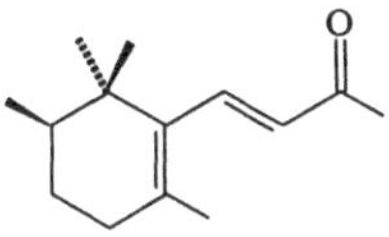 (5)

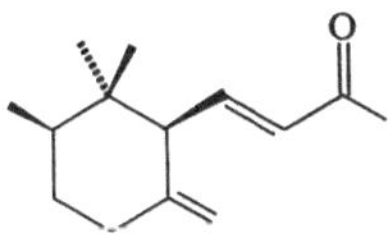

(6) 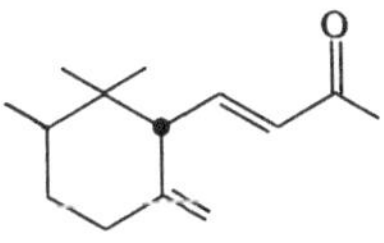 (7)

In contrast to the ionones that are apparently widespread in nature (*11*), the irones so far have been found only in orris root oil and more recently in oak moss extracts (*12*). Looking at the structural characteristics of irones as compared to ionones their rare occurrence is not very surprising. The ionones have a normal terpenoid skeleton and can be thought of as oxidative cleavage products of carotenoids, which indeed often are sources of these compounds. The structure of the irones, in contrast, does not follow the isoprene rule. In particular, the methyl group at C(2) of the ring represents a feature which requires an explanation. Nevertheless the terpenoid nature of the irones is beyond doubt. To our knowledge, however, nobody ever asked how the irones originate in the plant but for one attempt to find ring-methylated carotenoids in rhizome extracts which, however, failed (13).

   Certainly the C_{14}-ketones are not the original secondary metabolites of the plants. The producers of orris oil have long known that freshly harvested rhizomes do not contain any irones at all and that the scent develops gradually reaching its maximum after 3 to 4 years of storage of the peeled root-stocks (3); yet the reason for this development – whether enzymatic or by other means – has remained obscure. It was not until a few years ago that systematic attempts were made to approach the question of the nature and development of these ketones. Several observations make it unlikely that the formation of the irones during storage of the root stocks is due to an enzymatic reaction; these include the extreme slowness of the process and unsuccessful attempts to form irones by treatment of lipid or aqueous extracts of iris rhizomes with acids, bases or hydrolytic enzymes such as glucosi-dases (14). Instead, non-enzymatic oxidative degradation of some precursors is inferred as the source of the fragrance from the fact that only decorticated rhizomes give maximal yields of the desired compounds (4). Indeed, formation of irones was demonstrated by gas chromatography when lipid extracts of root stocks of *Iris pallida* and *Iris florentina* were treated with various oxidizing agents such as Jones reagent or pyridine chlorochromate (PCC) (15). Preparations from *I. germanica* yielded the less intensively fragrant *cis*-dihydroirones (8) and (9) as volatile cleavage products (16). Hence it was clear that the extracts contain precursor molecules which can be oxidatively cleaved to the aroma carriers in the plant tissues, and it became possible to track down these substances and to elucidate their structure and chemical relationships.

(8)   (9)

II. Isolation and Structure Determination

For identification of the postulated irone precursors the shredded rhizomes were mashed with an equal weight of water, and the slurry was extracted with methanol-chloroform (2:1 v/v). After concentration the essential oil was dispersed in methanol/water and extracted with several charges of ether. Then the organic solvent was evaporated and the oily residue (10 to 30 g per kg rhizome) fractionated coarsely on silicagel by a petrolether/chloroform/acetone/methanol gradient. The fractionation was first monitored by oxidizing a small aliquot of each fraction with PCC in dichloromethane and checking for the presence of irones or dihydroirones. It soon turned out that the irone precursors could be detected more easily by nmr spectroscopy since they all possess an aldehyde function the proton of which gives a distinct signal near 10 ppm. During chromatography the compounds appeared in two polarity zones: the less polar one was due to esters, mostly with saturated C_{10} to C_{16} fatty acids, but also with unsaturated C_{18} acids. The more polar zone contained the corresponding free alcohols. Final purification of the precursors was achieved by reversed phase medium pressure chromatography (RP-18 MPLC; methanol/water 6:4 to 9:1 v/v gradient; monitored by uv absorption). The pure compounds were concentrated to colourless, resinous residues. The yield of individual purified compounds was 0.2 to 1% of the weight of the fresh tissue for the main constituents and down to milligrams per kg fresh tissue for the minor compounds. All substances turned out to be mono- or bicyclic triterpenes, derived from the same basic structure skeleton which we named "iridal" (10) and number as shown below.

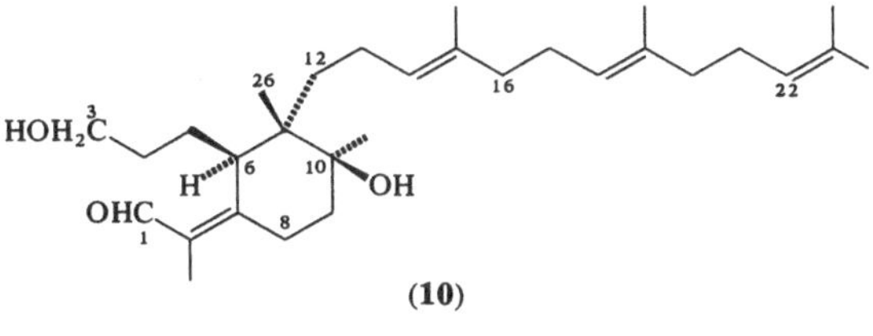

(10)

The main tools for structural elucidation were mass and nmr spectroscopy, supported and controlled by chemical degradation and comparison of the fragments with synthetic material. As an example, the procedure will be described for the precursor of γ-dihydroirone (9) from *I. germanica*.

This iridal derivative forms colourless needles from methanol (mp 74–75° C), is optically active ($[\alpha]_{578}^{20} + 10°$; CH_2Cl_2, c 14.4) and absorbs uv light ($\varepsilon_{256} = 14100$ in ethanol). The compound is stable to air but on oxidation with $KMnO_4$ in benzene/dicyclohexyl-18-crown-6 it yields 30% γ-dihydroirone (**9**). Use of PCC/CH_2Cl_2 gave (**9**) only in 2% yield.

The EI mass-spectrum (70 eV) shows a molecular ion at m/z 472. Fragment ions at m/z 457, 454, 439, 436 and 421 point to the successive

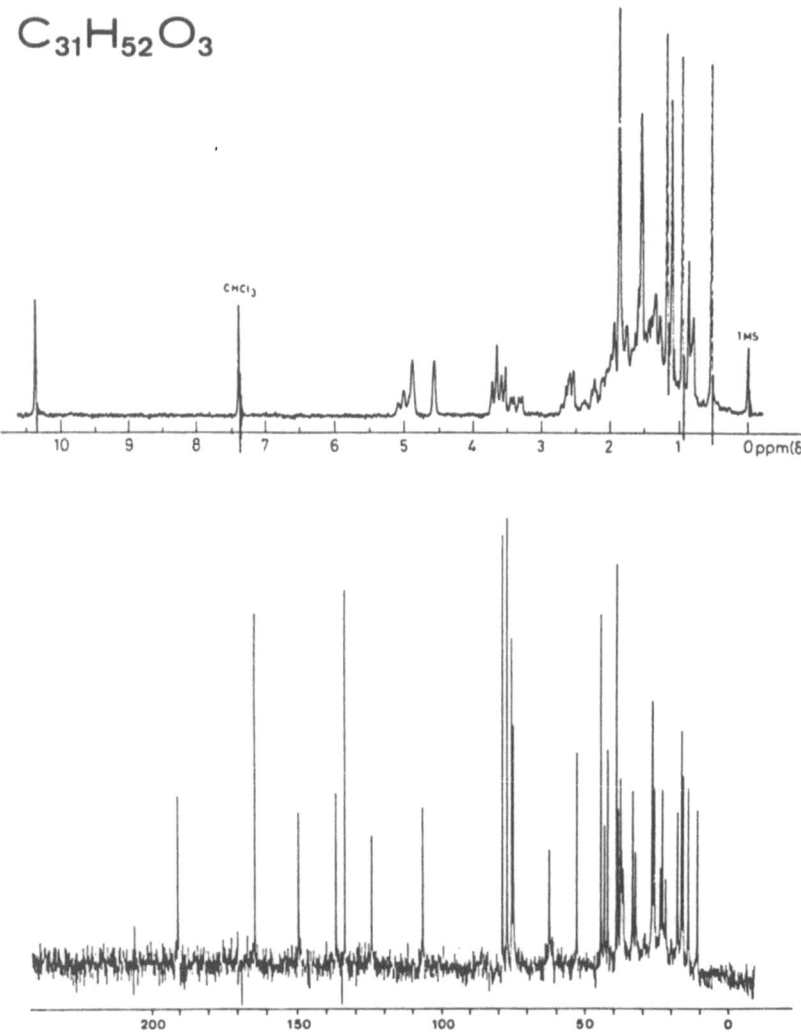

$C_{31}H_{52}O_3$

Fig. 1. 90 MHz ^{1}H- (top) and ^{13}C-nmr-spectra of (+)-(**12**)

loss of two molecules of water and one methyl group. From the exact mass of the molecular ion the formula $C_{31}H_{52}O_3$ (calc. 472.3916, found 472.3911) was calculated showing that a methylated triterpenoid was present.

The terpenoid nature of the compound was also evident from its nmr spectrum (Fig. 1). The ^{1}H-nmr spectrum shows 7 methyl signals in the range 0.55 to 1.82 ppm, 22 protons in the range 1.0 to 2.2 ppm, 3 olefinic protons between 4.5 and 4.94 ppm, one tertiary proton at 3.3 ppm, one methylene group at 2.5 ppm and another one at 3.55 ppm. This, together with the aldehyde proton at 10.3 ppm accounts for all 52 hydrogens. The presence of a γ-dihydroirone moiety could be ascertained by its characteristic signals on comparison with the authentic spectrum.

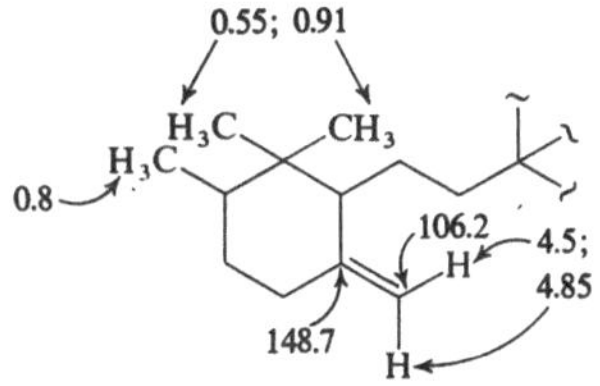

Other elements of the structure were identified by ^{13}C-nmr spectrometry, spin-spin decoupling and resolution enhancement experiments at 400 MHz:

(i) The ir band, v_{max} 1690 cm^{-1}, gives evidence for an α,β-unsaturated aldehyde. In the ^{1}H-nmr spectrum the aldehyde proton at 10.3 ppm shows only allylic coupling to a CH_3 at 1.84 ppm; CHO- and CH_3-groups therefore have to be attached to the carbons of a completely substituted double bond. In the ^{13}C-nmr spectrum the resonance of an olefinic carbon at 162.8 ppm was attributed to the carbon in β-position to the aldehyde function. Presumably this function is part of a ring system, since both the exocyclic nature of the double bond and the influence of the carbonyl group would explain the downfield shift of this resonance. The methine group, the proton of which appears at 3.3 ppm, must be attached to this same carbon. It interacts with two non-equivalent protons of a CH_2-group found near 2.1 ppm, thus giving support to the structural element shown below:

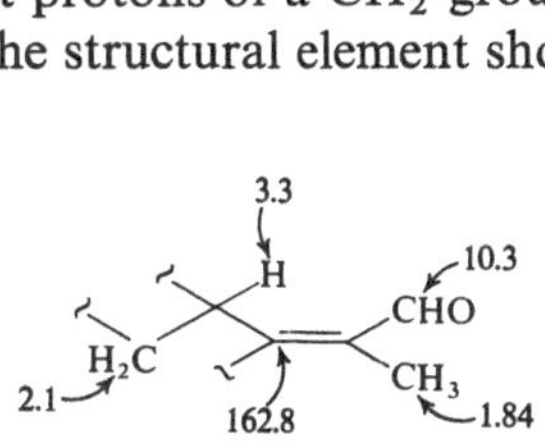

This was confirmed by reduction of the aldehyde with NaBH$_4$/alumina to the corresponding primary alcohol which resulted in a shift of the methine-proton to 2.6 ppm.

(*ii*) The CH$_2$ protons at 3.6 ppm are coupled to another methylene group at 1.35 ppm, indicating the presence of a CH$_2$-CH$_2$-OH moiety. The carbon resonance of this alcohol is found at 63.0 ppm. There is yet another signal for a quaternary carbon at 75.0 ppm, indicating the presence of a tertiary alcohol. This agrees perfectly with the mass spectrometric results.

(*iii*) The olefinic proton at 4.96 ppm exhibits allylic coupling to a CH$_2$ at 1.9 ppm and homoallylic coupling to the methyl group at 1.55 ppm which indicates the typical terpenoid partial structure:

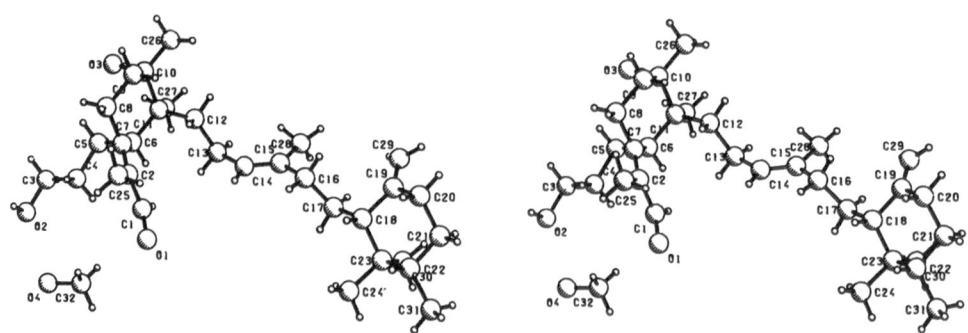

(*iv*) The unassigned two methyl-groups at 1.11 and 1.19 ppm are not coupled to any protons and therefore have to be connected to quaternary carbons.

Several structures satisfy the evidence presented so far; a complete solution of the problem became possible by x-ray analysis, after crystals of the substance were finally obtained from methanol. The crystals contain one mole of methanol, but tend to lose the solvent on exposure to air and become cloudy and useless for diffraction measurements. Therefore the analyses were carried out at $-100°$ C. The computer-generated drawing of the molecule (Fig. 2) clearly reveals a bicyclic triterpenoid structure with the cyclohexane rings in an almost perfect

Fig. 2. Stereoscopic view of $(+)$-(**12**)

chain configuration and tilted around the connecting six carbon chain by about 55°.

Thus the γ-dihydroirone precursor was identified as (+)-22-methyl-γ-cycloiridal (+)-(12) (*16*). X-ray analysis gave only the relative stereochemistry. The absolute configuration had to be either 6 R, 10 S, 11 S, 18 R, 22 S or its opposite. After oxidative cleavage of (+)-(12) to γ-dihydroirone (9) C(18) and C(22) become C(6) and C(2) of the latter. Because the absolute stereochemistry of (+)-γ-irone (+)-(6) from Italian iris oil had been determined by OHLOFF as 2 R, 6 S (*9*) and since it is possible to selectively reduce the bond conjugated double to the keto group, the stereochemical problem of (+)-(12) could be approached easily. Surprisingly the hydrogenation product of authentic (+)-(6) and the dihydroirone obtained by oxidation of (+)-(12) had exactly opposite optical rotations, namely $[\alpha]_{578}^{20} = +57°$ for the former and $[\alpha]_{578}^{20} = -53°$ for the latter, thus proving that the oxidation product is the 2 S, 6 R enantiomer (−)-(9); consequently, the precursor (+)-(12) has to have the configuration 6 R, 10 S, 11 S, 18 R, 22 S (*17*).

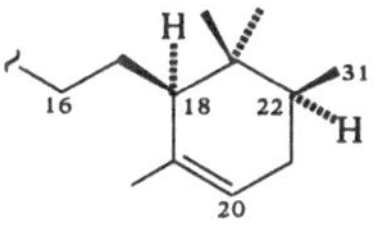

(11)

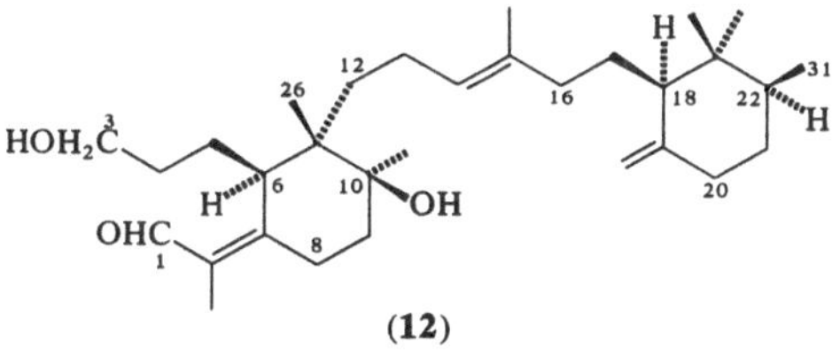

(12)

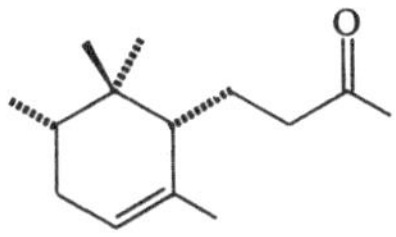

(−)-(8)

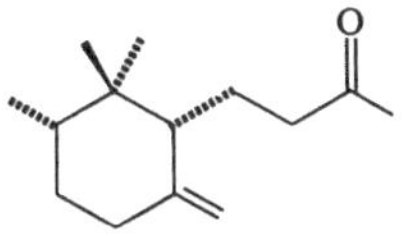

(−)-(9)

The γ-isomer (+)-(12) was accompanied by (+)-22-methyl-α-cycloiridal (+)-(11) (*16*), the precursor of (−)-α-dihydroirone (−)-(8). Its [1]H- and ^{13}C-nmr spectra are almost identical with the spectra of (+)-(12) except for small differences consistent with its isomeric structure.

III. Structure of the Irone Precursors

On the basis of these results it was also possible to elucidate the structural features of the other triterpenoids isolated from rhizomes

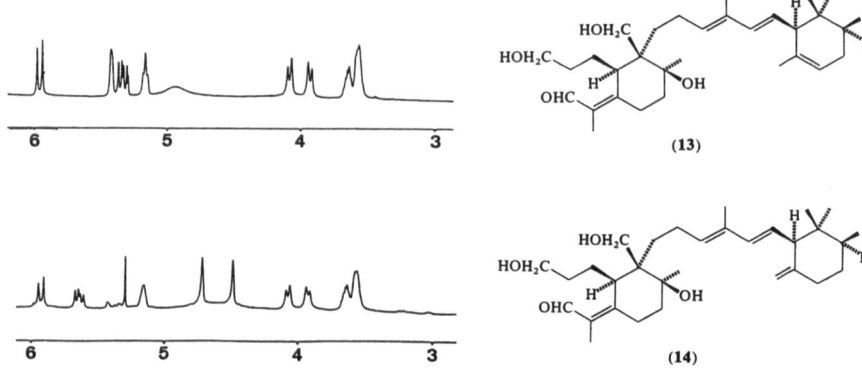

Fig. 3. 400 MHz ^{1}H-nmr spectra of (−)-(13) and (+)-(14)

of different indigenous and cultivated *Iris* species. Of particular interest
were the precursors of the irones proper. Oxidation experiments had
shown that the precursor fraction from extracts of *I. pallida* consisted
of almost pure α-irone precursor, whereas that of *I. florentina* contained
a mixture of both isomeric precursors, with the γ-isomer predominant.
Both compounds were isolated by the procedure outlined above to
give (−)-(13) ($[\alpha]^{20}_{578} = -7°$) and (+)-(14) ($[\alpha]^{20}_{578} = +48°$). The molecular composition for both compounds was deduced as $C_{31}H_{50}O_4$ from
^{13}C-nmr and mass spectrometric analyses which requires the presence
of an additional double bond, conjugated and located at C(16) to
account for the ^{1}H-nmr spectrum and the generation of irones on
oxidation. Additionally a hydroxy group, not contained in the dihydroirone precursors, has to be present in (−)-(13) and (+)-(14). The
1H-nmr spectra (Fig. 3) show an additional AB system for two protons
at 4.07 (4.08) and 3.93 (3.92) ppm respectively and instead lacks a
signal for one of the tertiary methyl groups at 1.11 or 1.19 ppm. Since
the CH_3-group of the acrolein side chain is still present, either C(26)
or C(27) must carry the hydroxyl-group, but since the molecule resists
glycol cleavage, the hydroxyl group is on C(26). Thus compound
(−)-(13) is 26-hydroxy-22-methyl-α-cycloirid-16-enal, and compound
(+)-(14) is 26-hydroxy-22-methyl-γ-cycloirid-16-enal (*15*). The additional hydroxy group at C(26) may explain the increased oxygen sensitivity of these compounds; simple exposure to air eventually gives rise
to the irones on storage of the dried root-stocks as already mentioned.
Further evidence for this deduction is the stability to air of a third
component in the extracts of *I. pallida*, which was identified as 22-
methyl-α-cycloirid-16-enal (15) which has the C(26) methyl group un-
oxidized as seen from the spectral data (Fig. 4).

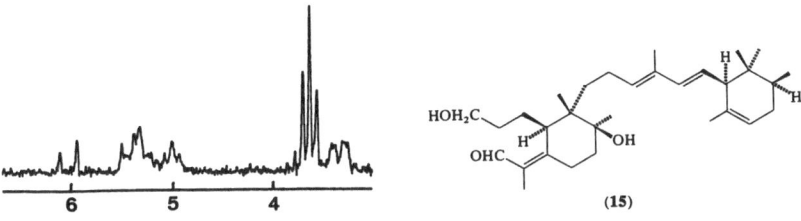

Fig. 4. 90 MHz ^{1}H-nmr spectrum of (+)-(15)

IV. Stereochemistry of the Iridals and the Irones Derived from Them

In contrast to the isolates used by RUZICKA (8) and OHLOFF (9) for their structure determinations of the irones, the oxidation products of (−)-(13) and (+)-(14) are the enantiomeric (2S, 6R)-(−)-cis-α-irone (−)-(3) and (2S, 6R)-(−)-cis-γ-irone (−)-(6), respectively (17).

(−)-(3)　　　　　　　(−)-(6)

Table 1 contains a compilation of the optical rotations of irones from different sources which shows that not only the compounds derived by oxidation of (−)-(13) and (+)-(14) but also a cis-α-irone isolated from one commercial iris oil (P. Kaders, Hamburg) possess the (2S, 6R)-configuration

Table 1. *Optical Rotations of Irones*

Irone	$[\alpha]_{578}^{20}$	GC-purity (%)	Conc. (g/100 ml)
(−)-cis-α Kaders iris oil	−115°	94	0.65
oxidation of (−)-(13)	−111°	97	8.8
(+)-cis-α-Firmenich iris oil	+109°	*	
(−)-cis-γ oxidation of (+)-(14)	−1°	99	0.5
(+)-cis-γ Firmenich iris oil	+2°	*	

* Data for the (+)-irones taken from (9).

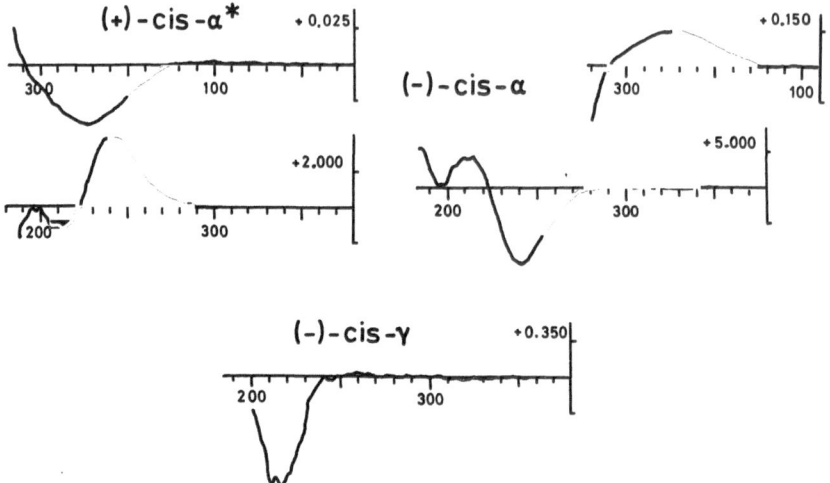

Fig. 5. CD-spectra of *cis*-irones. (*: spectrum of (+)-*cis*-α-irone courtesy of Prof. G. Snatzke, Bochum)

This is also evident from the CD-spectra. As seen from Fig. 5 and Table 2, the *cis*-irones formed from the precursors in fresh rhizomes of domestic cultures of *I. florentina* and *I. pallida* exhibit CD-spectra which are enantiomeric to the CD-spectra published for compounds from the absolute RUZICKA oil (*18*), hence they must possess the opposite configuration.

Table 2. *CD-data of α- and γ-irones*

Irone	λ_{max} (nm) ($\Delta \varepsilon$)
(+)-*cis*-α	318 (−0.23); 247 (+12.18); 216 (−4.0)
(−)-*cis*-α from (−)-(**13**)	327 (+0.14); 239 (−10.42); 212 (+4.28)
(+)-*cis*-γ	372 (+0.05); 354 (+0.15); 340 (+0.22); 326 (+0.21); 315 (+0.15); 223 (+3.86)
(−)-*cis*-γ from (+)-(**14**)	323 (−0.03); 318 (−0.03); 214 (−1.05)

Data taken for the (+)-irones from (*18*)

On closer examination, iris oils from different sources, revealed the interesting features shown in Table 3. As can be seen, the majority of the oils is dextrorotatory and all have a high content of *trans*-α-irone whereas the only laevorotatory oil contains only traces of this compound.

Table 3. *Optical Rotation and Composition (%) of Commercial Iris Oils*

Source	Origin	$[\alpha]^{20}_{578}$	*trans*-α	*cis*-α	*cis*-γ	β-irone
P. Kaders (Hamburg)	Morocco	−23°	1	61	37	1
F. Mülhens (Köln)		+37°	16	36	38	10
C. Georgie (Böblingen)		+27°	10	16	74	−
Haarmann & Reimer (Holzminden)		+2°	26	34	40	tr.
Firmenich (Geneva)	Italy	+28°	16	39	43	2

tr. = <1%

The *trans*-α-irone is unlikely to be derived from *cis*-α-irone by isomerization. Such a reaction only occurs at C(6) of the molecule as has been shown by appropriate experiments (9). Therefore the differing stereochemistry must be due to a specific step in the biosynthesis of the E-ring of the respective precursors. However, the β-irone present is probably produced by isomerization of either α- or γ-irone. Intriguing also is the circumstance that the laevorotatory oil and at least one of the dextrorotatory oils derive from different geographical regions. This is in perfect agreement with earlier literature. Iris oils from Italy have been found to be dextrorotatory whereas oils from Morocco were laevorotatory. Interestingly, these oils are reported to derive from Italian *I. pallida* and from Moroccan *I. germanica,* respectively (3). These apparent differences in the *Iris* species grown in various geographical areas led us to look for the irone precursors in rhizomes of *I. pallida dalmatica* grown in Italy for commercial purposes.

Table 4. *Optical Rotations of Irone-Precursors from Different Sources*

	(13)	(14)
I. pallida (Bonn, Germany)	−7°	−
I. florentina (Offstein, Germany)	−7°	+48°
I. pallida dalmatica (Tuscany, Italy)	+91°	+42°

The two triterpenoids (13) and (14) were isolated in 1:2 ratio. Their optical rotations already indicated that they were different from the compounds isolated earlier (Table 4); the difference is much more pronounced for the α-irone precursor (13). Indeed, the oxidation products of these two compounds turned out to be the (+)-*cis*-irones (+)-(3) and (+)-(6) (19). Although the chiralities of the three remaining centres at C(6), C(10) and C(11) in (13) and (14) from *I. pallida dalmatica* have not yet been proved, we take them to be identical for all irone precursors. The main argument for this assumption is that the proton and carbon signals for the B-ring system of the two isolates are identical with those of $(-)$-(13) and $(+)$-(14) described above. Presumably the biosynthesis of the triterpenoids – which no doubt starts from squalene as will be discussed below – is catalyzed by sets of enzymes with opposed streospecificity for the formation of the E-ring, depending on the genetic or environmental background of the *Iris* cultivar. It is the E-ring and part of the connecting chain which will form the irones on oxidation.

V. Minor Triterpenoids in *Iris* Extracts

The minor constituents of our *Iris* extracts and similar compounds found in less domesticated wild species greatly aided in unraveling the biogenetic relationships among the *Iris* triterpenes and with squalene. Interestingly, they all are triterpenoids with an open E-end, either C_{30}- or methylated C_{30} (C_{30+1}) compounds. The following basic structural details provide exemplaric proof for their constitution.

From *I. germanica* 21-hydroxy-iridal $(+)$-(16) was isolated (16), a compound accompanying the somewhat less polar (11) and (12). A molecular ion m/z 474 (M^-) seen only by negative chemical ionization established the formula as $C_{30}H_{50}O_4$. EI gave solely m/z 456 (M-H_2O). The water lost so easily could be traced to a hydroxyl group at C(21) of the open chain end attached to the highly substituted B-ring by its ^{1}H- (Fig. 6) and ^{13}C-nmr spectra and by spin-spin-decoupling and resolution enhancement experiments at 400 MHz. On mild oxidation the corresponding ketone was formed in whose spectrum the olefinic proton is shifted to 6.2 ppm and the CH_2-group appears as a singlet at 3.0 ppm.

Two minor components in *I. germanica* extracts were isomers of formula $C_{30}H_{50}O_3$. According to their spin resonance spectra they were the iridal $(+)$-(10) and 10-desoxy-21-hydroxy-iridal $(+)$-(17) (15). As seen from Fig. 6 the ^{1}H-nmr spectrum of $(+)$-(10) is very close to that of 21-hydroxy-iridal $(+)$-(16) but lacks the OH in the side chain. This establishes the structure. The second isomer possesses the hydroxyl group at C(21) of the side chain but has none at C(10) of the ring. Thus it has to be compound (17).

Another minor component identified in extracts of *I. florentina*, *I. pallida* and *I. versicolor* was the 16-hydroxy-iridal $(+)$-(18) (15). Its ^{1}H-nmr spectrum (Fig. 6) differs from that of (16) in the position of

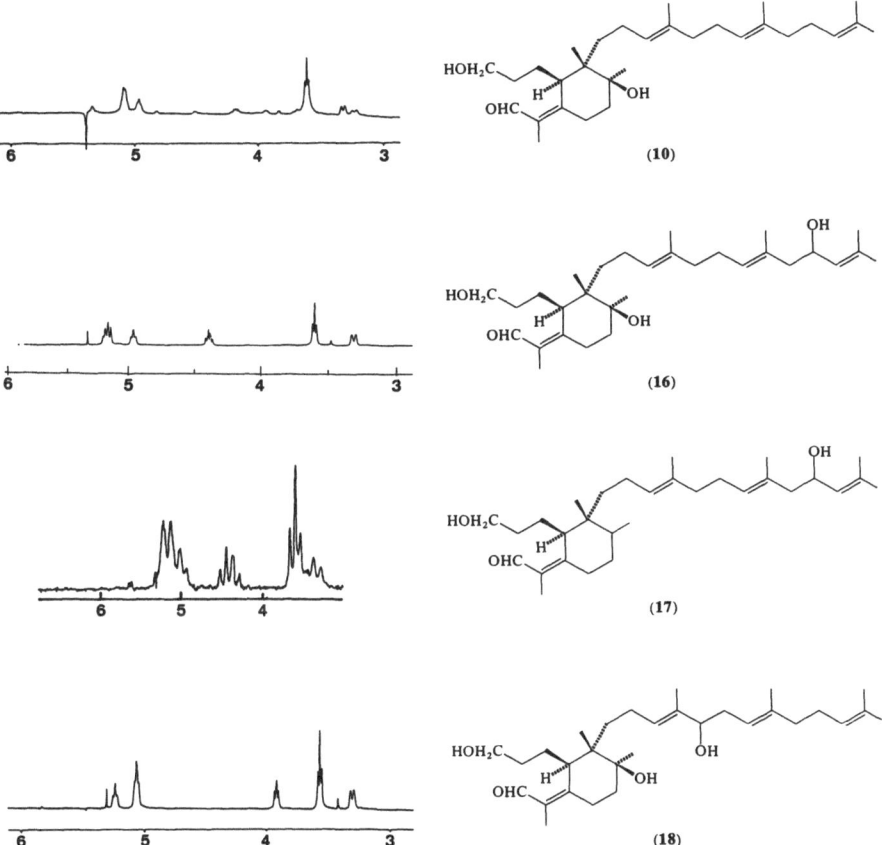

Fig. 6. 400 MHz ^{1}H-nmr spectra of (+)-(**10**), (+)-(**16**) and (+)-(**18**) and 90 MHz ^{1}H-nmr spectrum of (+)-(**17**)

the signal of the secondary hydroxyl group. On oxidation with KMnO$_4$ it gives 6-methyl-5-hepten-2-one (**19**) and 6,10-dimethyl-undeca-5,9-dien-2-on-3-ol (**20**). This proves that the hydroxyl group is on C(16), leading to (**18**) on the overall structure.

Of particular interest are two triterpenoids which represent about 0.2% of the fresh weight in extracts of rhizomes of *I. versicolor* and

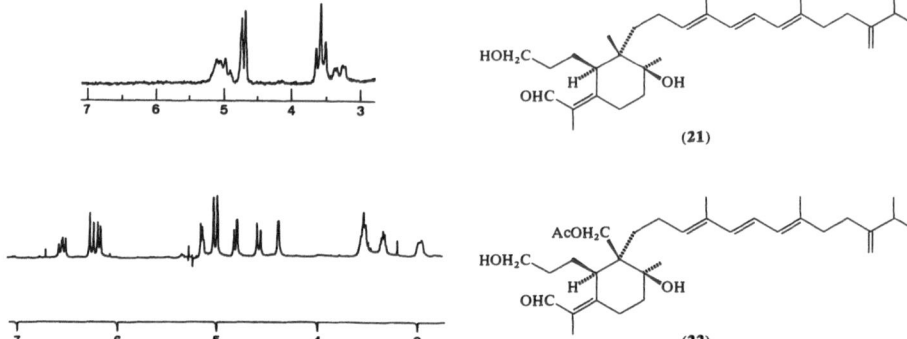

Fig. 7. 90 MHz ^{1}H-nmr spectrum of (21) (top) and 400 MHz ^{1}H nmr-spectrum of (22)

I. pseudacorus, respectively. The spectral data reveal in both compounds the characteristic signals of the polyfunctional ring B; the attached side chain, however, is not cyclized although carrying a methylene group.

In addition to the proton signals at 4.71 and 4.66 ppm (Fig. 7), the principal evidence of such a feature – at least for the constituent from *I. versicolor* – is its degradation to 6-methyl-heptane-2,5-dione (23), 5-methylene-6-methylheptan-2-one (24) and 6,10-dimethyl-9-methylene-5-undecen-2-one (25). Accordingly, the compound has to be 22-methylene-iridal (21) (*20*). The identity of the three degradation products (23)–(25) was confirmed by synthesis. Ketone (25) has been found naturally as ingredient of costus root oil (from *Saussurea lappa* Clarke) (*21*). The biogenetic origin of this compound, however, is unknown.

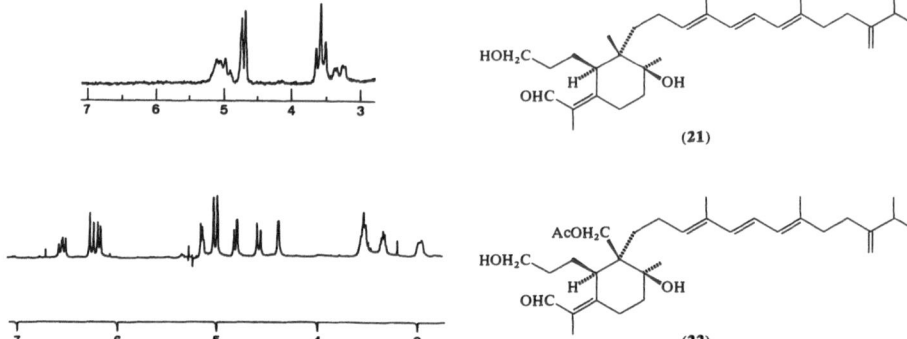

(23)   (24)   (25)

The component of *I. pseudacorus* oil is much more labile resembling in this aspect 26-hydroxylated irid-16-enals such as (13) or (14). Indeed, it was shown by spectral evidence to have a closely related but monocyclic structure. Again the B-ring is present in the same highly oxidized state as in (13) and (14). However, the carbinol at C(26) in this compound is acetylated causing a shift of the AB-system of the CH$_2$O-protons to 4.68 ppm (Fig. 7). The side chain is triply conjugated as indicated by the uv absorbance at $\lambda_{max} = 273$ nm and by the proton signals at 5.15 ppm and the three-spin system between 6.2 and 6.6 ppm. Again, a methylene group gives two signals at 5.02 and 4.99 ppm. Thus the structure of the compound may be that of a 26-acetoxy-22-methylene-irid-16-enal (**22**) (*22*).

It has been shown recently that such triterpenoids occur not only in *Iris* rhizomes of European or African origin. The two monocyclic iridals (**16**) and (**18**) have also been found in the North American species *I. missouriensis* (*23*). The data on the occurrence of different triterpenoids in rhizomes of several *Iris* species are summarized in Table 5.

Table 5. *Composition of Iridals in Extracts from Fresh Rhizomes*

Species	Main-components (>10%)	Minor components (1–10%)	Traces (<1%)
I. germanica	(+)-(**11**) (+)-(**12**) (+)-(**16**)	(+)-(**10**) (+)-(**17**)	(−)-(**13**) (+)-(**14**)
I. pallida	(−)-(**13**)	(+)-(**18**) (+)-(**11**)	(+)-(**15**)
I. florentina	(+)-(**14**)	(+)-(**18**)	(+)-(**10**) (−)-(**13**)
I. versicolor	(+)-(**21**)	(+)-(**18**)	(+)-(**10**)
I. pseudacorus	(**22**), (+)-(**18**)	not identified	
I. pallida dalmatica	(+)-(**13**) (+)-(**14**)	not identified	
I. missouriensis			(**16**), (**18**)

VI. Biogenesis of the Iridals

It is obvious that the eleven triterpenoids known at present in rhizomes of several *Iris* species are derived from all-*trans* squalene. In fact, radioactively labelled acetate, mevalonate and squalene injected into sterile, young *Iris* plantlets were incorporated at acceptable rates into the triterpenoid fraction (*24*). Several steps have to be involved in the formation of these compounds. The most intriguing question is the formation of the polyfunctional C(6)-C(11) ring system. As a model compound another squalenoid, the triterpene ambreine (26), may serve.

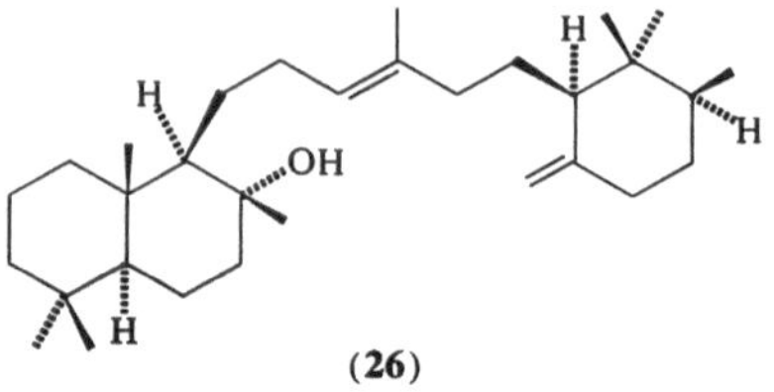

(26)

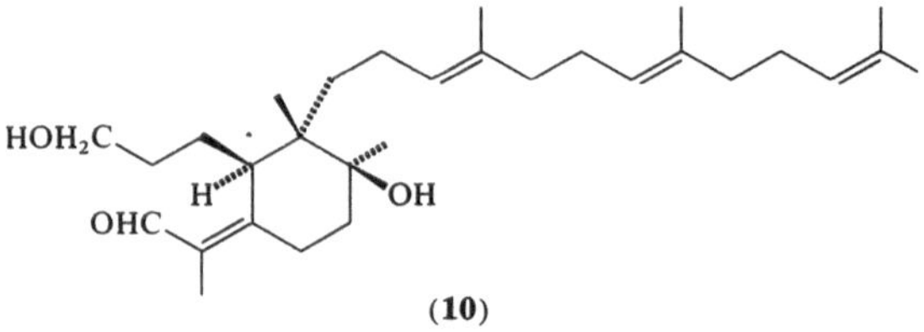

(10)

This constituent of ambergris from the sperm whale (25) gives γ-dihydroionone on oxidative degradation. Its skeleton is easily derived from all-*trans* squalene by proton-triggered cyclization from both ends. The formation of the basic structure (10) of iridal presumably follows a similar path. It is unlikely, however, that a saturated annellated bicyclic system similar to (26) is formed as an intermediate followed by oxidative cleavage of ring A. A more probable pathway is depicted in Fig. 8.

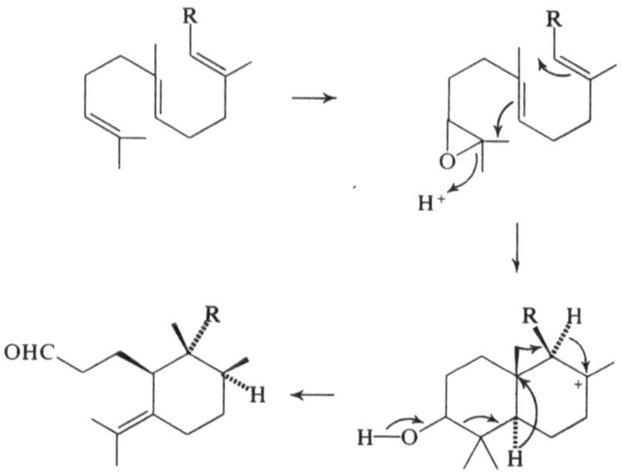

(27)

Fig. 8. Proposed biogenesis of the iridal skeleton

Epoxidation of squalene and subsequent cyclization leads to a bicyclic ionic intermediate which rearranges in the way shown to form the skeleton (27) of the *Iris* triterpenoids. By this scheme not only the unusual substitution pattern but also the stereochemistry of the compounds may be explained. C(1) and C(10) have to be oxidized – the latter with inversion of its configuration – and C(3) must be reduced to obtain the basic compound iridal (10).

A similar rearrangement has been shown to occur during the proton-initiated formation of pleuromutilin (28) from geranylgeraniol-pyrophosphate (26).

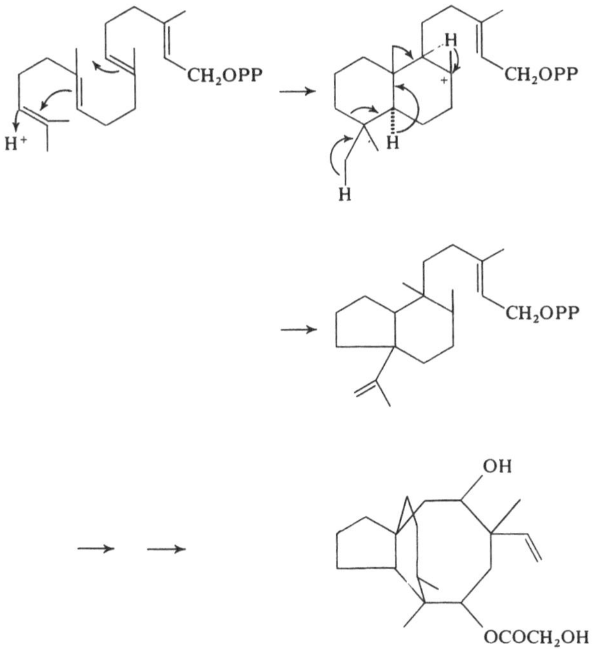

(28)

Preliminary experimental evidence for these intermediate steps has been obtained by feeding to the plants the synthetic (27) triterpene (29) labelled with ^{14}C in C(27). No incorporation of (29) into the precursor fraction could be observed. This behaviour is explained by the lack of a leaving group in (29). The additional hydroxy group in (30) – presently being synthesized – should make the rearrangement possible.

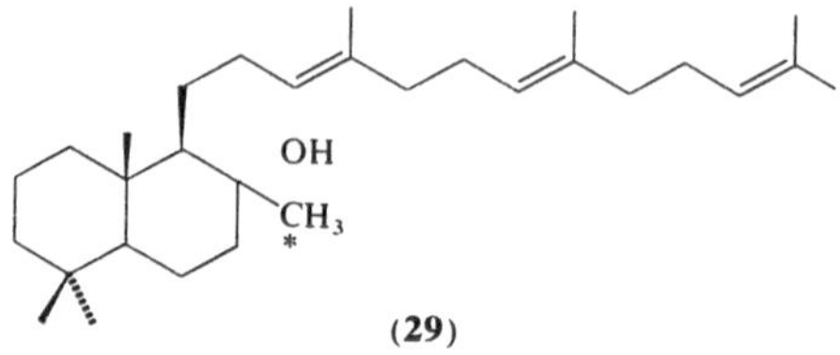

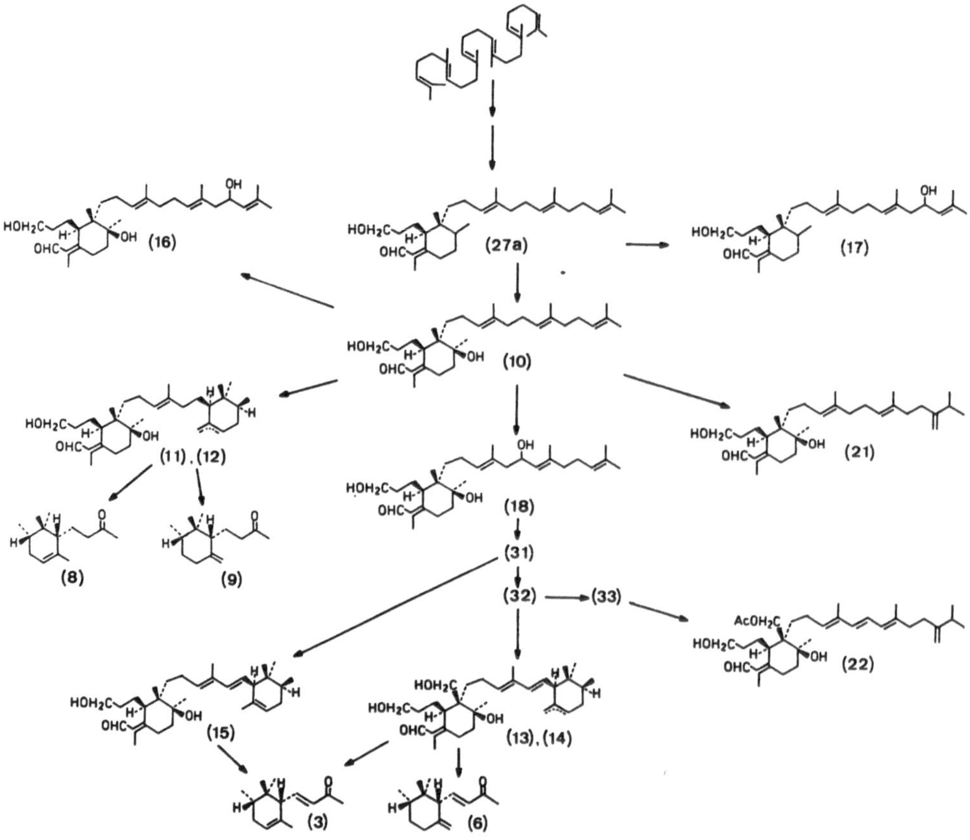

Fig. 9. Proposed biogenetic pathway of iridals

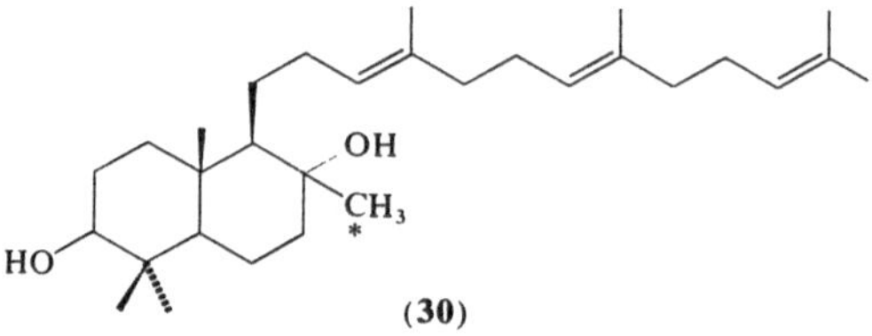

The proposed biogenetic pathway leading from all-*trans* squalene to the irone precursors is outlined in Fig. 9. The terpenoids differ in their sum formula either by additional OH-groups, newly introduced double bonds, or a methylene carbon. Formation of the highly substituted B-ring – on the proposed pathway outlined above – should eventually result in the (postulated) compound (27a) which on oxidation at C(21) yields the 10-desoxy-21-hydroxy-iridal (17); on oxidation of C(10) the iridal (10). Iridal (10) again can be oxidized at C(21) to give 21-hydroxy-iridal (16).

Iridal (10) on the other hand may be the acceptor of an activated methyl group to form the open chain 22-methylene-iridal (21). The methylated cyclization products of (10) are the α- and γ-cycloiridals (11) and (12), from which the dihydroirones (8) and (9) are formed on oxidative cleavage.

A further reaction sequence could begin with hydroxylation of (10) to 16-hydroxy-iridal (18) followed by removal of water to give the postulated conjugated triene irid-16-enal (31); this would yield 26-hydroxy-irid-16-enal (32) by oxidation of the quaternary 26-methyl group. This intermediate, however, has not yet been found and also the 26-hydroxy-22-methylene-irid-16-enal (33) is still missing. However, its 26-acetylated derivative as (22) is present in the lipid fraction of *I. pseudacorus* extracts.

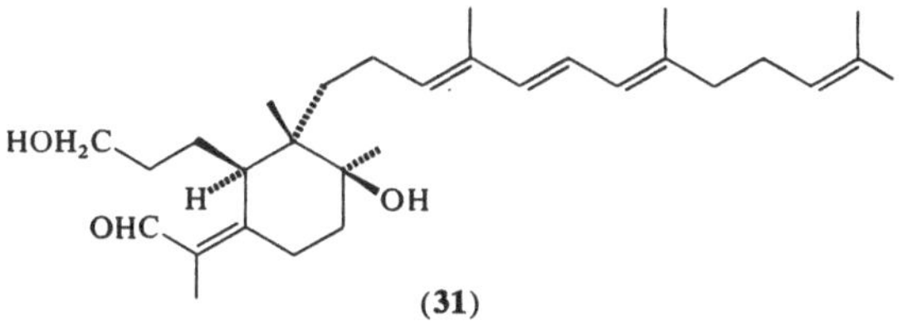

(31)

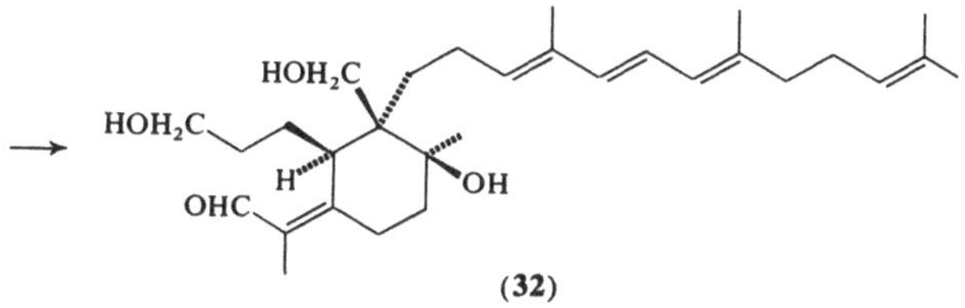

(32)

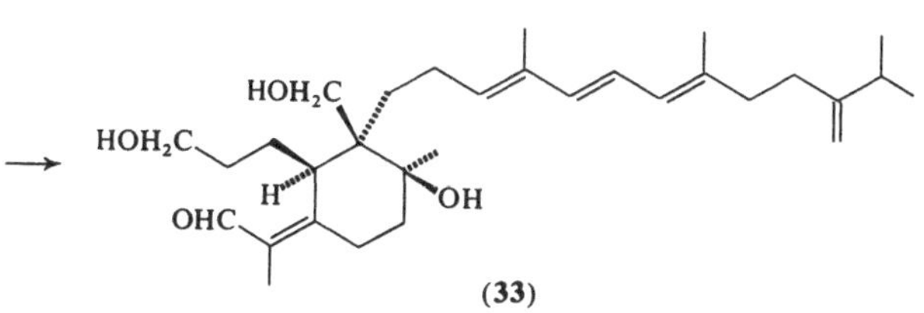

(33)

Presumably, some of the intermediates are so quickly converted to other consecutive products that they do not accumulate sufficiently in the rhizomes. Cyclization instead of methylene-formation would lead from (32) to (13) and (14), both of which are found in *I. pallida* and *I. florentina* as the supplier of the irones. The more stable irone precursor (15) in *I. pallida* root stocks may be formed by direct methylation and cyclization of (31).

It should be stressed again that the occurrence and mixture of the different triterpenoids can vary considerably. For example it is remarkable that a domestic cultivar of *I. germanica* is almost free of (13) and (14) but is rich in (11) and (12) and therefore produces nothing but weakly aromatic dihydroirones on vigorous oxidation whereas the same species in Morocco is a source of an iris oil with a high content of α- and γ-irone. An explanation may be that some of the species distinguished by a Latin name may be hybrid instead of true-bred forms (*28*). Even *Iris*-specialists often disagree on the classification of species (*29*), and it may turn out that the content and composition of these so easily analyzed triterpenoids provides a chemotaxonomic tool for ordering the genus.

VII. Mechanism of Irone Ring Formation

Finally the question posed at the beginning has to be answered: when and how is the extra methyl group introduced into the unsaturated C(22) position of the iridals? Incorporation studies with labelled compounds into *Iris* plantlets established that the source of this methyl group is the one-carbon fragment of methionine *via* S-adenosyl methionine (SAM) (*24*). The absence of C_{30}-triterpenoids with a cyclized E-end makes it unlikely that ring closure occurs before transfer of the methyl group from SAM. Rather, cyclization starts by addition of the methyl group to the C(22)-C(23) double bond. Insertion of methyl groups into unsaturated systems by means of SAM is well known, for example in the methylation of steroid or triterpenoid side chains. Even closer to the formation of the irone precursors is the cyclization of the methyl-hopanoids, which, indeed, recently has been reported to be initiated by transmethylation of SAM to squalene (*30*).

The most plausible mechanism for the addition would be one in which a cyclopropane cation is formed as depicted in Fig. 10. The cyclization reaction may start immediately; alternatively, the cyclopropane cation may open to leave the positive charge either at C(22) or C(23). The former will give 22-methylene open chain compounds

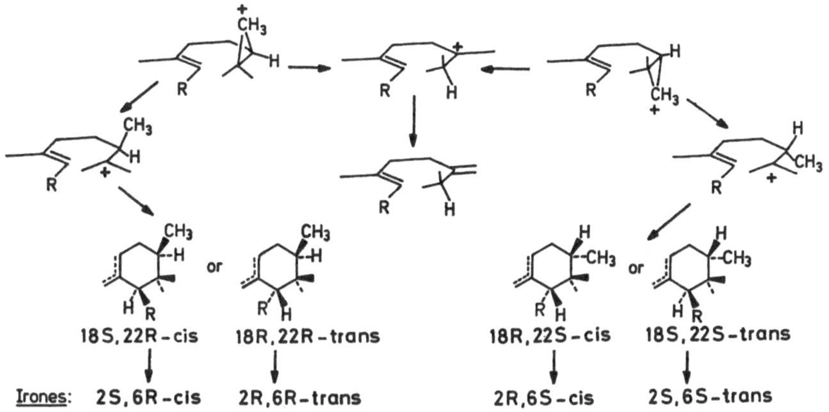

Fig. 10. Proposed mechanism of methylation of iridals

as end products, such as have actually been found as mentioned above. The latter, on the other hand, will lead to the cyclic compounds. The stereochemistry at C(22) of the cyclic products probably depends solely on the side from which SAM attacks the molecule. Attack from above leads to the 22 R-molecule; attack from below to the 22 S-isomer. Depending on the direction of the ring formation the 18 S- or 18 R-products will be formed. Thus the formation of all possible *cis*- or *trans*-irone precursors is explained as shown in Fig. 10. The simultaneous occurrence of both 2 R,6 S-*cis*- and 2 S,6 S-*trans*-irones within the same iris-oil (9) implies, that the formation of the cyclopropane cation is not stereospecific – though one product seems to be preferred. The direction of ring closure, however, and the twisting of the side chain R is unidirectional. This may depend on the enzymatic 'environment' during the cyclization *e.g.* the position of a stabilizing negative charge or/and steric effects.

This scheme would lead to the prediction that the *trans*-γ-irone (7) accompanying the 2 R,6 S-*cis*-irones (10) has to have 2 S,6 S-configuration whereas the *trans*-irones – if ever found – accompanying the 2 S,6 R-*cis*-irones would have 2 R,6 R-geometry.

VIII. Biological Significance of the Compounds

The new and unusual terpenoid structure of the iridals raises the question of their role in the life of the plant. So far, however, no answer has been found. Neither the extracts nor the compounds showed significant insecticidal (31; 32) activity nor are they effective as antifeeding agents against the corn beetle (*Sitophilus granarius*) (33) as test animal. They may be just another example of the bitter-tasting or saponine-like protectants so frequent in ecological plant/animal interactions. One may only speculate on their possible role as membrane plasticizers which is suggested by their amphiphilic structure and the high content in the rhizomes. It will be a challenge to natural products chemists to find the answers to the numerous biological and biochemical questions raised and to give full insight into the biogenesis and function of these newly discovered polyprenoids.

References

1. FROHNE, G., and H.J. PFÄNDER: Giftpflanzen, 2nd ed., p. 137. Stuttgart: Wiss. Verl.-GmbH. 1983.
2. ROTH, L., M. DAUNDERER, and K. KORMANN: Giftpflanzen – Pflanzengifte, p. IV-1-I-5. Landsburg-München: Ecomed Verl. GmbH. 1984.
3. CRABALONA, J.: Élaboration des irones dans le rhizome d'iris au cours de sa conversation. Fr. Ses Parfumes 13, 22 (1959).
4. TSCHIRSCH, A.: Rhizoma Iridis. In: Handbuch der Pharmakognosie, pp. 1143–1156. Leipzig: C.H. Tauchnitz Verlag. 1917.
5. TIEMANN, F., and P. KRÜGER: Über Veilchenaroma. Ber. dtsch. chem. Ges. 26, 2675 (1893).
   TIEMANN, F., and P. KRÜGER: Zum Nachweis von Ionon und Iron. Ber. dtsch. chem. Ges. 28, 1754 (1895).
   TIEMANN, F.: Über die Veilchenketone und die in Beziehung dazu stehenden Verbindungen der Citral-(Geranial-)reihe. Ber. dtsch. chem. Ges. 31, 808 (1898).
6. UHDE, G., and G. OHLOFF: Parmon, eine Phantomverbindung im Veilchenblütenöl. Helv. Chim. Acta 55, 2621 (1972).
7. RUZICKA, L., C.F. SEIDEL, and H. SCHINZ: Veilchenriechstoffe III. Über die Bruttoformel und einige Umsetzungen des Irons. Helv. Chim. Acta 16, 1143 (1933).
8. NAVES, Y.R., A.V. GRAMPOLOFF, P. BACHMANN: Etudes sur les matières végétales volatiles L. Etudes dans les séries des méthyl-3-linalols, des méthyl-3-citrals et des méthyl-6-ionones. Helv. Chim. Acta 30, 1599 (1947).
   RUZICKA, L., C.F. SEIDEL, H. SCHINZ, and M. PFEFFER: Veilchenriechstoffe. Die Konstitution des Irons. Helv. Chim. Acta 30, 1807 (1947).
9. RAUTENSTRAUCH, V., and G. OHLOFF: Die Stereochemie der Irone. Helv. Chim. Acta 54, 1776 (1971).
10. RAUTENSTRAUCH, V., B. WILLHALM, W. THOMMEN, and G. OHLOFF: On the stereochemistry of the irones. Helv. Chim. Acta 67, 325 (1984).
11. BEDOUKIAN, P.Z.: Violet fragrance compounds. In: Fragrance Chemistry, ed. by E.T. Theimer. New York: Academic Press. 1982.
12. TER HEIDE, R., N. PROVATOROFF, P.C. TRAS, P.J. DE VALOIS, N. VAN DER PLASSE, H.J. WOBBEN, and R. TIMMER: Qualitative analysis of the odoriferous fraction of oakmoss (*Evernia prunastri* (L.) Ach.). J. Agric. Food Chem. 23, 950 (1975).

13. BUCHHECKER, R., and S. LIAAEN-JENSEN: The carotenoid pattern in *Iris germanica*. Phytochem. **14**, 851 (1975).

14. MARNER, F.-J.: Unpublished results.

15. KRICK, W., F.-J. MARNER, and L. JAENICKE: Isolation and structure determination of the precursors of α- and γ-irone and homologous compounds from *Iris pallida* and *Iris florentina*. Z. Naturforsch. **38c**, 179 (1983).

16. MARNER, F.-J., W. KRICK, B. GELLRICH, L. JAENICKE, and W. WINTER: Irigermanal and Iridogermanal: Two new triterpenoids from rhizomes of *Iris germanica* L. J. Org. Chem. **47**, 2531 (1982).

17. KRICK, W., F.-J. MARNER, and L. JAENICKE: On the stereochemistry of natural irones, dihydroirones and their precursors. Helv. Chim. Acta **67**, 318 (1984).

18. OHLOFF, G., E. OTTO, V. RAUTENSTRAUCH, and G. SNATZKE: Chiroptische Eigenschaften einiger Trimethyl-cyclohexen-Derivate: Die Ionone, Irone und Abszisinsäuren. Helv. Chim. Acta **56**, 193 (1973).

19. MARNER, F.-J., and L. JAENICKE: In preparation.

20. KRICK, W., F.-J. MARNER, and L. JAENICKE: Isolation and structural determination of a new methylated triterpenoid from rhizomes of *Iris versicolor* L. Z. Naturforsch. **38c**, 689 (1983).

21. MAURER, B., and G. OHLOFF: (E)-9-Isopropyl-6-methyl-5,9-decadien-2-one, a terpenoid C_{14}-ketone with a novel skeleton. Helv. Chim. Acta **60**, 2191 (1977).

22. AROLD, R., F.-J. MARNER, L. JAENICKE, and K. SEFERIADIS: In preparation.

23. FARNSWORTH, N.R., Chicago: Pers. communication.

24. GLADTKE, D., F.-J. MARNER, and L. JAENICKE: In preparation.

25. RUZICKA, L., and F. LARDON: Zur Kenntnis der Triterpene. Über das Ambrein, einen Bestandteil der grauen Ambra. Helv. Chim. Acta **29**, 912 (1946).
    LEDERER, E., F. MARX, D. MERCIER, and G. PEROT: Sur les constituents de l'ambre gris II. Ambréine et Coprostanone. Helv. Chim. Acta **29**, 1354 (1946).

26. BIRCH, A.J., C.W. HOLZAPFEL, and R.W. RICKARDS: The structure and some aspects of the biosynthesis of Pleuromutilin. Tetrahedron **22**, 359 (1966).
    ARIGONI, D.: Some Studies in the biosynthesis of terpenes and related compounds. Pure Appl. Chem. **17**, 331 (1968).

27. GLADTKE, D., and L. JAENICKE: In preparation.

28. MATHEW, B.: The Iris. London: B.T. Batford Ltd. 1981.

29. PASCHE, E., Wuppertal: Pers. communication.

30. ZUNDEL, M., and M. ROHMER: Procaryotic triterpenoids. 3. The biosynthesis of 2β-methylhopanoids and 3β-methylhopanoids of *Methylobacterium organophilum* and *Acetobacter pasteurianus* ssp. *pasteurianus*. Eur. J. Biochem. **150**, 35 (1985).

31. ANKE, T., Kaiserslautern: Pers. communication.

32. HOLST, H., Gießen: Pers. communication.

33. LEVINSON, H.Z., Seewiesen: Pers. communication.

(Received February 10, 1986)

The Condylocarpine Group of Indole Alkaloids

By M. LOUNASMAA and P. SOMERSALO, Laboratory for Organic and Bioorganic Chemistry, Department of Chemistry, Technical University of Helsinki, Finland

With 12 Figures

Contents

1. Introduction

In this review we consider 24 alkaloids having the precondylocarpine-type pentacyclic skeleton and a two-carbon unit (carbons 18 and 19 in the "biogenetic numbering" of LE MEN and TAYLOR (*1*)). All can be derived from precondylocarpine (**1**) *via* condylocarpine (**2**). To date, no comprehensive review of these alkaloids has appeared. However, considerable information can be found in M. HESSE's (*2*) Tables of Indole Alkaloids under the aspidospermatine-type alkaloids. HUSSON (*3*) has recently reviewed *Strychnos* alkaloids discovered or studied in the seventies and up to 1982. Besides the chemical and structural properties he notes plant sources of condylocarpine derivatives. VAN BEEK *et al.* (*4*) have listed the alkaloids isolated from the genus *Tabernaemontana*.

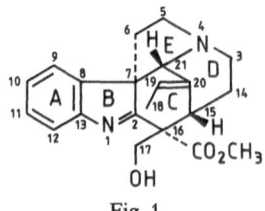

Fig. 1

Some of the condylocarpine-type alkaloids have been prepared by simple reactions from other alkaloids. The five-ring skeleton with an ethyl side chain has been synthesized and condylocarpine itself has been prepared by total synthesis.

The first condylocarpine-type alkaloid to be isolated was aspidospermatine (**16**). It was discovered by O. HESSE, the publication appearing in 1882 (*5*), during his search for the substances responsible for the activity of the drug "quebracho", which was obtained from the bark of *Aspidosperma quebracho-blanco* Schlecht and was then being used against dyspnoea and as a febrifuge. Aspidospermatine, like other condylocarpine-type alkaloids, does not have strong physiological effects. Condylocarpine was isolated from *Diplorrhynchus condylocarpon* (Muell. Arg.) Pichon ssp. *mossambicensis* (Benth.) Duvign. in 1961 (*6*). Its structure was deduced from mass spectra (*7, 8*) and confirmed in 1962 by synthesis from stemmadenine (*7*). The absolute stereochemistry was established in 1963 by chemically relating it to the known structure of strychnine (*9*) and was confirmed in 1977 by X-ray analysis of the borane complex prepared from stemmadenine (*10*).

2. Structures

The *E* geometry of the C_{19}–C_{20} double bond in condylocarpine-type alkaloids follows from the greater stability relative to the *Z* isomer of the iminium intermediate (see section 4 and, e.g. (*11*)) during the formation of precondylocarpine (**1**). The absolute stereochemistry of the cage part of the molecules has in all cases studied been that of condylocarpine (**2**) by correlation with compounds whose structure is known through X-ray crystallography.

2.1. Precondylocarpine

Precondylocarpine (**1**) is the only alkaloid in the condylocarpine group that has retained C-17. The chirality at C-16 was elucidated by KUTNEY and FULLER (*12*) in their synthesis of 16-epistemmadenine.

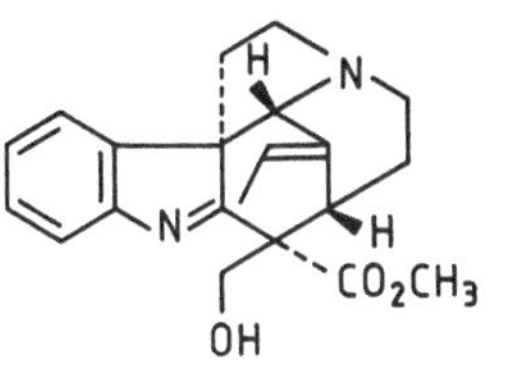

(**1**) Precondylocarpine

Fig. 2

2.2. Condylocarpines

This group consists of condylocarpine (**2**) and condylocarpine N-oxide (**3**). The latter, like all the N-oxide alkaloids, may be an artefact formed during isolation.

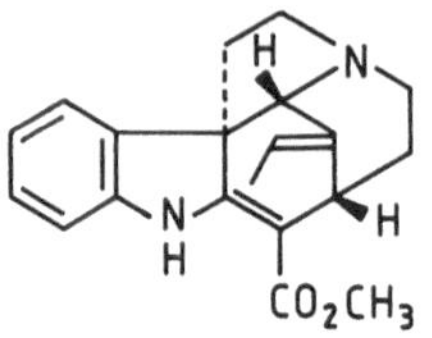

(2) Condylocarpine
(3) Condylocarpine N-oxide

Fig. 3

2.3. Dihydrocondylocarpines

The chirality at C-20 in 19,20-dihydrocondylocarpines has been a matter of controversy. None of the isolated 19,20-dihydrocondylocarpines has been shown to differ in chirality from tubotaiwine (4) isolated from *Pleiocarpa pycnantha* (K. Schum.) Stapf var. *tubicina* (Stapf) Pichon. Chemical Abstracts describes tubotaiwine (4) [6711-69-9] as a 19,20-dihydrocondylocarpine with *R*-chirality at C-20. However, catalytic hydrogenation of condylocarpine (2) to tubotaiwine (4) suggests the *S*-chirality at C-20 (see part 5.1.2) and proton nmr studies confirm this (see section 6.3). The chirality at C-19 and C-20 in 19,20-dihydro-19-hydroxycondylocarpines (6 and 7) has not been established. C-20 in 11-methoxytubotaiwine (8) has the same chirality as tubotaiwine (4) (see the ^{13}C-nmr spectra, section 6.2).

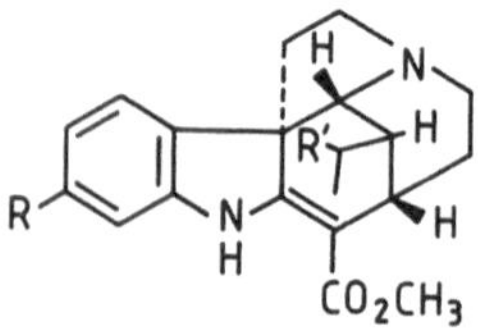

	R	R′		
(4)	H	H	Tubotaiwine	
(5)	H	H	Tubotaiwine N-oxide	
(6)	H	OH	19,20-Dihydro-19-hydroxycondylocarpine	
(7)	H	OH	19,20-Dihydro-19-hydroxycondylocarpine	C-19 epimers
(8)	OH	H	11-Methoxytubotaiwine	

Fig. 4

2.4. Tubotaiwinal

Tubotaiwinal (9) has the same C-20 stereochemistry as tubotaiwine (4).

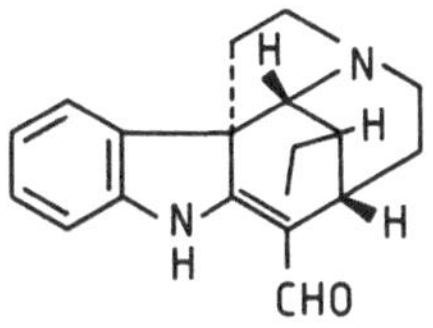

(9) Tubotaiwinal

Fig. 5

2.5. Aspidospermatidines

Aspidospermatidines lack the carbomethoxy group of condylocarpine (2), and have a single bond between C-2 and C-16.

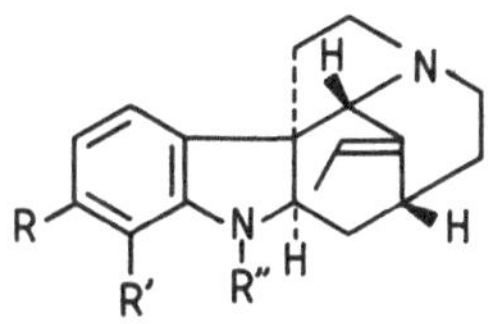

	R	R'	R''	
(10)	H	H	H	Aspidospermatidine
(11)	H	H	Me	N-Methylaspidospermatidine
(12)	H	H	Ac	N-Acetylaspidospermatidine
(13)	OH	H	Ac	N-Acetyl-11-hydroxyaspidospermatidine
(14)	H	OH	Ac	Limatinine (≡12-hydroxy-N-acetylaspidospermatidine)
(15)	H	OMe	H	Deacetylaspidospermatine (≡12-methoxyaspidospermatidine)
(16)	H	OMe	Ac	Aspidospermatine (≡N-acetyl-12-methoxyaspidospermatidine)
(17)	H	OH	EtCO	Limatine (≡12-hydroxy N-propionylaspidospermatidine)
(18)	OH	OH	Ac	N-Acetyl-11,12-dihydroxyaspidospermatidine
(19)	OMe	OH	Ac	11-Methoxylimatinine
				(≡N-acetyl-12-hydroxy-11-methoxyaspidospermatidine)
(20)	OMe	OH	EtCO	11-Methoxylimatine
				(≡12-hydroxy-11-methoxy N-propionylaspidospermatidine)

Fig. 6

2.6. 19,20-Dihydroaspidospermatine

The stereochemistry of C-20 in 19,20-dihydroaspidospermatine (21) has not been definitely established. Hydrogenation of aspidospermatine (16) gives a compound whose mass spectrum is identical with that of the alkaloid (21) (13).

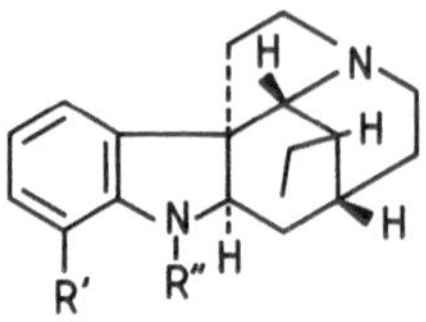

R   R'
(21) OMe Ac 19,20-Dihydroaspidospermatine
              (≡N-acetyl-19,20-dihydro-12-methoxyaspidospermatidine)

Fig. 7

2.7. Geissovelline

Geissovelline (22) has the five-ring structure of condylocarpine (2) under mildly acidic conditions as a result of bonding between N-4 and C-21 (see part 5.1.5). The alkaloid has been fully characterized save for the absolute configuration, which has not been established.

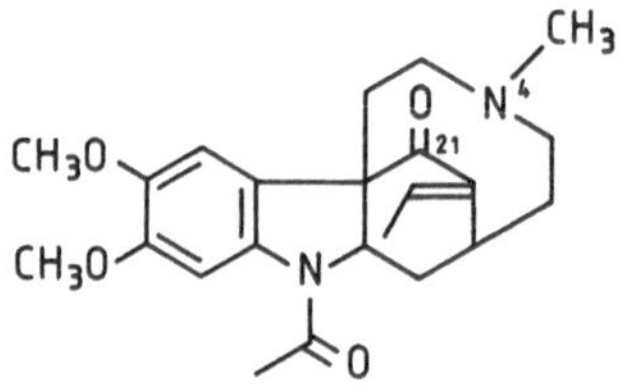

(22) Geissovelline

Fig. 8

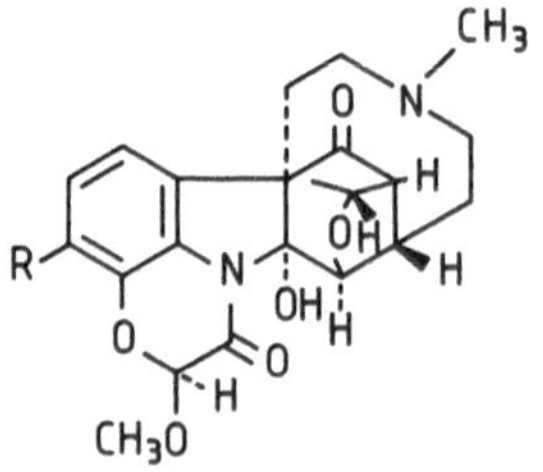

R
(23) H   Dichotine
(24) OMe 11-Methoxydichotine

Fig. 9

2.8. Dichotines

As demonstrated by the X-ray analysis of dichotine (**23**), dichotine and 11-methoxydichotine (**24**), when protonated, have seven rings, including the pentacyclic structure of condylocarpine (**2**).

3. Isolation and Occurrence

3.1. Isolation

The isolation of alkaloids proceeds as follows (see e. g. (*14*)): Dried and ground plant material is extracted with a polar solvent (usually ethanol under basic conditions). Most of the nonalkaloid material is precipitated when the extracts are kept cold. After filtration and concentration the raw alkaloids may be fractionated by extracting them with organic solvents from pH-buffered solutions where the basicity of the buffer solution is gradually increased. Chromatography is usually needed to further separate and purify the individual alkaloids. BIEMANN *et al.* (*13*) separated six aspidospermatidine-type alkaloids from many others by gas chromatography, but more often liquid chromatography with silica gel or alumina is used. Very polar alkaloids (quaternary ammonium compounds) require ion-pair or reversed phase chromatography for separation (*15*). Other than the N-oxides, condylocarpine-type alkaloids with a quaternary nitrogen have not yet been detected.

Owing to its rigid and open position in the cage structure, N-4 has a strong tendency to form quaternary salts. If the alkaloids are purified with chlorinated hydrocarbons, material is lost. For example, BESSELIÈVRE *et al.* (*16*) could not find free tubotaiwine (**4**) when they extracted alkaloids with dichloromethane because of the formation of tubotaiwine-dichloromethane adducts. Also oxygen, heat, light and strongly basic or acidid conditions should be avoided. Condyfoline (**37**), a sensitive compound obtainable from tubotaiwine (**4**) and structurally closely related to tubifoline (**36**), has not been found in plant material.

3.2. Occurrence

Evolution and hybridization of species lead to wide variety among plant specimens and their proper classification is often difficult. Sources of condylocarpine-type alkaloids are listed in Table 1 on the basis of

Table 1

FAMILY
SUBFAMILY
TRIBE

Genus and species	alkaloid source reference

APOCYNACEAE

PLUMERIOIDEAE

CARISSEAE

Melodinus aeneus Baill.	(4) L/TW (*19*), (5) L/TW (*19*)
M. australius (Muell.) Pierre	(2) SB (*20*)
Hunteria zeylanica var. *africana* Pichon	(4) SB (*21*)
Pleiocarpa pycnantha (K. Schum.) Stapf, var. *tubicina* (Stapf) Pichon	(4) L (*22*)

TABERNAEMONTANEAE

Ervatamia heyneana (Wall.) Cooke see *Tabernaemontana heyneana*

Tabernaemontana amblyocarpa Urb.	(4) S (*23*)
T. attenuata (Miers) Urb.	(4) L (*24*)
T. chippii (Stapf) Pichon	(4) RB (*25*)
T. divaricata (L.) R.BR. ex Roem. et Schult	(4) C (*26*)
T. echinata Aubl.	(4) L (*27*)
T. eglandulosa Stapf	(4) L, TW (*28*)
T. eusepala Aug. DC.	(4) SB (*29*)
T. heyneana Wall.	(4) S/SB (*30*)
T. holstii K. Schum. see *T. pachysiphon*	
T. humboltii (Baill.) Pichon	(4) RB, SB (*31*)
T. mauritiana Poir.	(4) L, RB, SB (*32*)
T. minutiflora Pichon	(4) L (*33*)
T. mocquerysii Aug. DC.	(4) (*34*)
T. olivacea Muell. Arg.	(3) TW (*35*)
T. pachysiphon Stapf	(4) RB, SB (*36*), (5) R, RB, SB (*36, 37*)
T. siphilitica (L.F.) Leeuwenberg	(4) L (*38*)
T. stapfiana Britten	(4) RB (*39*), (5) RB (*39*)

Peschiera echinata A. DC. see *Tabernaemontana echinata*

Tabernanthe iboga Baill.	(4) C (*26*)

Stemmadenia glabra Benth.	(4) L (*40*)
S. tomentosa var. *palmeri* Woodson	(2) C (*41*), (4) C (*41*)

Conopharyngia johnstonii Stapf see *Tabernaemontana stapfiana*

Pandaca boiteaui MGF. see *Tabernaemontana mocquerysii*
P. eusepala (Aug. DC.) MGF. see *T. eusepala*
P. mauritiana (Poir.) MGF. et Boiteau see *T. mauritiana*
P. minutiflora (Pichon) MGF. see *T. minutiflora*
P. ochrascens (Pichon) MGF. see *T. humboltii*

Anartia meyeri (G. Don) Miers see *Tabernaemontana attenuata*

Bonafousia tetrastachya (H.B.K.) MGF. see *Tabernaemontana siphilitica*

References, pp. 52–56

Table 1 (continued)

ALSTONIEAE	
Craspidospermum verticillatum BOJ. var. petiolare A.DC.	(2) L (42), (4) L/SB (16)
Strempeliopsis strempelioides K. Schum.	(2) L, SB (43), (4) L, R, SB (43, 44), (5) R (44)
Alstonia angustiloba Miq.	(6) SB (45), (7) SB (45)
A. quaternata Heurck et Muell. Arg.	(4) L/S, SB (46)
A. scholaris R. BR.	(4) R, S (47, 48)
Diplorrhynchus condylocarpon (Muell. Arg.) Pichon ssp. mossambicensis (Benth.) Duvign.	(2) RB (6)
Aspidosperma album (Vahl) R. Ben. ex Pich.	(2) SE (85), (4) SE (85)
A. compactinervium Kuhlmann	(13) B (49)
A. excelsum Benth.	(8) RB (50)
A. limae Woods.	(4) RB (51), (14) SB (52), (17) SB (52, 53), (19) SB (52), (20) SB (52)
A. peroba Allem. ex Sald, see A. polyneuron	
A. polyneuron Muell. Arg.	(16) B (54)
A. populifolium A. DC.	(8) B (49)
A. quebracho-blanco Schlecht.	(10) B (13, 55), (11) B (13, 55), (12) B (13, 55), (15) B (13, 55), (16) B, F (5, 13, 54, 55), (21) B (13, 55)
A. spec. no 9610	(18) (2)
A. subincanum Mart.	(14) (56)
A. tomentosum Mart.	(14) (56)
Geissospermum vellosii Allem.	(22) B (57)
Amsonia tabernaemontana Walt.	(4) L (58)
Catharanthus ovalis MGF.	(4) A (59, 60)
C. roseus G. Don	(4) C (61)
RAUVOLFIEAE	
Vallesia dichotoma Ruiz et Pav	(1) L (62), (2) L (62), (4) L (62), (12) L (62), (23) B (63), (24) B (63)
LOGANIACEAE	
Strychnos angolensis Gilg	(4) RB/SB (64)
S. dolichotyrsa Gilg ex Onochie et Hepper	(2) SB (65), (4) SB (65)
S. ngouniensis Pellegr.	(9) RB, SB (66)
S. nux-vomica L.	(2) SE (67)

A = aerial parts, B = bark, C = cell culture, F = fruits, L = leaves, R – roots, RB = root bark, S – stems, SB = stem bark, SE = seedlings, TW = twigs

the species listed in the references. The system used is that of M. HESSE (17). The genus Tabernaemontana is arranged as in reference (4). An alternative classification has been presented by ALLORGE (18) for the subfamily Tabernaemontanoideae. The alkaloid content varies according to plant part, growing conditions and variety of species. Some of the alkaloids are barely detectable.

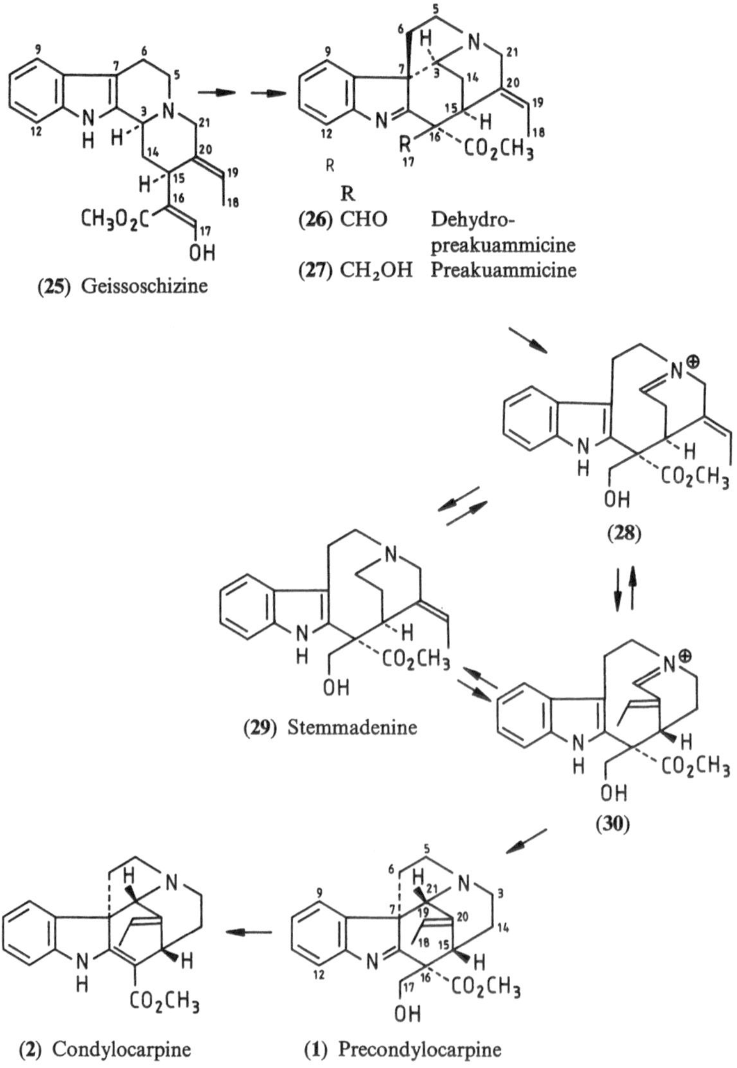

(25) Geissoschizine

R	
(26) CHO	Dehydro-preakuammicine
(27) CH$_2$OH	Preakuammicine

(28)

(29) Stemmadenine

(30)

(2) Condylocarpine

(1) Precondylocarpine

Scheme 1. Plausible biosynthesis of condylocarpine

4. Biogenesis

A plausible biosynthetic scheme for condylocarpine (**2**), supported, for example, by the radioactive feeding experiments of SCOTT *et al.* (*67–69*), is depicted in Scheme 1. Geissoschizine (**25**) is formed from strictosidine, which is a condensation product of secologanin and tryptamine, and reacts further, presumably *via* an oxindole derivative and dehydropreakuammicine (**26**), to preakuammicine (**27**), a precursor of the *Strychnos* alkaloids (*3*). Opening of the C_3–C_7 bond and shifting of the iminium double bond gives an intermediate (**30**) which readily closes to precondylocarpine (**1**). Stemmadenine (**29**), obtainable from preakuammicine (**27**) and precondylocarpine (**1**) by reduction, gives again, by oxidation, preakuammicine and precondylocarpine. Retroaldol reaction of precondylocarpine (**1**) yields condylocarpine (**2**).

Tubotaiwine (**4**) is formed by reduction of condylocarpine (**2**), and aspidospermatidine (**10**) by decarboxylation and reduction.

LING and DJERASSI (*63*) have presented a plausible biogenetic sequence from condylocarpine (**2**) to dichotine (**23**) *via* (**31**), (**32**) and (**33**) (Scheme 2). The order of the reactions is not definitive. Compound (**32**) was selected as an intermediary, since the correct stereochemistry of the alcohol (**33**) results when the epoxide ring opens. Other activating groups may be necessary. Condylocarpine (**2**) and N-acetylaspidospermatidine (**12**) have been isolated from the same plant as dichotine (**23**).

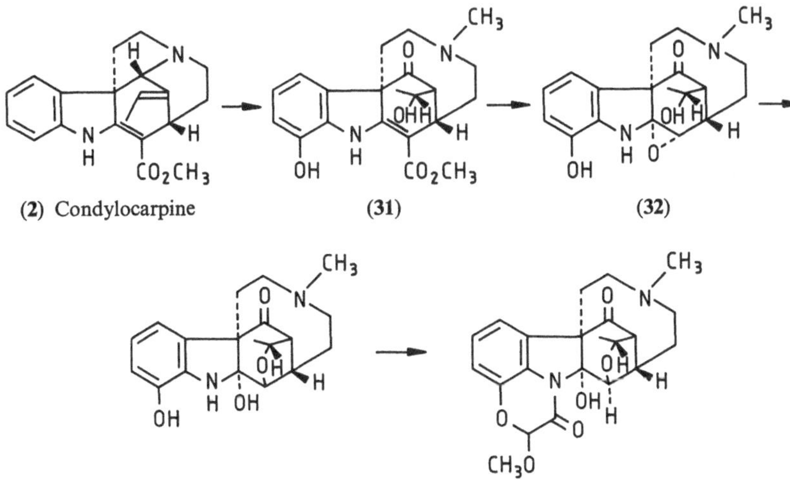

(**2**) Condylocarpine   (**31**)   (**32**)

(**33**)   (**23**) Dichotine

Scheme 2. Plausible biogenetic sequence from condylocarpine to dichotine

5. Chemistry

5.1. Correlation of Alkaloid Structures

5.1.1. Precondylocarpine and Condylocarpine

Condylocarpine (2) can be obtained from stemmadenine (29) *via* precondylocarpine (1). SANDOVAL *et al.* (7) treated stemmadenine hydrochloride with aqueous $KMnO_4$ and heated the resulting crude product in nitrogen at 90° C. After evolution of formaldehyde a crystalline product identical with authentic condylocarpine (2) was isolated. SCOTT and WEI (70, 71) acylated stemmadenine (29) and oxygenated the product with plantinum catalyst to give precondylocarpine acetate in about 20% yield. Reaction of this with NaOMe by the procedure of WALSER and DJERASSI (62) gave condylocarpine (2). The borane adduct of condylocarpine characterized by WANG and PAUL (10) was made by treating stemmadenine (29) first with mercuric acetate and then with borohydride.

5.1.2. Tubotaiwine

SCHUMANN and SCHMID (9) have correlated both condylocarpine (2) and tubotaiwine (4) with akuammicine (34) (Scheme 3).

19,20-Dihydrocondylocarpine, which is identical with authentic tubotaiwine (4) (mixed melting point of picrate; also ^{13}C NMR spectrum, unpublished results of our laboratory), has been obtained as the only product from catalytic hydrogenation of condylocarpine (2) (9). Hydrogen adds from the less hindered side forcing the ethyl side chain close to the aromatic ring and producing the 20 S configuration. Heating tubotaiwine (4) in hydrochloric acid gave condyfoline (37) in 93% yield. When condyfoline was heated in the absence of acid, unreacted condyfoline (37), 20-epi-condyfoline (38) and tubifoline (36) were isolated. Condyfoline (37) and tubifoline (36) gave the same indole product (39) when treated with potassium borohydride, whereas 20-epicondyfoline (38) gave the isomeric product (40). Heating of 20-epicondyfoline (38) yielded the same three products as heating of condyfoline (37); tubifoline (36) did not isomerize. The reaction from condyfoline to tubifoline has been proposed to proceed through intermediates (41) and (42) (Scheme 4). The driving force of the isomerization might be steric repulsion between the side chain and the aromatic part. Treatment of indole (39) with oxygen gave condyfoline (37) and tubifoline (36) (Scheme 3).

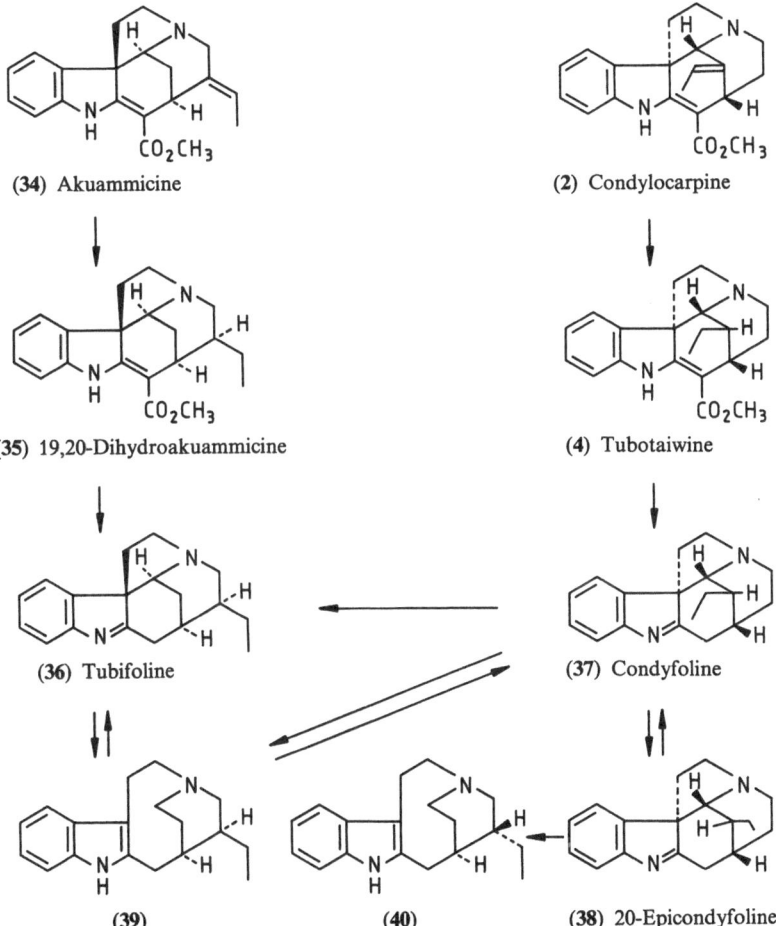

(34) Akuammicine

(2) Condylocarpine

(35) 19,20-Dihydroakuammicine

(4) Tubotaiwine

(36) Tubifoline

(37) Condyfoline

(39)

(40)

(38) 20-Epicondyfoline

Scheme 3. Correlation of condylocarpine and tubotaiwine with akuammicine

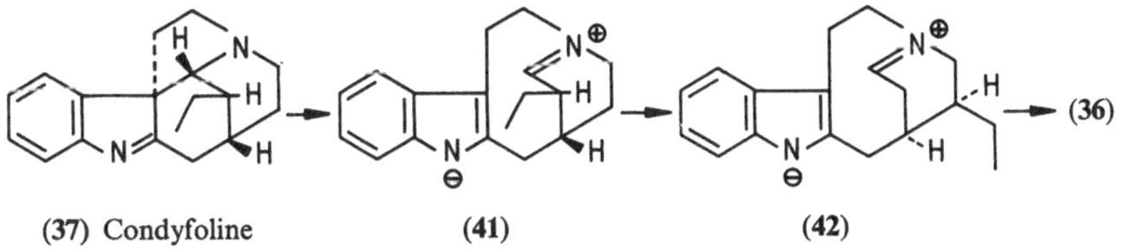

(37) Condyfoline

(41)

(42)

Scheme 4. Proposed isomerization of condyfoline to tubifoline

5.1.3. N-oxides

The N-oxides of condylocarpine (**2**) and tubotaiwine (**4**) are easily prepared. ACHENBACH and RAFFELSBERGER (*35*) treated natural condylocarpine (**2**) with hydrogen peroxide, obtaining a product identical with condylocarpine N-oxide (**3**). Reduction of the N-oxide with iron(II) sulphate gave condylocarpine (**2**), and hydrogenation of the N-oxide gave tubotaiwine (**4**).

PINAR *et al.* (*39*) have oxidized tubotaiwine (**4**) with m-chloroperbenzoic acid to tubotaiwine N-oxide (**5**) and reduced the N-oxide with sulfur dioxide back to tubotaiwine (**4**). BAASSOU *et al.* (*19*) reduced tubotaiwine N-oxide (**5**) with iron in acetic acid to tubotaiwine (**4**), and oxygenated it back to the N-oxide with hydrogen peroxide.

5.1.4. Aspidospermatidines

WALSER and DJERASSI (*62*) have deacetylated N-acetylaspidospermatidine (**12**) using refluxing HCl/MeOH, and BIEMANN *et al.* (*8*) have decarboxylated (20% HCl, 115° C, 2 h) and reduced (LAH) condylocarpine (**2**). In both cases the products resembled aspidospermatidine (**10**). They exhibited similar mass spectra but different melting points and high resolution NMR spectra would have clarified the situation.

To study the structure of the indoline alkaloids KLYNE *et al.* (*72*) prepared N-acetyl-19,20-dihydroaspidospermatidine (**43**) by catalytic hydrogenation of N-acetylaspidospermatidine (**12**) and by reduction and acetylation of condyfoline (**37**). The products were identical in all respects.

ARNDT *et al.* (*56*) methylated limatinine (**14**) to aspidospermatine (**16**), which BIEMANN *et al.* (*13*) further deacetylated to deacetylaspidospermatine (**15**) and hydrogenated to 19,20-dihydroaspidospermatine (**21**).

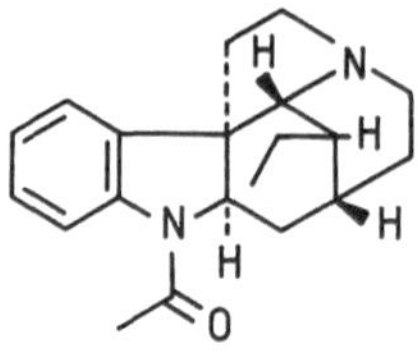

(43) 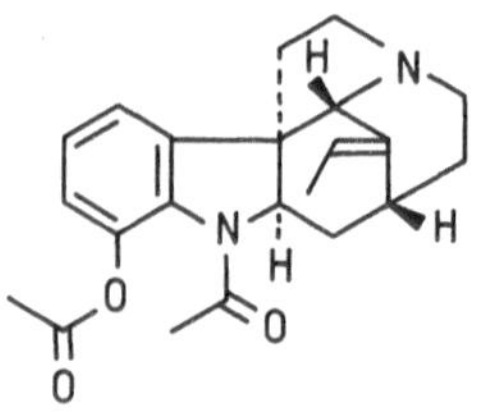   (44)

Fig. 10

PINAR and SCHMID (52) have correlated limatinine (14) with limatine (17) by preparing the acetylated limatinine (44). 11-Methoxylimatinine (19) was similarily correlated with 11-methoxylimatine (20).

5.1.5. Geissovelline and Dichotine

MOORE and RAPOPORT (57) have studied the chemistry of geissovelline (22) and worked out its structure save for the absolute stereochemistry. The acetyl group gives rise to two different rotamers and to simplify the spectra geissovelline (22) was deacetylated. The structure of deacetylgeissovelline (45 and 46) depends on the solvent used (Scheme 5). The ^{13}C NMR spectra (see part 6.2.) of these two forms strongly support the established structures. The partial bond between C-21 and N-4 makes reduction at C-21 difficult.

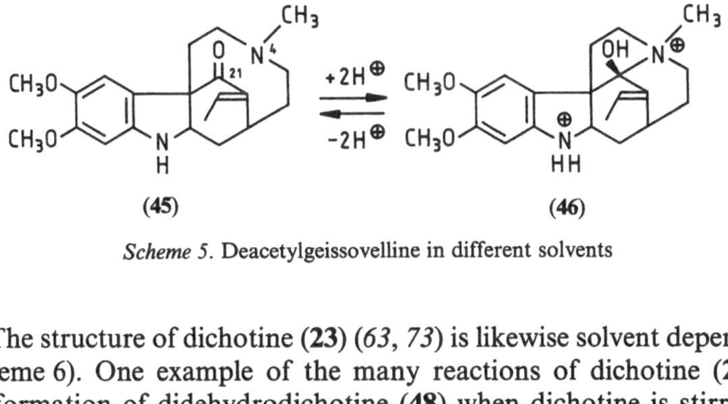

Scheme 5. Deacetylgeissovelline in different solvents

The structure of dichotine (23) (63, 73) is likewise solvent dependent (Scheme 6). One example of the many reactions of dichotine (23) is the formation of didehydrodichotine (48) when dichotine is stirred at room temperature with palladium-on-carbon in absolute ethanol under nitrogen.

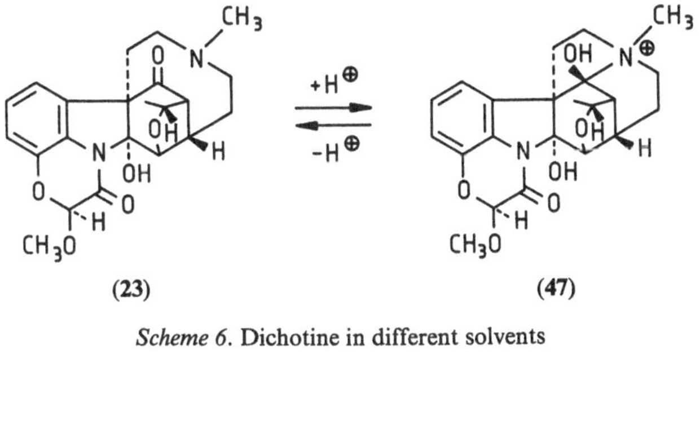

Scheme 6. Dichotine in different solvents

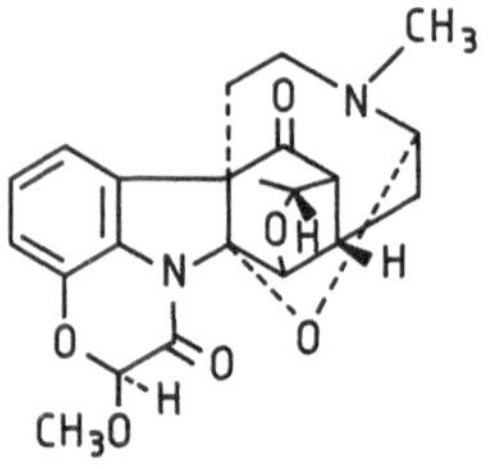

(48)

Fig. 11

5.1.6. Further Alkaloids from Precondylocarpine

When precondylocarpine acetate (49) was heated in ethyl acetate at 100° C, a racemic compound was isolated in 28% yield which was identical with andranginine (51) (3, 74). The reaction most likely proceeds *via* intermediate (50) and Diels-Alder reaction (Scheme 7). Andranginine is isolated from *Craspidospermum verticillatum* BOJ., with other alkaloids such as condylocarpine (2) and stemmadenine (42).

When Scott and Wei (70, 71) hydrogenated precondylocarpine acetate (49) and heated the raw product they detected racemic tabersonine (54) and vincadifformine (55) (Scheme 8).

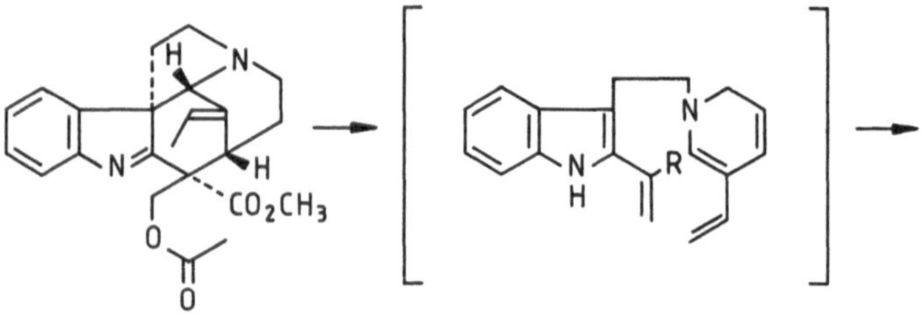

(49)   (50)

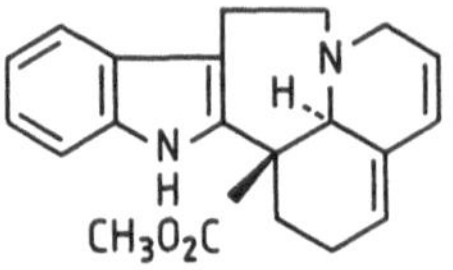

(51) Andranginine

Scheme 7. Formation of andranginine from precondylocarpine acetate

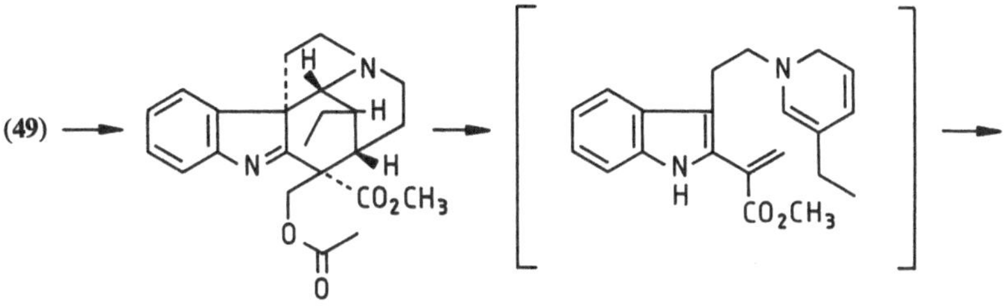

(52)   (53) Dehydrosecodine

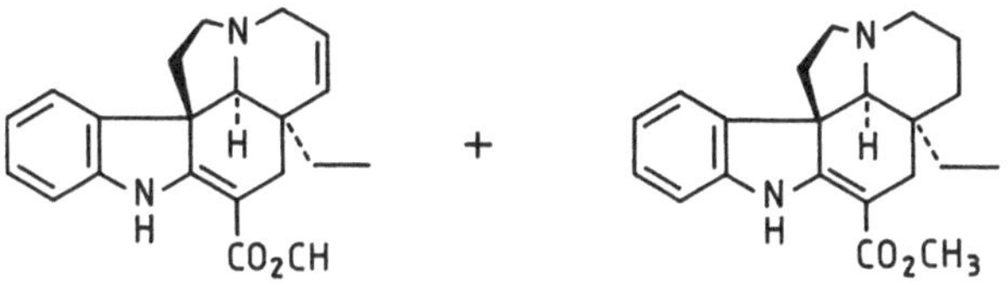

(54) (±)-Tabersonine   (55) (±)-Vincadifformine

Scheme 8. Formation of (±)-tabersonine and (±)-vincadifformine from precondylocarpine acetate

5.2. Total Syntheses

Since all total syntheses thus far have produced racemates, synthetic challenges still remain.

5.2.1. Condyfoline

DADSON et al. (75) carried out the first total synthesis of the condylocarpine (2) skeleton and also the first synthesis of the stemmadenine skeleton (39) in preparing condyfoline (37) (Scheme 9). Acid anhydride treatment of the hexahydroindolizinoindole (56) gave the amido-ester (57). Cautious alkaline hydrolysis, oxidation of the alcohol and base-induced cyclization of the ketone gave the tetracyclic amide (58), whereafter reduction of the keto- and amido-groups produced the amine (39). This was oxidized to a mixture of condyfoline (37) and tubifoline (36) by a procedure described by SCHUMANN and SCHMID (9) (part 5.1.2).

The different approach of WU and SNIECKUS (76) to amine (39) is depicted in Scheme 10. Here the ethyl side chain was introduced

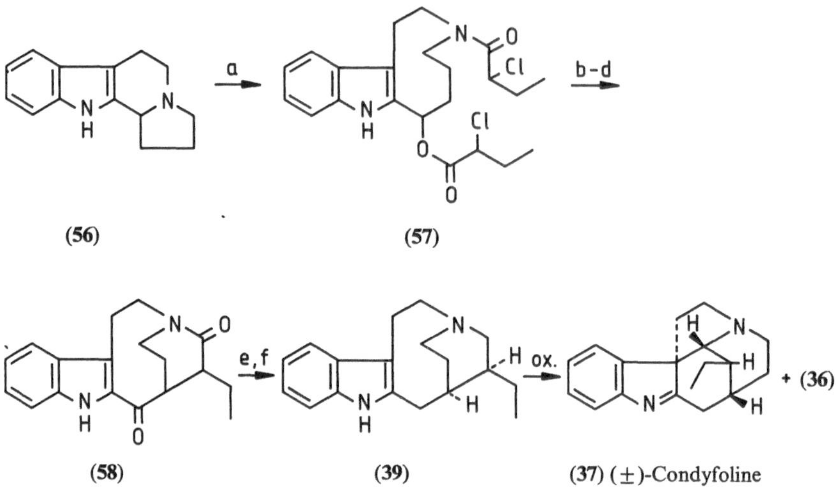

(56)   (57)

(58)   (39)   (37) (±)-Condyfoline

Scheme 9. Synthesis of condyfoline and tubifoline by DADSON et al. (75).
Reagents: a) O(COCHClEt)$_2$, b) OH$^\ominus$, c) MnO$_2$, d) sodium t-pentoxide, e) Wolff-Kish-
ner, f) LiAlH$_4$

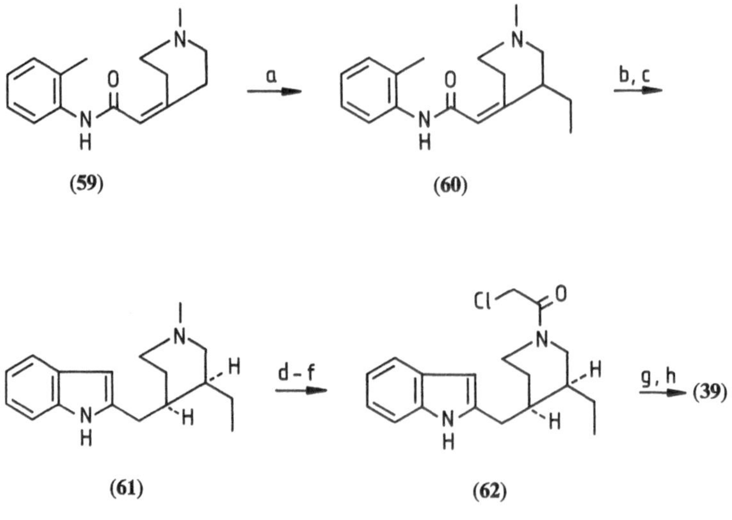

(59)   (60)

(61)   (62)

Scheme 10. WU and SNIECKUS approach to condyfoline (76).
Reagents: a) 1. BuLi 2 equiv., 2. EtBr 1 equiv., b) H$_2$/Pd, c) t-BuOK, 340° C,
d) ClCO$_2$CH$_2$CCl$_3$, e) Zn, AcOH, f) ClCOCH$_2$Cl, g) hv, h) LiAlH$_4$

by regioselective γ-alkylation to the α, β-unsaturated amide (59) via a dianion route. Hydrogenation of (60) proceeded with high stereoselectivity and Madelung reaction gave the indole (61). The methyl group was replaced by the chloroacetyl group to give the amide (62) and photocyclization and reduction of the amide afforded the amine (39).

TAKANO et al. (77) exploited the thio-Claisen rearrangement in their route to condyfoline (37) (Scheme 11). Methyl γ-bromocrotonate reacted with thioamide (63) to the salt (64), which gave, by base-induced thio-Claisen rearrangement, thioamide (65). Phosphorus oxychloride and borohydride treatment furnished further the indole compound (66) with an E-double bond. The allylic ester group was reduced with DIBAL to alcohol, which when methanesulfonylated formed the quaternary ammonium salt (67). This was reduced with sodium in liquid ammonia to the amine (68), which was reduced and oxidized to condyfoline (37).When the amine (68) was directly oxidized the *Strychnos* framework was produced and not the condylocarpine skeleton.

A synthesis of condyfoline (37) by BAN et al. (78, 79), which is part of a larger undertaking is depicted in Scheme 12. Ketoacid (70) was prepared and this reacted in Fischer's indole synthesis to give the amide (71). Amide (72) was obtained by Curtius rearrangement. The amine was liberated, and photoreaction yielded the nine-membered lactam (73). This was reduced and the obtained amine reacted with

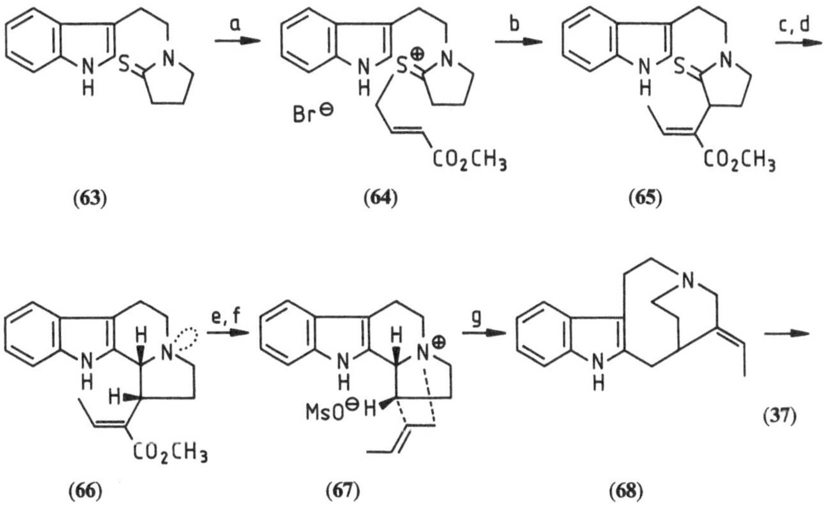

Scheme 11. Synthesis of condyfoline by TAKANO et al. (77).
Reagents: a) CH₂BrCHCHCO₂Me, b) NaOMe, c) POCl₃, d) NaBH₄, DIBAL, f) MsCl, g) Na, NH₃

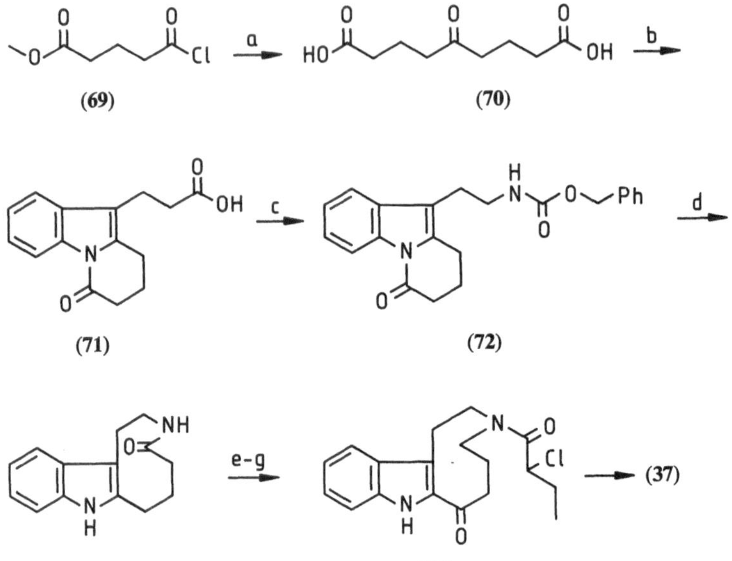

(69) (70) (71) (72) (73) (74)

Scheme 12. Synthesis of condyfoline by BAN et al. (*78, 79*).
Reagents: a) 1. Et_3N, 2. KOH, b) 1. $PhNHNH_2$, 2. H_2SO_4, c) 1. $(COCl)_2$, 2. NaN_3, 3. Δ, 4. $PhCH_2OH$, d) 1. H_2/Pd, 2. hν, e) $LiAlH_4$, f) EtCHClCOCl, g) I_2O_5

2-chlorobutanoic acid chloride. Oxidation of the product with I_2O_5 gave keto-amide (**74**). The subsequent reactions are those used in ref. (*75*).

BOSCH and AMAT (*80*) prepared 20-deethyl-6-methylthiocondyfoline (**77**) (Scheme 13) by first constructing rings A, B, C and D (**75**). The piperidine ring was alkylated with bromoacetaldehyde diethylacetal,

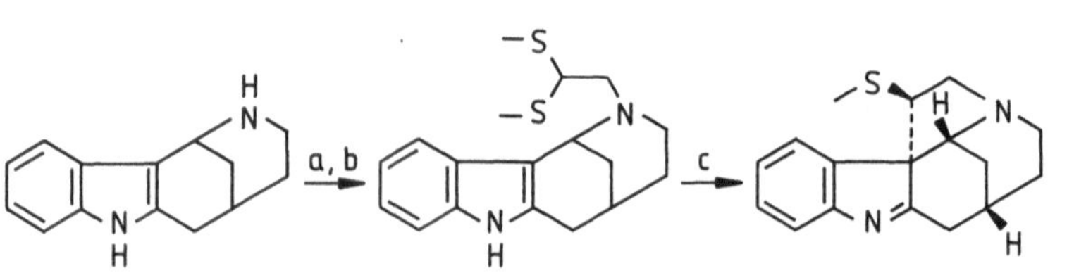

(75) (76) (77) 20-Deethyl-6-methylthiocondyfoline

Scheme 13. Synthesis of 20-deethyl-6-methylthiocondyfoline by BOSCH and AMAT (*80*).
Reagents: a) $BrCH_2CH(OEt)_2$, Na_2CO_3, dioxane, b) MeSH, CH_2Cl_2, $BF_3 \cdot OEt_2$, c) dimethyl(methylthio)sulfonium fluoroborate, CH_2Cl_2

and the ethoxy groups were replaced by methylthio groups to yield the dithioacetal (76). From (76), under mild conditions, dimethyl(methylthio)sulfonium tetrafluoroborate (DMTSF) generated a thionium ion which cyclized to the pentacycle (77).

5.2.2. Tubotaiwine

The tubotaiwine synthesis of DADSON and HARLEY-MASON (81) (Scheme 14) is based on their synthesis of condyfoline (37) (part 5.2.1). The ketone was reduced to alcohol which was converted by normal manipulation to ester (78). Upon heating of the ester with phosphorus oxychloride racemic tubotaiwine (4) could be isolated. This last step involves a change in the oxidation level and is low yielding.

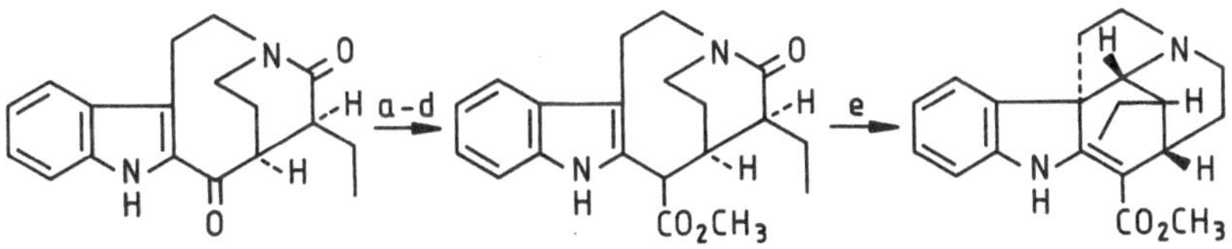

Scheme 14. Synthesis of (±)-tubotaiwine by DADSON and HARLEY-MASON (81).
Reagents: a) $NaBH_4$, b) acetylation, c) NaCN, DMSO, d) H_2SO_4, MeOH, e) $POCl_3$

5.2.3. Condylocarpine

Like the synthesis of condyfoline (37) by DADSON et al., the condylocarpine (2) synthesis of HARLEY-MASON (82, 83) is based on the ring enlargement reaction of the tetracyclic indole (56) (Scheme 15). Acid anhydride (79) reacted with amine (56), as expected, to give the amidoester (80). This was hydrolyzed, oxidized and, when treated with base, cyclized with double bond formation to ketoamide (81) and its Z-isomer in 34% overall yield from (56). The carbomethoxy group was added through Wittig reaction and nitrile formation. Addition of Meerwein's reagent to the amide (82) afforded the iminium-ether salt (83) which was then reduced with borohydride to an amino-ether with the same oxidation level as condylocarpine (2). Further reduction is hindered through steric strain. Finally, Lewis acid treatment gave condylocarpine (2).

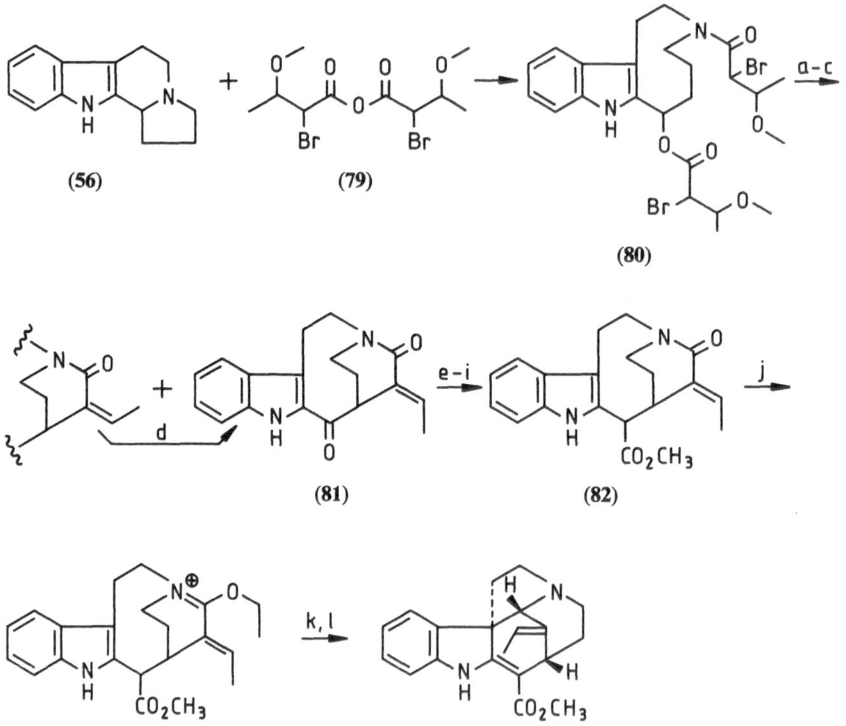

(56) (79) (80) (81) (82) (83) (2) (±)-Condylocarpine

Scheme 15. Synthesis of (±)-condylocarpine by HARLEY-MASON (*82, 83*).
Reagents: a) $OH^{\ominus}$, b) $Pb(OAc)_4$, c) sodium t-pentoxide, d) NaOMe, e) $Ph_3PCHOMe$, f) $H^{\oplus}$, g) NH_2OH, h) $TiCl_4$, i) methanolysis, j) F_4BOEt_3, k) $NaBH_4$, l) BF_3

6. Physical Properties

6.1. X-ray Crystallography

X-ray structures are available only for the condylocarpine-borane complex (*10*) and dichotine hydrobromide (*73*).

6.2. ^{13}C NMR Spectroscopy

^{13}C NMR spectra of alkaloids are highly informative and nowadays easily recorded. Unfortunately ^{13}C spectroscopy was not available when many of the alkaloids were first characterized and spectra of the aspidospermatidines, for example, have not been published. The spectra of some related alkaloids are useful, e. g. the shifts of the aromatic carbons of the *Aspidosperma* alkaloids studied by WENKERT *et al.* (*84*) are practically the same as those of the condylocarpine-type alkaloids. The published spectra of alkaloids (**2, 4, 8, 9, 45, 46**) are compiled in Table 2. Without further information some of the signal assignments are only suggestive. (Incautious interpretation has led to wrong conclusions about the stereochemistry of tubotaiwine (**4**) (*66*). The spectral values for the cage carbons (C-3, C-5, C-6, C-14, C-15, C-18, C-19,

Table 2

Atom	Alkaloid					
	(**2**)(*85*)	(**4**)(*85*)	(**8**)(*50*)	(**9**)(*66*)	(**45**)[d](*57*)	(**46**)[d](*57*)
2	167.3[b]	170.3[b]	168.9[b]	172.1	58.7	66.3
3	45.2[c]	43.6[c]	45.3[c]	44.8[c]	55.1	61.5
5	52.9	53.7	53.9	54.5	49.3	59.6
6	45.9[c]	45.2[c]	44.0[c]	44.1[c]	44.8	25.4
7	59.9	54.8	54.5	55.9	57.6	59.6
8	137.4	136.7	129.5	137.4	142.3	128.1
9	120.5	121.2	119.9	122.3	110.6	103.0
10	119.6	119.6	105.3	120.0	143.0	150.8
11	127.0	127.4	159.5	127.6	149.3	150.8
12	109.2	109.8	97.1	110.7	95.6	110.8
13	144.0	143.6	144.8	143.1	139.7	128.1
14	28.4	28.5[a]	23.9[a]	29.8[a]	36.1	36.6
15	29.0	30.6	30.9	32.1	29.4	29.4
16	101.0	95.7	95.8	106.5	29.8	32.9
18	12.9	11.6	11.6	11.1	12.9	12.9
19	116.5	23.6[a]	28.5[a]	23.7[a]	125.0	128.8
20	135.0	40.9	41.3	41.3	121.9	126.8
21	68.6	65.5	65.4	66.3	186.4	104.5
CO_2Me	168.7[b]	168.8[b]	170.9[b]			
CO_2*Me*	50.8	51.2	51.1			
C_{10}-O*Me*					55.9	56.8
C_{11}-O*Me*			55.4		56.5	56.7
CHO				190.1		
N-*Me*					40.2	43.6

[a,b,c] The assignments may be interchanged.
[d] Methylene and quaternary carbon assignments are tentative.

The following values without assignments have been published (*57*) for geissovelline (**22**): 12.8, 15.1, 22.9, 24.0, 29.0, 29.2, 29.4, 29.6, 29,9, 30.9, 31.6, 33.4, 40.0, 40.4, 44.9, 50.0, 55.5, 55.8, 56.0, 56.2, 61.7, 62.2, 100.4, 102.1, 108.7, 110.2, 124,3, 125.5, 126.1, 127.1, 132.8, 134.0, 138.7, 145.9, 148.3, 167.1, 184.3.

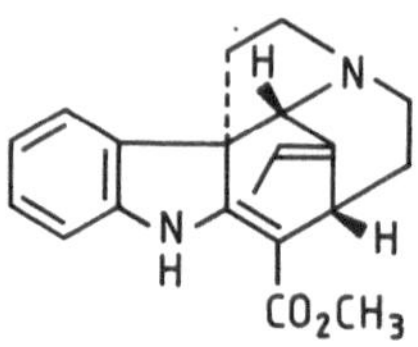

(2)

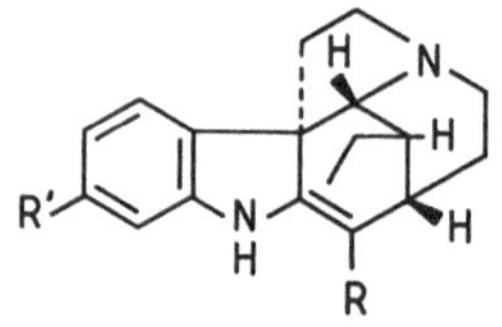

(4) R=CO_2CH_3, R'=H
(8) R=CO_2CH_3, R'=OH
(9) R=CHO, R'=H

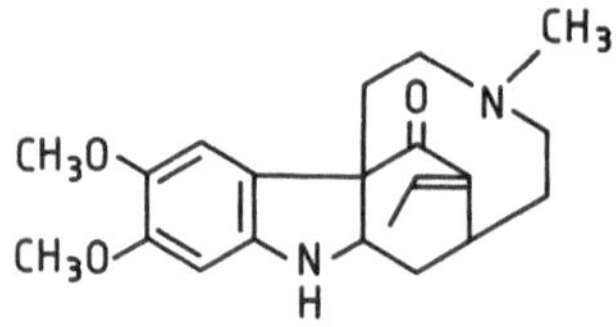

(45) 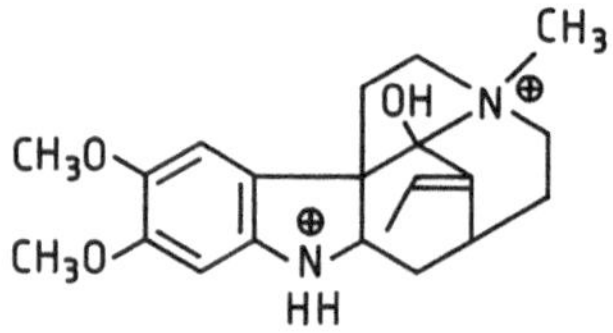   (46)

Fig. 12

C-20 and C-21) of tubotaiwine (4), 11-methoxytubotaiwine (8) and tubotaiwinal (9) are nearly identical, so the orientation of the ethyl side chain must be the same in these compounds.

6.3. ^{1}H NMR Spectroscopy

Proton nmr spectroscopy provided a simple way for distinguishing condylocarpine (2) from its isomer akuammicine (34): in condylocarpine the hydrogens at C-15 and C-21, and the C-18 methyl group, are in the same plane as the double bond, so there is no homoallylic coupling and the C-18 hydrogens give a clear doublet (53). LOUNASMAA et al. (86) have studied the configuration of tubotaiwine (4) at 500 MHz. All the hydrogens and the connectivities were elucidated using the COSY method. Using the NOESY method it was shown that there was interaction between hydrogens 20 and 14β but not between hydrogens 19 and 14β. The ethyl group is thus oriented close to the aromatic ring and C-20 has S-chirality.

6.4. Mass Spectrometry

The molecular peaks of the condylocarpine-type alkaloids, like those of aromatic compounds in general, are quite strong. BIEMANN

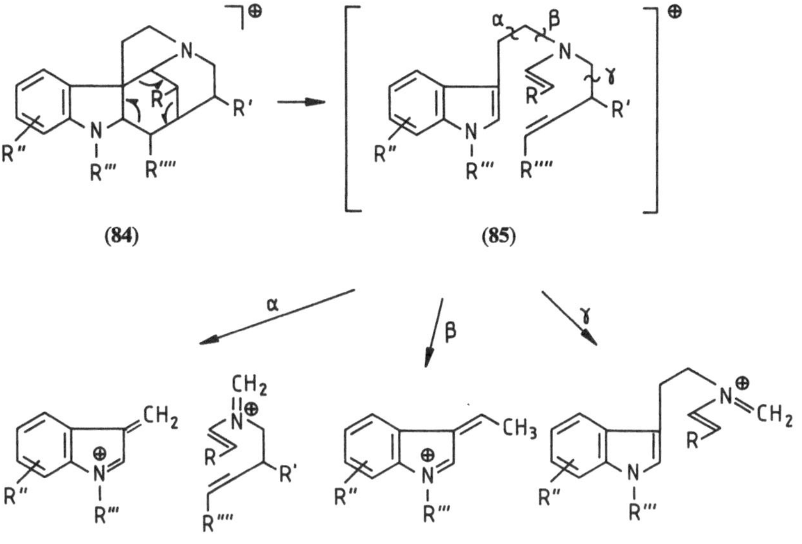

Scheme 16. Mass spectral fragmentation pattern of hydrogenated condylocarpine (and akuammicine) derivatives

et al. (13) and BUDZIKIEWICZ (87) have explained the fragmentation pattern of hydrogenated condylocarpine (R = Et, R' = H) [and akuam-micine (R = H, R' = Et)] derivatives. Collapse of ring C in (84) leads to intermediate (85) where several bond fissions can take place leading to ions (86), (87), (88) (which can also be written as a spiro structure) and (89) (Scheme 16). The fragment (89) helps to distinguish condylo-carpine derivatives from akuammicine derivatives.

6.5. Other Spectra

UV spectra of model compounds (e. g. the methoxycarbazoles stud-ied by CHALMERS et al. (88)) have proved useful in determining the substitution of the aromatic ring.

KLYNE et al. (72, 89) measured ORD spectra of condylocarpine-type alkaloids and found the curves to be of opposite sign to those of strych-nine derivatives; confirming thereby the proposed absolute stereochem-istry of these alkaloids.

7. Pharmacology

The condylocarpine-type alkaloids show little pharmacological activity. KINGSTON et al. (37) have shown that tubotaiwine N-oxide (5) has an ED_{50} of 1.8 µg/ml in the P-388 cell culture, but is inactive in the 9KB system. According to BOHLIN et al. (64) tubotaiwine (4) induces weak clonic convulsions. GUNASEKERA et al. (30) measured the activity of tubotaiwine against P-388 lymphocytic leukemia in vitro and found the ED_{50} to be 23 µg/ml (a compound is considered active if it displays an $ED_{50} \leq 4.0$ µg/ml).

VERPOORTE et al. (50) found 11-methoxytubotaiwine (8) to be active against Bacillus subtilis and Staphylococcus aureus which are Gram-positive but not against Gram-negative bacteria and fungi. The minimum inhibitory concentration (MIC) using B. subtilis was 230 µg/ml.

References

1. LE MEN, J., and W.I. TAYLOR: A Uniform Numbering System for Indole Alkaloids. Experientia 21, 505 (1965).

2. a) HESSE, M.: Indolalkaloide in Tabellen, p. 41. Berlin-Göttingen-Heidelberg: Springer. 1964.
   b) HESSE, M.: Indolalkaloide in Tabellen, Ergänzungswerk, p. 90. Berlin-Heidelberg-New York: Springer. 1968.

3. HUSSON, H.-P.: The Strychnos Alkaloids. In: Indoles, Part Four, The Monoterpenoid Indole Alkaloids (Saxton, J.E., ed.), in "The Chemistry of Heterocyclic Compounds" (Weissberger, A., and Taylor, E.C., eds.), p. 293. New York: John Wiley and Sons. 1983.

4. VAN BEEK, T.A., R. VERPOORTE, A. BAERHEIM SVENDSEN, A.J.M. LEEUWENBERG, and N.G. BISSET: Tabernaemontana L. (Apocynaceae): A Review of Its Taxonomy, Phytochemistry, Ethnobotany and Pharmacology. J. Ethnopharmacol. 10, 1 (1984).

5. HESSE, O.: Studien über argentinische Quebracho-drogen. Liebigs Ann. 211, 249 (1882).

6. STAUFFACHER, D.: Alkaloide aus Diplorrhynchus condylocarpon (Muell. Arg.) Pichon ssp. mossambicensis (Benth.) Duvign. Helv. Chim. Acta 44, 2006 (1961).

7. SANDOVAL, A., F. WALLS, J.N. SHOOLERY, J.M. WILSON, H. BUDZIKIEWICZ, and C. DJERASSI: Alkaloid Studies. The Structures of Stemmadenine and Condylocarpine. Tetrahedron Lett. 1962, 409.

8. BIEMANN, K., A.L. BURLINGAME, and D. STAUFFACHER: Application of Mass Spectrometry to Structure Problems: Condylocarpine. Tetrahedron Lett. 1962, 527.

9. SCHUMANN, D., and H. SCHMID: Chemische Korrelation von Condylocarpin mit Akuammicin. Helv. Chim. Acta 46, 1996 (1963).

10. WANG, A.H.-J., and I.C. PAUL: The Borine Adduct of Condylocarpine: A Case of Partially Mistaken Identity. Acta Crystallogr., Sect. B B 33, 2977 (1977).

11. USKOKOVIĆ, M.R., R.L. LEWIS, J.J. PARTRIDGE, C.W. DESPREAUX, and D.L. PRUESS: Asymmetric Synthesis of allo-Heteroyohimbine Alkaloids. J. Am. Chem. Soc. 101, 6742 (1979).

12. KUTNEY, J.P., and G.B. FULLER: The Total Synthesis of Akuammicine and 16-Epi-

stemmadenine. The Absolute Configuration of Stemmadenine. Heterocycles 3, 197 (1975).

13. BIEMANN, K., M. SPITELLER-FRIEDMANN, and G. SPITELLER: Application of Mass Spectrometry to Structure Problems. X. Alkaloids of the Bark of *Aspidosperma quebracho-blanco*. J. Am. Chem. Soc. **85**, 631 (1963).

14. RAPOPORT, H., T.P. ONAK, N.A. HUGHES, and M.G. REINECKE: Alkaloids of *Geissospermum vellosii*. J. Am. Chem. Soc. **80**, 1601 (1985).

15. VERPOORTE, R., and A. BAERHEIM SVENDSEN: Chromatography of Alkaloids, Pt. B: Gas-Liquid Chromatography and High-Performance Liquid Chromatography. In: Journal of Chromatography Library, vol. 23, p. 425. Amsterdam: Elsevier. 1984.

16. a) BESSELIÈVRE, R., N. LANGLOIS, and P. POTIER: Chlorure de méthylène, solvant ou réactif? Bull. Soc. Chim. Fr. **1972**, 1477. See also: MILLS, J.E., C.A. MARYANOFF, R.M. COSGROVE, L. SCOTT, and D.F. McCOMSEY, Org. Prep. Proc. Int. **16**, 99 (1984).

   b) KAN-FAN, C., R. BESSELIÈVRE, A. CAVÉ, B.C. DAS, and P. POTIER: Nouveaux alcaloides du *Craspidospermum verticillatum* Boj. *ex* DC (Apocynacées): Δ^{14}-vincine et Δ^{14}-épi-16 vincine. C.R. Acad.Sci., Sér. C **272**, 1431 (1971).

17. HESSE, M.: Alkaloid Chemistry. New York: Wiley-Interscience. 1981.

18. ALLORGE, L.: Monographie des Apocynacées – Tabernaemontanoidées Américaines Morphologie, Systématique, Chimio-taxonomie. In: Mémoires du Muséum national d'Histoire naturelle, nouvelle série, Série B, Botanique, vol. 30. Paris: Muséum national d'Histoire naturelle. 1985.

19. BAASSOU, S., H. MEHRI, and M. PLAT: Alcaloides de *Melodinus aeneus*. Phytochem. **17**, 1449 (1978).

20. LINDE, H.A.: Die Alkaloide aus *Melodinus australis* (F. Mueller) Pierre (Apocynaceae). Helv. Chim. Acta **48**, 1822 (1965).

21. LAVAUD, C. G. MASSIOT, J. VERCAUTEREN, and L. LE MEN-OLIVIER: Alkaloids of *Hunteria zeylanica*. Phytochem. **21**, 445 (1982).

22. KUMP, W.G., M.B. PATEL, J.M. ROWSON, and H. SCHMID: Indolalkaloide aus Blättern von *Pleiocarpa pycnantha* (K. Schum.) Stapf, *var. tubicina* (Stapf) Pichon. Helv. Chim. Acta **47**, 1497 (1964).

23. PÉREZ, I., and P. SIERRA: Alcaloides de *Tabernaemontana amblyocarpa* Urb. Rev. Latinoam. Quim. **11**, 132 (1980).

24. LADHAR, F., M. DAMAK, A. AHOND, C. POUPAT, P. POTIER, and C. MORETTI: Contribution à l'étude des Tabernaemontanées américaines. III. Alcaloides de *Anartia* cf. *meyeri*. J. Nat. Prod. **44**, 459 (1981).

25. VAN BEEK, T.A., R. VERPOORTE, and A. BAERHEIM SVENDSEN: Antimicrobially Active Alkaloids from *Tabernaemontana chippii*. J. Nat. Prod. **48**, 400 (1985).

26. PAWELKA, K.-H., and J. STÖCKIGT: Indole Alkaloids from Cell Suspension Cultures of *Tabernaemontana divaricata* and *Tabernanthe iboga*. Plant Cell Rep. **2**, 105 (1983).

27. GHORBEL, N., M. DAMAK, A. AHOND, E. PHILOGÈNE, C. POUPAT, P. POTIER, and H. JACQUEMIN: Contribution à l'étude des Tabernaemontanées américaines. IV. Alcaloides de *Peschiera echinata*. J. Nat. Prod. **44**, 717 (1981).

28. VAN BEEK, T.A., R. VERPOORTE, and A. BAERHEIM SVENDSEN: Alkaloids of *Tabernaemontana eglandulosa*. Tetrahedron **40**, 737 (1984).

29. QUIRIN, F., M.-M. DEBRAY, C. SIGAUT, P. THÉPENIER, L. LE MEN-OLIVIER, and J. LE MEN: Alcaloides du *Pandaca eusepala*. Phytochem. **14**, 812 (1975).

30. GUNASEKERA, S.P., G.A. CORDELL, and N.R. FARNSWORTH: Anticancer Indole Alkaloids of *Ervatamia heyneana*. Phytochem. **19**, 1213 (1980).

31. PANAS, J.M., B. RICHARD, C. SIGAUT, M.-M. DEBRAY, L. LE MEN-OLIVIER, and J. LE MEN: Alcaloides du *Pandaca ochrascens*. Phytochem. **13**, 1969 (1974).

32. PICOT, F., F. LALLEMAND, P. BOITEAU, and P. POTIER: Alcaloides indoliques de *Pandaca mauritiana*. Phytochem. **13**, 660 (1974).

33. Petitfrère, N., A.M. Morfaux, M.-M. Debray, L. Le Men-Olivier, and J. Le Men: Alcaloides des feuilles du *Pandaca minutiflora*. Phytochem. **14**, 1648 (1975).

34. Andriantsiferana, M., F. Picot, P. Boiteau, and H.-P. Husson: Alcaloides de *Pandaca boiteaui* (Apocynaceae). Phytochem. **18**, 911 (1979).

35. Achenbach, H., and B. Raffelsberger: Alkaloide in *Tabernaemontana*-Arten, XII, Untersuchung der Alkaloide von *Tabernaemontana olivacea* – Condylocarpin-N-oxid, ein neues Alkaloid aus *T. olivacea*. Z. Naturforsch., B: Anorg. Chem., Org. Chem. **35 B**, 885 (1980).

36. van Beek, T.A., F.L.C. Kuijlaars, P.H.A.M. Thomassen, R. Verpoorte, and A. Baerheim Svendsen: Antimicrobially Active Alkaloids from *Tabernaemontana pachysiphon*. Phytochem. **23**, 1771 (1984).

37. Kingston, D.G.I., F. Ionescu, and B.T. Li: Plant Anticancer Agents IV: Identification of Tubotaiwine-N-oxide as a Cytotoxic Constituent of *Tabernaemontana holstii*. Lloydia **40**, 215 (1977).

38. Damak, M., A. Ahond, and P. Potier: Contribution à l'étude des Tabernaemontanées américaines. II. Nouveaux alcaloides de *Bonafousia tetrastachya* (Humboldt, Bonpland et Kunth) Markgraf (Apocynacées). Bull. Soc. Chim. Fr. **1981**, II-213.

39. Pinar, M., U. Renner, M. Hesse, and H. Schmid: Tubotaiwin-N-oxid aus Wurzelrinde von *Conopharyngia johnstonii* Stapf. Helv. Chim. Acta **55**, 2972 (1972).

40. Ciccio, J.F., C.H. Herrera, V.H. Castro, and M. Ralitsch: Aislamiento y caracterizacion de alcaloides de *Stemmadenia glabra* Benth. (Apocynaceae). Rev. Latinoam. Quim. **10**, 67 (1979).

41. Stöckigt, J., K.-H. Pawelka, A. Rother, and B. Deus: Indole Alkaloids from Cell Suspension Cultures of *Stemmadenia tomentosa* and *Voacanga africana*. Z. Naturforsch., C. Biosci. **37 C**, 857 (1982).

42. Kan-Fan, C., B.C. Das, H.-P. Husson, and P. Potier: Plantes malgaches. XV. – Alcaloides de *Craspidospermum verticillatum* var. *petiolare* (Apocynacées): andrangine ou (+) époxy-14,15 nor-1 vallésamidine. Bull. Soc. Chim. Fr. **1974**, 2839.

43. Laguna, A., L. Dolejš, and L. Novotný: Alkaloids from *Strempeliopsis strempelioides* K. Schum. Collect. Czech. Chem. Commun. **45**, 1419 (1980).

44. Laguna, A., L. Novotný, L. Dolejš, and M. Buděšinský: Alkaloids from Roots of *Strempeliopsis strempelioides* – Structures of Strempeliopine and Strempeliopidine. Planta Med. **51**, 285 (1984).

45. Zèches, M., T. Ravao, B. Richard, G. Massiot, L. Le Men-Olivier, J. Guilhem, and C. Pascard: Structure de l'échitamidine, d'un stéréoisomère et de deux régioisomères. Tetrahedron Lett. **25**, 659 (1984).

46. Mamatas-Kalamars, S., T. Sèvenet, C. Thal, and P. Potier: Alcaloides d'*Alstonia quaternata*. Phytochem. **14**, 1849 (1975).

47. Boonchuay, W., and W.E. Court: Minor Alkaloids of *Alstonia scholaris* root. Phytochem. **15**, 821 (1976).

48. Boonchuay, W., and W.E. Court: Alkaloids of *Alstonia scholaris* from Thailand. Planta Med. **29**, 380 (1976).

49. Gilbert, B., A.P. Duarte, Y. Nakagawa, J.A. Joule, S.E. Flores, J. Aguayo Brissolese, J. Campello, E.P. Carrazzoni, R.J. Owellen, E.C. Blossey, K.C. Brown, Jr. and C. Djerassi: Alkaloid Studies-L. The Alkaloids of Twelve *Aspidosperma* species. Tetrahedron **21**, 1141 (1965).

50. Verpoorte, R., E. Kos-Kuyck, A. Tjin A Tsoi, C.L.M. Ruigrok, G. de Jong, and A. Baerheim Svendsen: Medicinal Plants of Surinam, III. Antimicrobially Active Alkaloids from *Aspidosperma excelsum*. Planta Med. **48**, 283 (1983).

51. Pinar, M., and H. Schmidt: 3'-Methoxy-limaspermin, Limapodin, 3'-Methoxylimapodin und Tubotaiwin aus *Aspidosperma limae* Woodson. Liebigs Ann. Chem. **668**, 97 (1963).

52. PINAR, M., and H. SCHMID: Weitere Alkaloide aus *Aspidosperma limae* Woods. Helv. Chim. Acta **50**, 89 (1967).

53. PINAR, M., B.W. BYCROFT, J. SEIBL, and H. SCHMID: Notiz über Limatin aus *Aspidosperma limae* Woods. Helv. Chim. Acta **48**, 822 (1965).

54. BISSET, N.G.: The Occurrence of Alkaloids in the Apocynaceae. Ann. Bogor. **3**, 105 (1958).

55. BIEMANN, K., M. FRIEDMANN-SPITELLER, and G. SPITELLER: An Investigation by Mass Spectrometry of the Alkaloids of *Aspidosperma quebracho-blanco*. Tetrahedron Lett. **1961**, 485.

56. ARNDT, R.R., S.H. BROWN, N.C. LING, P. ROLLER, C. DJERASSI, J.M. FERREIRA, F.B. GILBERT, E.C. MIRANDA, S.E. FLORES, A.P. DUARTE, and E.P. CARRAZZONI: Alkaloid Studies-LVIII. The Alkaloids of Six *Aspidosperma* Species. Phytochem. **5**, 1653 (1967).

57. MOORE, R.E., and H. RAPOPORT: Geissovelline, a New Alkaloid from *Geissospermum vellossi*. J. Org. Chem. **38**, 215 (1973).

58. PANAS, J.-M., A.-M. MORFAUX, L. OLIVIER, and J. LE MEN: Alcaloides des feuilles de l'*Amsonia tabernaemontana* Walt., Apocynacées. Ann. Pharm. Fr. **30**, 273 (1972).

59. DIATTA, Li, Y. LANGLOIS, N. LANGLOIS, and P. POTIER: Alcaloides de *Catharanthus ovalis* Mgf.: réactivité d'alcaloides du type vindoline (et corrélation avec la désacétyl-cathovaline). Bull. Soc. Chim. Fr. **1975**, 671.

60. LANGLOIS, N., L. DIATTA, and R.Z. ANDRIAMIALISOA: Alcaloides monoindoliques de *Catharanthus ovalis*. Phytochem. **18**, 467 (1979).

61. PÉTIARD, V., D. COURTOIS, F. GUÉRITTE, N. LANGLOIS, and B. MOMPON: New Alkaloids in Plant Tissue Cultures. Plant Tissue Cult., Proc. Int. Congr. Plant Tissue Cell Cult., 5th. 1982, 309. Ref. Chem. Abstr. **99**, 172856 u (1983).

62. WALSER, A., and C. DJERASSI: Alkaloid-Studien LII. Die Alkaloide aus *Vallesia dichotoma* Ruiz et Pav. Helv. Chim. Acta **48**, 391 (1965).

63. LING, N.C., and C. DJERASSI: Alkaloid Studies. LXIII. The Constitution and Chemistry of Dichotine and 11-Methoxydichotine. J. Am. Chem. Soc. **92**, 6019 (1970).

64. BOHLIN, L., W. ROLFSEN, J. STRÖMBOM, and R. VERPOORTE: Alkaloids and Biological Activity of *Strychnos angolensis*. Planta Med. **35**, 19 (1979).

65. VERPOORTE, R., M.J. VERZIJL, and A. BAERHEIM SVENDSEN: Further Alkaloids from *Strychnos dolichothyrsa*. Planta Med. **44**, 21 (1982).

66. MASSIOT, G., P. THÉPENIER, M.-J. JACQUIER, J. LOUNKOKOBI, C. MIRAND, M. ZÈCHES, L. LE MEN-OLIVIER, and C. DELAUDE: Further Alkaloids from *Strychnos ngouniensis*. Tetrahedron **39**, 3645 (1983).

67. HEIMBERGER, S.I., and A.I. SCOTT: Biosynthesis of Strychnine. J. Chem. Soc., Chem. Commun. **1973**, 217.

68. SCOTT, A.I., and A.A. QURESHI: Biogenesis of *Strychnos*, *Aspidosperma*, and *Iboga* Alkaloids. The Structure and Reactions of Preakuammicine. J. Am. Chem. Soc. **91**, 5874 (1969).

69. SCOTT, A.I.: Biosynthesis of the Indole Alkaloids. Acc. Chem. Res. **3**, 151 (1970).

70. SCOTT, A.I., and C.C. WEI: Regio- and Stereospecific Models for the Biosynthesis of the Indole Alkaloids. The *Corynanthe-Aspidosperma* Relationship. J. Am. Chem. Soc. **94**, 8264 (1972).

71. SCOTT, A.I., and C.C. WEI: Regio- and Stereospecific Models for the Biosynthesis of the Indole Alkaloids-II. Biogenetic Type Synthesis of *Aspidosperma* and *Iboga* alkaloids from a *Corynanthe* Precursor. Tetrahedron **30**, 3003 (1974).

72. KLYNE, W., R.J. SWAN, B.W. BYCROFT, and 'H. SCHMID: Ermittlung der absoluten Konfiguration von Indolinalkaloiden durch Vergleiche der Optischen Rotations-dispersionen ihrer N(a)-Acylderivate. Helv. Chim. Acta **49**, 833 (1966).

73. LING, N.C., C. DJERASSI, and P.G. SIMPSON: Alkaloid Studies. LXII. X-Ray Crystallo-

graphic Structure Determination of Dichotine hydrobromide. J. Am. Chem. Soc. **92**, 222 (1970).

74. KAN-FAN, C., G. MASSIOT, A. AHOND, B.C. DAS, H.-P. HUSSON, P. POTIER, A.I. SCOTT, and C.-C. WEI: Structure and Biogenetic-type Synthesis of Andranginine: an Indole Alkaloid of a New Type. J. Chem. Soc., Chem. Commun. **1974**, 164.

75. DADSON, B.A., J. HARLEY-MASON, and G.H. FOSTER: Total Synthesis of (±)-Tubifoline, (±)-Tubifolidine and (±)-Condyfoline. J. Chem. Soc., Chem. Commun. **1968**, 1233.

76. WU, A., and V. SNIECKUS: A New Synthesis of a Stemmadenine Model. Tetrahedron Lett. **1975**, 2057.

77. TAKANO, S., M. HIRAMA, and K. OGASAWARA: A New Entry into the Synthesis of the *Strychnos* Indole Alkaloids Containing 19,20-Double Bond *via* the Thio-Claisen Rearrangement. Tetrahedron Lett. **23**, 881 (1982).

78. BAN, Y., K. YOSHIDA, J. GOTO, and T. OISHI: Novel Photoisomerization of 1-Acylindoles to 3-Acylindolenines: General Entry to the Total Synthesis of *Strychnos* and *Aspidosperma* Alkaloids. J. Am. Chem. Soc. **103**, 6990 (1981).

79. BAN, Y., K. YOSHIDA, J. GOTO, T. OISHI, and E. TAKEDA: A Synthetic Road to the Forest of *Strychnos, Aspidosperma, Schizozygane* and *Eburnamine* Alkaloids by Way of the Novel Photoisomerization. Tetrahedron **39**, 3657 (1983).

80. BOSCH, J., M. AMAT: A New Synthetic Entry to Pentacyclic *Strychnos* Indole Alkaloids. Tetrahedron Lett. **26**, 4951 (1985).

81. a) DADSON, B.A., and J. HARLEY-MASON: Total Synthesis of (±)-Geissoschizoline. J. Chem. Soc., Chem. Commun. **1969**, 665.
    b) DADSON, B.A., and J. HARLEY-MASON: Total Synthesis of (±)-Tubotaiwine (Dihydrocondylocarpine). J. Chem. Soc., Chem. Commun. **1969**, 665.

82. CRAWLEY, G.C., and J. HARLEY-MASON: Total Synthesis of (±)-Fluorocurarine, the Racemate of a Calabash-curare Alkaloid. J. Chem. Soc., Chem. Commun. **1971**, 685.

83. HARLEY-MASON, J.: Synthetic Studies in the *Strychnos*-type Alkaloid Field. Pure Appl. Chem. **41**, 167 (1975).

84. WENKERT, E., D.W. COCHRAN, E.W. HAGAMAN, F.M. SCHELL, N. NEUSS, A.S. KATNER, P. POTIER, C. KAN, M. PLAT, M. KOCH, H. MEHRI, J. POISSON, N. KUNESCH, and Y. ROLLAND: Carbon-13 Nuclear Magnetic Resonance Spectroscopy of Naturally Occurring Substances. XIX. *Aspidosperma* Alkaloids. J. Am. Chem. Soc. **95**, 4990 (1973).

85. URREA, M., A. AHOND, H. JACQUEMIN, S.-K. KAN, C. POUPAT, P. POTIER, and M.-M. JANOT: Nouveaux alcaloides extraits des graines de *Aspidosperma album* (Vahl) R. Bent. (Apocynacées). C.R. Hebd. Séances Acad. Sci., Sér. C **287**, 63 (1978).

86. LOUNASMAA, M., A. KOSKINEN, and J. O'CONNELL: NMR Studies of Alkaloids. Assignment of the Stereochemistry at C(20) in Tubotaiwine (Dihydrocondylocarpine). Helv. Chim. Acta **69**, 1343 (1986).

87. BUDZIKIEWICZ, H., J.M. WILSON, C. DJERASSI, J. LÉVY, J. LE MEN, and M. JANOT: Mass Spectrometry in Structural and Stereochemical Problems-XIX. Akuammicine and Related Alkaloids. Tetrahedron **19**, 1265 (1963).

88. CHALMERS, J.R., H.T. OPENSHAW, and G.F. SMITH: The Constitution of Aspidospermine. Part II. Ultraviolet Absorption of the Bz-Methoxy-tetra- and -hexa-hydrocarbazoles. J. Chem. Soc. **1957**, 1115.

89. KLYNE, W., R.J. SWAN, B.W. BYCROFT, D. SCHUMANN, and H. SCHMID: Absolute Konfiguration von Alkaloiden der Aspidospermin-Gruppe. Helv. Chim. Acta **48**, 443 (1965).

(Received January 22, 1986)

The Antibiotics of the Pluramycin Group (4*H*-Anthra[1,2-*b*]pyran Antibiotics)

By U. SÉQUIN, Institut für Organische Chemie der Universität, Basel, Switzerland

With 9 Figures

Contents

1. Introduction

The pluramycin-like antibiotics are a group of highly substituted
4H-anthra[1,2-b]pyran-4,7,12-triones. They show antitumor and antimicrobial activity and have thus mainly been investigated for their
biological activities. Their chemical instability and photolability make
their handling difficult. This review stresses the chemical aspects of
the pluramycin antibiotics; their biochemical properties will, however,
also be summarized.

In 1974, BÉRDY (7) published an elaborate classification of antibiotics, which was based on their chemical structures. In the chapter 'anthraquinone derivatives', the griseophagins, hedamycin, the indomycins, the iyomycins, kidamycin F, neopluramycin, the pluramycins, the
plurallin chromophore, rubiflavin, tumimycin and the antibiotics
4418-I and -II were grouped together and called pluramycin type compounds after the antibiotic that was first described from this class.
Since not a single structure had been fully elucidated at that time,
the classification was made on the basis of the physical and chemical
properties of the compounds that were published in the respective original papers.

Today, the compounds of this group of antibiotics are either said
to be pluramycins or to belong to the pluramycin group according
to BÉRDY (7), or they are referred to as anthra[1,2-b]pyran antibiotics
according to the suggestion of HAUSER and RHEE (37). A short review
of this group of antibiotics was published in 1981 by ECKARDT (19).
The compounds or families of compounds that have been described
to date are compiled in Table 1. As can be seen, only a few of them

Table 1. *The Pluramycin Antibiotics*[a]

Antibiotic or group of antibiotics	Producing organism	First described by	Constitution determined by
Pluramycins	*Streptomyces pluricolorescens*	MAEDA et al., 1956 (57)	KONDO et al., 1977 (50)
Iyomycins	*S. phaeoverticillatus*	HATA et al., 1963 (35, 78)	–
Indomycins	*S. spec.* Ind. 927	SCHNELL and BROCKMANN, 1963 (9, 80)	–
Anthracidins	*S. spec.* no. 190	YOSHIDA and KATAGIRI, 1964 (104)	
Rubiflavins	*S. spec.* SC 3728	ASZALOS et al., 1965 (1)	NADIG and SÉQUIN, 1980 (65, 67)
Plurallin A chromophore	*S. pluricolorescens*	OGAWARA et al., 1966 (72)	–
Hedamycin	*S. griseoruber* C-1150	SCHMITZ et al., 1967 (79)	SÉQUIN et al., 1978 (82, 105)
4418's	*Actinomyces griseorubiginosus*	KUDINOVA et al., 1968 (53)	–
Griseophagins	*S. griseus*	TOTH-SARUDY et al., 1970 (92)	
Neopluramycin	*S. pluricolorescens*	KONDO et al., 1970 (51)	KONDO et al., 1977 (50)
Largomycin FII chromophore constituents	*S. pluricolorescens*	YAMAGUCHI et al., 1970 (103)	GONDA et al., 1984 (11, 33)
Kidamycin	*S. phaeoverticillatus*	KANDA, 1971 (44)	FURUKAWA and IITAKA, 1974 (28, 29)
Tumimycin	*S.* ATCC 21501	ASZALOS et al., 1972 (2)	–
Griseorubins	*S. fimicarius*	DORNBERGER et al., 1980 (18)	–
PD 121,222	*S. spec.* WP 0123	SÉQUIN, FRENCH et al., 1985 (68)	SÉQUIN, FRENCH et al., 1985 (68)
Chromoxymycin	*S. rubropurpureus*	SETO et al., 1985 (47)	SETO et al., 1985 (47)

[a] Antibiotics listed by BÉRDY (7), updated.

have been structurally elucidated. In this paper, the focus will be on those compounds that unambiguously belong to this group, *i.e.* those whose constitution has been fully or at least partly elucidated. Thus, the iyomycins, the anthracidins, the plurallin chromophore, the 4418's, the griseophagins and tumimycin will not be discussed further. They were included in Table 1, since their spectroscopic data – as far as they are published – closely resemble those of typical pluramycin antibiotics and because they were in BÉRDY's compilation.

2. General Structural Characteristics and Nomenclature

The general structure of the pluramycins is shown in Fig. 1. All pluramycin antibiotics possess a 4*H*-anthra[1,2-*b*]pyran-4,7,12-trione moiety. The correct numbering for this is given in Fig. 1. However, in earlier papers, carbon 12b was sometimes designated as carbon 1a. The rings of the anthraquinone part are known as rings B to D, the pyrone ring is called A. Position 5 is always methylated and C(2) carries a side chain whose constitution varies between the different pluramycin antibiotics. This side chain either contains four carbons and is derived from the 1-methylpropyl group or it is a six carbon fragment related to the 1-methylpentyl residue. The carbon atoms of this side chain are customarily not numbered according to IUPAC standards: With the methyl group at C(5) of the aromatic nucleus being given the number 13, the side chain carbons are numbered from 14 to 17 or from 14 to 19, depending on the length of this side chain. Positions 8 and 10 of the 4*H*-anthra[1,2-*b*]pyran skeleton are substituted by deoxyaminosugars bound *C*-glycosidically to the chromophore. The substituent in position 8, referenced as ring E, is angolosamine (2,3,6-trideoxy-3-(dimethylamino)-D-*arabino*-hexose), a sugar that was found to be a constituent of the macrolide antibiotic angolamycin (*10, 14, 49*). In contrast, the substituent in position 10, known as ring F, is *N,N*-dimethylvancosamine (2,3,6-trideoxy-3-(dimethylamino)-3-*C*-methyl-L-*lyxo*-hexose), the dimethyl derivative of vancosamine (*43, 98*), which was found to be part of the glycopeptide antibiotic vancomycin (*62*). In the naturally occurring pluramycins, the hydroxyl group of ring E has so far never been found to be acetylated; in contrast, the ring F hydroxyl group may or may not be acetylated. The atoms of rings E and F are numbered starting with the oxygen atom, using primed numbers for ring E and doubly primed numbers for ring F.

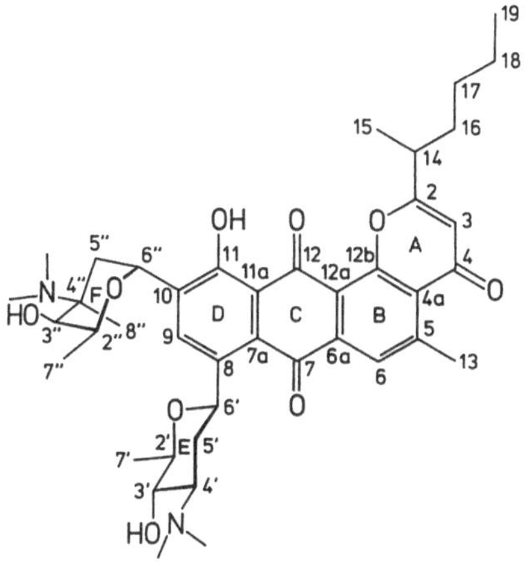

Fig. 1. General formula of a pluramycin antibiotic

This is consistent with the numbering of tetrahydropyrans. In contrast, the carbohydrate numbering is used for these two rings in the Chemical Abstracts entries (see below).

Trivial names together with the numbering scheme outlined above and indicated in Fig. 1 are usually used for the pluramycin antibiotics, since their systematic names are far too cumbersome. Hedamycin, *e.g.*, is listed in Chemical Abstracts as 2-(3,3′-dimethyl[2,2′-bioxiran]-3-yl)-11-hydroxy-5-methyl-8-[2,3,6-trideoxy-3-(dimethylamino)-β-D-*arabino*-hexopyranosyl]-10-[2,3,6-trideoxy-3-(dimethylamino)-3-*C*-methyl-α-L-*lyxo*-hexopyranosyl]-4*H*-anthra[1,2-*b*]pyran-4,7,12-trione. Note that the numbering scheme for the C(2)-side chain in this systematic name follows the IUPAC rules and for rings E and F is according to the carbohydrate nomenclature. In this review, the numbering given in Fig. 1 will be used exclusively, since all papers dealing with pluramycins have adopted it.

Although chromoxymycin does not contain the 4*H*-anthra[1,2-*b*]pyran-4,7,12-trione chromophore, and thus – strictly speaking – is not a pluramycin antibiotic, its relation to this class of compounds is so obvious that chromoxymycin is included in this review.

Pluramycins lacking the two sugar rings are called pluramycinones or are referred to as the 'aglycones' of the corresponding antibiotics. Strictly speaking, these 'aglycones' are not part of the antibiotic family, since they have no antimicrobial activity. But as they are closely related to the pluramycins, they are also included in this review. Unfortunately, the pluramycinones, which sometimes can be found in the fermentation broths besides the glycosylated compounds, have usually not been in-

Table 2. *Some Naturally Occurring Pluramycin Antibiotics and Selected Derivatives*

Compounds with a four-carbon side chain

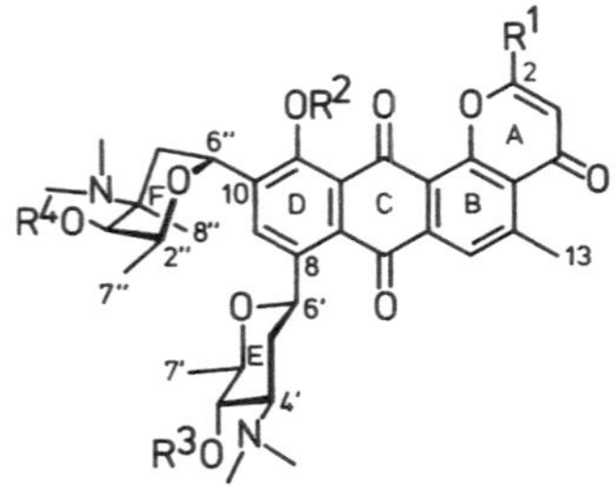

Trivial names	R¹	R²	R³	R⁴	Relative configurations within the side chain ᵃ	Molecular formula and molecular weight	Natural product	References
Kidamycin (1) Rubiflavin B	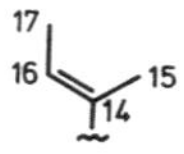 [see structure]	H	H	H	ok	$C_{39}H_{48}N_2O_9$ 688.82	yes	(28, 44) (65)
Neopluramycin (2)		H	H	Ac	ok	$C_{41}H_{50}N_2O_{10}$ 730.86	yes	(50, 51)
11,3''-Diacetylkidamycin (3) Kidamycin 11,3''-diacetate		Ac	H	Ac	ok	$C_{43}H_{52}N_2O_{11}$ 772.89	no	(28)
3',3''-Diacetylkidamycin (4) Kidamycin 3',3''-diacetate		H	Ac	Ac	ok	$C_{43}H_{52}N_2O_{11}$ 772.89	no	(28)
11,3',3''-Triacetylkidamycin (5) Kidamycin 11,3',3''-triacetate Acetylkidamycin Diacetylneopluramycin		Ac	Ac	Ac	ok	$C_{45}H_{54}N_2O_{12}$ 814.93	no	(28, 45) (50)
Epoxykidamycin (6) Largomycin FII chromophor constituent 4	 [see structure]	H	H	H	ok	$C_{39}H_{48}N_2O_{10}$ 704.82	yes	(11)

ᵃ The configurations of the side chain carbon atoms relative to those in rings E and F have only been determined for kidamycin (1) and are thus known only for this compound and its direct derivatives.

Table 2 (*continued*)

Compounds with a six-carbon side chain

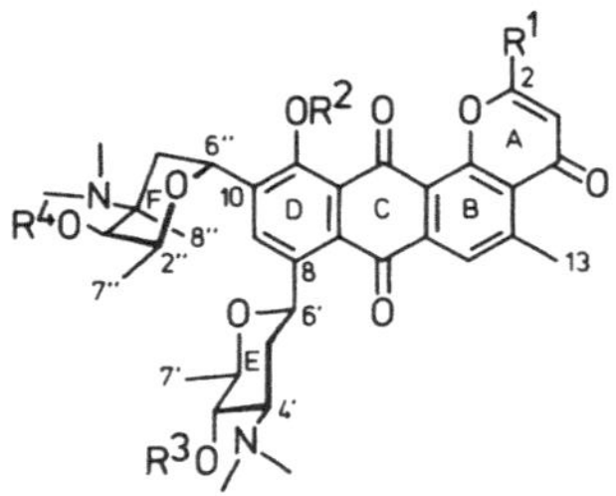

Trivial names	R¹	R²	R³	R⁴	Relative configurations within the side chain[b]	Molecular formula and molecular weight	Natural product	References
Rubiflavin C-1 (7)	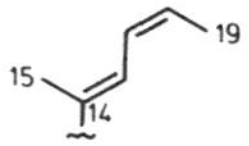	H	H	H	ok	$C_{41}H_{50}N_2O_9$ 714.86	yes	(65)
Rubiflavin C-2 (8)	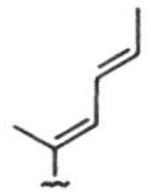	H	H	H	ok	$C_{41}H_{50}N_2O_9$ 714.86	?	(65)
Rubiflavin D (9)	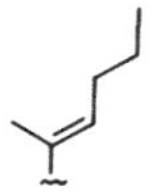	H	H	H	ok	$C_{41}H_{52}N_2O_9$ 716.87	yes	(65)

Table 2 (continued)

Trivial names	R^1	R^2	R^3	R^4	Relative configurations within the side chain[b]	Molecular formula and molecular weight	Natural product	References
Rubiflavin A (10) Desacetylpluramycin A Largomycin FII chromophore component 7	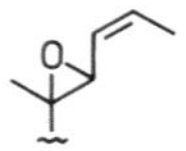 (structure)	H	H	H	ok	$C_{41}H_{50}N_2O_{10}$ 730.86	yes	(33, 67)
Pluramycin A (11) Largomycin FII chromophore component 8	(structure)	H	H	Ac	ok	$C_{43}H_{52}N_2O_{11}$ 772.89	yes	(13, 50) (33)
11,3'-Diacetylpluramycin A (12) 11,3',3''-Triacetylrubiflavin A	(structure)	Ac	Ac	Ac	ok	$C_{47}H_{56}N_2O_{13}$ 856.97	no	(50) (65)
Rubiflavin E (13)	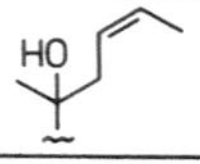 (structure)	H	H	H	ok	$C_{41}H_{52}N_2O_{10}$ 732.87	yes	(65)
PD 121,222 (14)	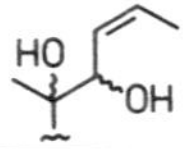 (structure)	H	H	H	ok	$C_{41}H_{52}N_2O_{11}$ 748.87	yes	(68)
11,16,3',3''-Tetraacetyl-PD 121,222 (15)	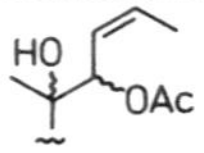 (structure)	Ac	Ac	Ac	ok	$C_{49}H_{60}N_2O_{15}$ 917.02	no	(68)
11,15,16,3',3''-Pentaacetyl-PD 121,222 (16)	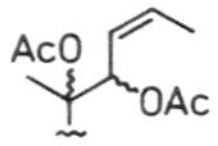 (structure)	Ac	Ac	Ac	ok	$C_{51}H_{62}N_2O_{16}$ 959.05	no	(68)
Hedamycin (17)	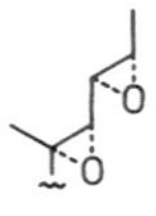 (structure)	H	H	H	ok	$C_{41}H_{50}N_2O_{11}$ 746.85	yes	(79, 82) (105)
11,3',3''-Triacetylhedamycin (18)	(structure)	Ac	Ac	Ac	ok	$C_{47}H_{56}N_2O_{14}$ 872.97	no	(105)

Table 2 (*continued*)

Trivial names	R^1	R^2	R^3	R^4	Relative configurations within the side chain[b]	Molecular formula and molecular weight	Natural product	References
Chromoxymycin (19)	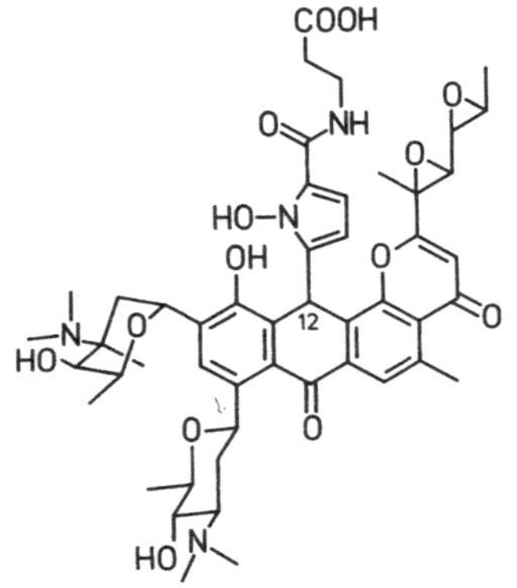				?	C$_{49}$H$_{60}$N$_4$O$_{14}$ 929.02	yes	(47)

[b] The configurations of the side chain carbon atoms relative to those in rings E and F have only been determined for hedamycin (17) and are thus known only for this compound and its direct derivatives.

Table 3. *Pluramycinones*

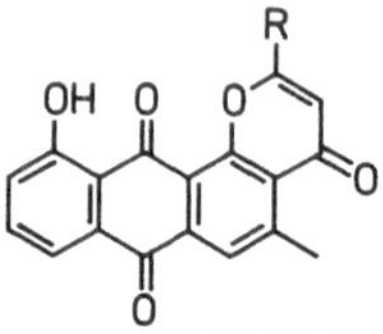

Trivial names	R	Relative configurations within the side chain	Molecular formula and molecular weight	Natural product	References
Rubiflavinone C-1 (**20**)	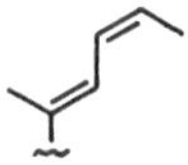	ok	$C_{24}H_{18}O_5$ 386.40	yes	(*65*)
Rubiflavinone C-2 (**21**)	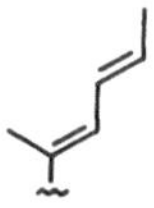	ok	$C_{24}H_{18}O_5$ 386.40	?	(*65*)
α-Indomycinone (**22**)	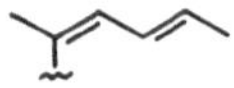	?	$C_{24}H_{18}O_5$ 386.40	yes	(*9, 26*)
Dihydro-α-indo-mycinone (**23**)	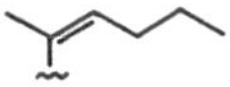	?	$C_{24}H_{20}O_5$ 388.42	no	(*9, 26*)
Tetrahydro-α-indo-mycinone (**24**)	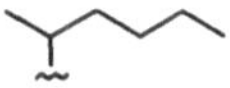		$C_{24}H_{22}O_5$ 390.44	no	(*9, 26*)
β-Indomycinone (**25**)	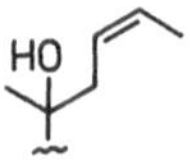	ok	$C_{24}H_{20}O_6$ 404.42	yes	(*15, 26*)

vestigated due to their biological inactivity. Structural elucidations in the pluramycin group are, however, much facilitated when these compounds are also investigated: their NMR spectra are simpler and thus easier to interpret than the spectra of the parent antibiotics. The side chain resonances are not obscured by the resonances of the many sugar protons.

The structures of the most important naturally occurring pluramycins and some of their derivatives, as well as some pluramycinones are presented in Tables 2 and 3.

3. The Families of Pluramycin Antibiotics

3.1. Pluramycin A and Neopluramycin

The pluramycins A and B were the first antibiotics of this class of anthra[1,2-*b*]pyrans to be described. In 1956 MAEDA and coworkers found the culture filtrate of the species *Streptomyces pluricolorescens* obtained from a Japanese soil sample to be highly active against ascites type of Ehrlich carcinoma and HeLa cells (*57*). Two active fractions were obtained by countercurrent distribution. The one called pluramycin A carried most of the activity and was therefore purified and investigated further. The other portion, pluramycin B, was less active and was not investigated further. Pluramycin A (**11**) has since become some sort of a standard. It was included in many comparative studies of antitumor active and antimicrobial compounds so as to represent the class of anthrapyran intercalative substances.

Pluramycin A (**11**) has a six carbon side chain with an epoxide adjacent to the pyrone ring and an (*Z*)-double bond between C(17) and C(18). The hydroxyl group of ring F is acetylated. KONDO *et al.* (*50*) determined the constitution, but assigned wrong configurations to the side chain carbons; these were corrected later by CERONI *et al.* (*85*).

Fifteen years after the detection of pluramycin A, KONDO, MAEDA and coworkers isolated another antitumor antibiotic from a strain of Streptomyces pluricolorescens which was different from the strain producing pluramycin A. The new compound was named neopluramycin (**2**) (*51*) and proved to be the 3″-*O*-acetyl derivative of kidamycin (**1**).

No derivatives were prepared from pluramycin A and neopluramycin with the exception of the peracetates which were used for the structural analysis.

3.2. Kidamycin and Isokidamycin

Kidamycin (**1**) is a secondary metabolite of *Streptomyces phaeoverticillatus* (*44*). It is one of the best investigated puramycins and the first compound of this family of antibiotics whose structure was fully elucidated. The constitution and configuration were determined from chemical and NMR spectroscopical investigations (*28*) in conjunction with X-ray analyses of the bis(*m*-bromobenzoyl) and bis(*m*-iodobenzoyl) derivatives (**28**) and (**29**) of isokidamycin (**26**) (*29*) (see Scheme 1) as well as of the rather peculiar kidamycin derivative (**30**), where C(12) is in fact the acetylated methyl hemiketal of the parent quinone (*30*). The fact that the structure of kidamycin had been determined unambig-

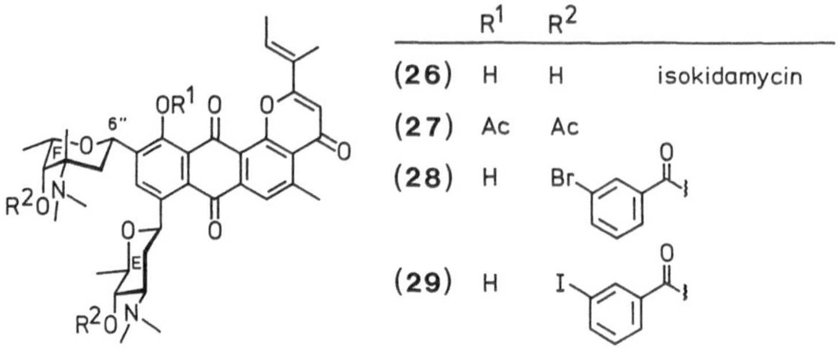

Scheme 1. Isokidamycin and derivatives

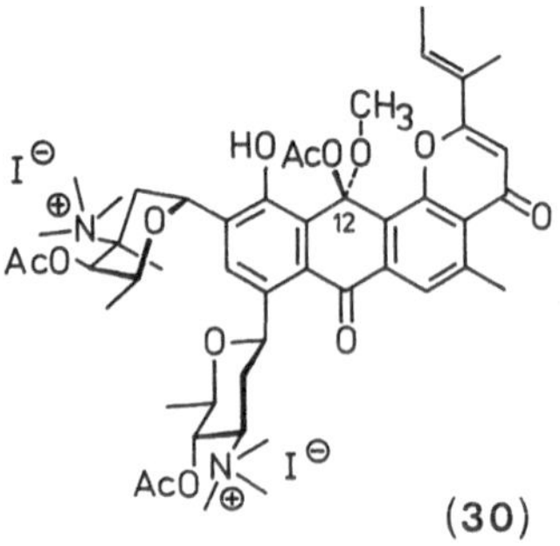

(30)

uously was of great use in all subsequent structure determinations of pluramycin antibiotics. Thus the constitution and configuration of hedamycin (17) could not have been established without a thorough spectral comparison with kidamycin (82) (105).

Kidamycin (1) is the only pluramycin antibiotic where an effort was made to reduce its toxicity and thus get a better therapeutic index by derivatization. Two partially acetylated compounds, (3) and (4), as well as the peracetate (5) were synthesized. It is noteworthy, however, that the derivative in which only the hydroxyl group of ring F was acetylated was never obtained. This particular substance was later found as a natural product, neopluramycin (2) (50).

Isokidamycin (26), the C(6″) epimer of kidamycin, was formed when kidamycin (1) was treated with p-toluenesulfonic acid in refluxing chloroform. The compound proved to be rather interesting in connection with the structure determination of kidamycin (29), the conformational analysis of ring F (86), and some first experiments towards an understanding of the structure-activity relationship with the pluramycins (24).

3.3. Hedamycin

Hedamycin (**17**) was isolated from the fermentation broth of *Streptomyces griseoruber* strain C-1150 (ATCC 15422). Only a few purification steps were necessary to obtain relatively large amounts of pure, crystalline hedamycin (*79*). The rather large quantities of hedamycin made thus available are certainly responsible for the fact that this compound was extensively investigated. Its constitution was determined from a detailed spectral comparison with kidamycin (**1**) (*82*). Finally, an X-ray structure determination served to establish the configurations and to corroborate the constitution. In contrast to kidamycin, where the X-ray structure determinations were carried out using heavy atom derivatives, crystals of hedamycin itself could finally be grown that were suitable for this analysis (*105*).

With hedamycin (**17**), the photodegradation of the pluramycin antibiotics was studied; after irradiation of this compound in solution with

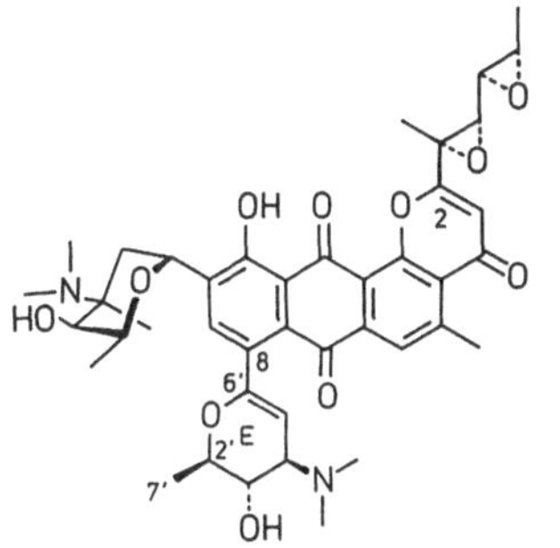

photohedamycin A   (**31**)

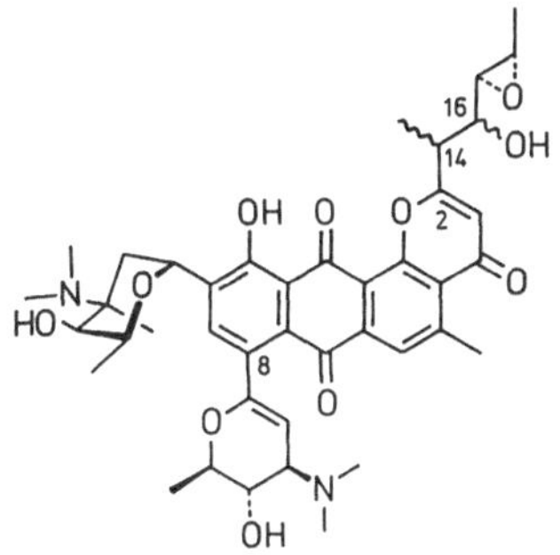

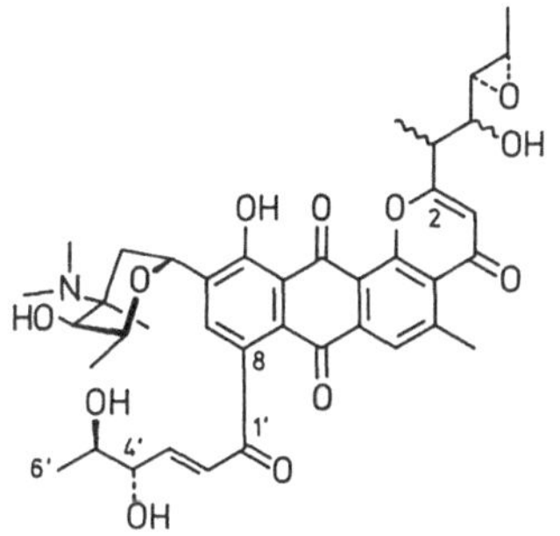

photohedamycin B   (**32**)
photohedamycin C   (**33**)

photohedamycin D   (**34**)

Scheme 2. The photohedamycins

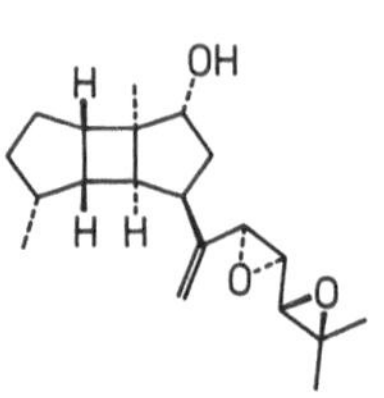

spatol   (35)

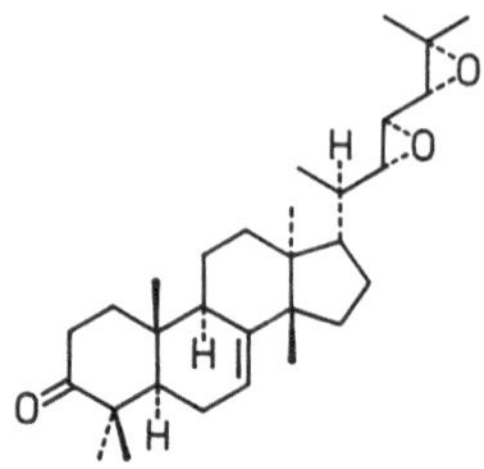

cneorin-NP$_{36}$ (36)

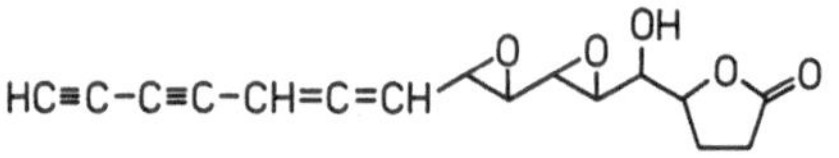

$HC{\equiv}C-C{\equiv}C-CH{=}C{=}CH$

cepacin B   (37)

Scheme 3. Natural products with open chain 1,2:3,4-diepoxides

UV light, several degradation products, the photohedamycins A (31), -B (32), -C (33), and -D (34), could be isolated (23). Their structures were determined (see Scheme 2), and a comparison of the cytotoxic activities of hedamycin and one of the photodegradation products, (31), shed additional light on structure-activity relationships in the pluramycin antibiotics (24) (see section 7.1.).

Hedamycin proved to have a 1,2:3,4 diepoxide (bioxiran) in the side chain. Such diepoxides have been found quite often in structures of natural products, but in most cases at least one of the epoxides was involved in an additional ring. By contrast, the open chain form of the bioxiran as in hedamycin had never before been observed in a natural product. Recently, this unique structural element was detected in four additional natural products (see Scheme 3): in spatol (35), a diterpenoid from the seaweed *Spatoglossum schmittii* (31, 32), in cneorin-NP$_{36}$ (36) from *Neochamaelea pulverulenta* (63), in cepacin B (37), an acetylenic antibiotic from *Pseudomonas cepacia* SC 11783 (76) and in chromoxymycin (19) (47) (see Table 2).

3.4. The Rubiflavins and Rubiflavinones

In 1965, the isolation of rubiflavin from a *Streptomyces* species (SC 3728) and some preliminary structural work on rubiflavin were published (1). Rubiflavin was obtained in amorphous form after an elaborate purification process involving precipitation as the hydrochloride and countercurrent distribution. It was thought to be uniform, and a molecular weight of ca. 400 was assigned from titrations, elemental analysis and ultracentrifugation.

Many years later, the investigation of a prepurified sample of rubiflavin by HPLC and NMR-spectroscopy clearly indicated that it was a complex mixture (67). So far, nine components have been identified from this mixture. Seven of them were pluramycin antibiotics and were called rubiflavins A (10), -B (1), -C-1 (7), -C-2 (8), -D (9), -E (13), and -F (26); none of them was acetylated in any position. The two remaining components were identified as the aglycones rubiflavinone C-1 (20) and rubiflavinone C-2 (21) (65).

The rubiflavins B and -F could be identified as kidamycin (1) (67) and isokidamycin (26) (65), respectively. It is not clear, whether the isokidamycin found is a genuine metabolite of the *Streptomyces* or whether it is an artifact generated during the isolation and workup procedure.

Rubiflavin A (10) was found to be almost identical with pluramycin A (11), the only difference was the non-acetylated C(3'')-hydroxy group in rubiflavin A. The chromophore, ring E and the side chain were clearly the same as in pluramycin A. The configurations of the epoxide and the double bond in the side chain were determined unambiguously from the ^{1}H- and ^{13}C-NMR spectra (67). Peracetylation of rubiflavin A (10) gave a compound whose ^{1}H-NMR spectrum was virtually the same as that of diacetyl pluramycin A (12) (65). This fact further corroborated the structural revision of pluramycin A, which was implicated from a comparison of the spectral data of pluramycin A and related model compounds (85).

The two rubiflavins C-1 (7) and -C-2 (8) are stereoisomers. They both contain a conjugated diene system in the side chain. The two rubiflavinones C-1 (20) and -C-2 (21) are the corresponding aglycones which were isolated from the same prepurified rubiflavin mixture as the glycosylated antibiotics (65). This is astonishing, since the antibiotic mixture was obtained from the fermentation broth by precipitation steps with HCl. We have no explanation for the appearance of the nitrogen-free aglycones after such treatment. In rubiflavin C-1 (7) and the corresponding aglycone (20) the terminal olefin has the (Z)-configuration, whereas in rubiflavin C-2 (8) and its aglycone (21) this double bond has the (E)-arrangement. Rubiflavinone C-1 shows in solution a slow transition into the more stable isomer rubiflavinone C-2. We assume that this same isomerisation takes also place with rubiflavin C-1, but the phenomenon was not observed (nor was it looked for). The question whether rubiflavin C-2 (8) and its aglycone (21) are actually artifacts formed later from the biologically produced rubiflavin C-1

(7) and rubiflavinone C-1 (20), respectively, cannot be answered yet (65). The rubiflavinones C-1 and C-2 are, of course, reminiscent of α-indomycinone (22) (9). Unfortunately, no NMR data were recorded for α-indomycinone, so that a comparison between these compounds was not possible.

Rubiflavin D (9) is a compound which is formally derived from the rubiflavins C by hydrogenation of the C(17)-C(18) double bond. This particular side chain had not yet been found in pluramycins. However, a related compound, dihydro-α-indomycinone (23), was obtained by partial hydrogenation of α-indomycinone (22) in BROCKMANN's laboratory (9, 16). The structure elucidation of rubiflavin D (9) was difficult inasmuch as the protons at the saturated carbon atoms of the side chain were hidden under the dimethylamino and sugar methyl resonances. Difference spectroscopy proved successful in revealing these resonances (65).

The last compound so far investigated from the rubiflavin mixture is rubiflavin E (13). Its side chain has a (Z)-double bond between C(17) and C(18) and a tertiary hydroxy group at C(14). This same side chain was found many years ago in β-indomycinone (25) (26). The NMR data of this partial structure are virtually identical for β-indomycinone (25) (15) and rubiflavin E (13) (65).

3.5. Largomycin FII Chromophore Constituents

Largomycin FII is a protein antibiotic that was isolated from *Streptomyces pluricolorescens* and exhibits antitumor activity (103). It possesses a non-protein chromophore, which is antitumor active as well and consists of a mixture of pluramycin antibiotics. In this respect largomycin FII resembles plurallin, also a protein antibiotic with a chromophore that was thought to belong to the pluramycins (72), but was never investigated more closely.

The largomycin FII chromophore could be separated from the protein and was then subjected to HPLC; eight major constituents were thus detected (33). Three of them, the constituents no. 4, 7, and 8, have so far been elucidated structurally. The component 4 proved to be the most interesting, since it was a new compound, epoxykidamycin (6) (11). Components 7 and 8, on the other hand, were recognized as rubiflavin A (10) and pluramycin A (11), respectively (33).

3.6. PD 121,222

Recently, a compound called PD 121,222 (14) was isolated from *Streptomyces* sp. WP 0123. This antibiotic formally corresponds to rubiflavin A (10) where the epoxide in the side chain has been opened hydrolytically to the corresponding diol. The diol structure was derived from NMR spectroscopic investigations as well as from acetylation (leading to (15) and (16)) and from cleavage with periodic acid (68). The configurations at C(14) and C(16) in the side chain of PD 121,222 could, however, not be determined.

3.7. Chromoxymycin

Chromoxymycin (19) is a new antibiotic that was recently isolated by HORI and coworkers from *Streptomyces libani* subsp. *rubropurpureus* obtained from a Japanese soil sample (39). Although the structure of chromoxymycin has not yet been fully elucidated – only the constitution was established, the configurations remain yet to be determined – the compound seems to be a very close relative of hedamycin (17) (47). The surgar rings E and F have the same constitution as in hedamycin; the detailed NMR spectral analysis carried out by SETO and coworkers revealed, however, for ring F a chair conformation in the solvent pyridine rather than the twist conformation found for the same ring of hedamycin in chloroform. The side chain contains the diepoxide as in hedamycin, whereas the chromophore is different from that of the true anthra[1,2-b]pyran antibiotics inasmuch as C(12) is not a carbonyl group but a methine carrying a substituent consisting of N-hydroxypyrrole-2-carboxylic acid and β-alanine. Chromoxymycin (19) is thus the first natural product related to the pluramycins where one of the anthraquinone carbonyls is reduced.

3.8. The Griseorubins

The antibiotic complex called the griseorubins was isolated from a *Streptomyces fimicarius* strain which was obtained from the shrimp *Crangon crangon* L. of the Baltic Sea (18). A very elaborate separation procedure lead to eight fractions named griseorubins A through H. TLC of griseorubins A through E shows that each fraction is still a mixture of five to ten compounds. Fraction E was the most active and was thus purified further, again yielding eight fractions, griseorubins E1 through E8. No structures have so far been fully elucidated. Griseorubin E1, however closely resembles kidamycin (1). It differs from this compound by its behaviour on TLC and by its biological properties (17, 18).

3.9. The Indomycins and Indomycinones

The results of all the research on the indomycines and indomycinones has, unfortunately, only been published in doctoral theses from BROCKMANN's laboratory (*26, 80*) and in an abstract of a lecture given by BROCKMANN in 1968 (*9*). This is particularly regrettable, since in this group the aglycones were also isolated and investigated.

Tedious chromatography of the crude material gave three fractions, called indomycins A, -B, and -C, respectively. The former two were amorphous, whereas the latter was crystalline. Two additional compounds could be isolated; they contained no nitrogen and were the α- and β-indomycinones (**22**) and (**25**) (*80*). The constitution of (**22**) could be derived by DAHM from the products obtained after ozonolysis of the dihydro and tetrahydro derivatives (**23**) and (**24**) (*9, 16*); the constitution of (**25**) was determined spectroscopically (*15*).

The structures of the indomycines could not be determined. However, hints as to the nature of several structural elements can be found in FRICKE's thesis, such as the possibility of deoxysugars bound C-glycosidically to the anthra[1,2-*b*]pyran skeleton (*26*). Strange is, however, the observation, that the indomycines were only slightly soluble in the usual solvents used for NMR-spectroscopy so that recording of a spectrum was not possible. This statement contrasts with one made elsewhere in the thesis cited, that the indomycins are readily soluble in chloroform and butanol when in the form of the free base (but not in the salt form).

4. Chemical Properties

4.1. Separation Techniques

It is quite rare that a microorganism produces just one single pluramycin antibiotic; usually these substances occur in families of structurally (and biogenetically) very closely related compounds, which differ in the constitutions of the side chains at C(2) and sometimes in the degree of acetylation. The whole trunk of the molecules is thus identical. Hence the structural differences amenable to separation techniques are rather small and the isolation of pure compounds is accordingly very difficult. Compounds, which earlier were thought to be uniform now prove to be mixtures when subjected to modern chromatographic techniques such as HPLC (*67*).

Hedamycin (**17**) (*79*) was an unusual case, where the antibiotic could be obtained even in crystalline form with comparatively small effort.

The rubiflavins (*65*), the griseorubins (*18*) and the indomycins (*26*) on the other hand are typical examples where separation of the compounds and isolation of more or less pure single constituents was extremely difficult. The separation problems are enhanced by the chemical instability and photolability of the puramycins.

4.1.1. Preparative Separations

The pluramycin antibiotics were usually obtained by extraction of the fermentation broth and/or the mycelium with butanol, chloroform or ethyl acetate. This organic phase was then extracted with acidic water, which was made neutral or slightly basic again as soon as possible. After that, the pluramycin bases were again extracted into an organic phase such as chloroform or ethyl acetate. This procedure of acid extraction followed by basification and extraction of the pluramycin bases into an organic solvent was sometimes repeated once or twice at different pH values (*18, 68*).

The mixture obtained after removal of the solvent then had to be fractionated further. At the time, when pluramycin A (**11**), hedamycin (**17**) and the rubiflavin mixture were first detected, countercurrent distribution between ethyl acetate and an aqueous phosphate buffer solution was frequently used (*1, 57, 79*). In more recent reports, chromatographic procedures prevail. HPLC proved to be a particularly valuable tool, but even with today's elaborate techniques several successive chromatographic steps are usually necessary.

Sephadex LH 20 proved to be quite useful for the purification of pluramycin antibiotics (*18, 68*). The column particularly retains polar components (which may either be genuine or degradation products). Fractionation can be achieved by elution with dichloromethane containing increasing amounts of methanol (*65*).

Crude separations can normally be achieved on silica gel either with regular columns or with preparative HPLC. However, the solvents often have to contain a base to make the pluramycins move on the acidic silica gel (*67*); an alternate method is the pretreatment of the adsorbent with $NaHCO_3$ (*18*). Alumina with chloroform-ether mixtures or chloroform alone was successfully used in the purification of hedamycin (**17**) (*79*) and kidamycin (**1**) (*44*), respectively, but proved to be less satisfactory when used with ethyl acetate for pluramycin A (**11**) (*57*).

The fractions so obtained can then be separated into their components by reversed phase HPLC (*11, 33*). Paired ion chromatography also proved to be very useful; its disadvantage is that the pluramycin bases have finally to be separated from the ion pairing salt, which necessitates another HPLC run on silica gel (*65*) or an extraction step (*33, 68*). It must be pointed out that some of these preparative separations can be very tedious. They often have to be carried out with analytical HPLC columns; to avoid overload of the column, which would degrade its resolving power, many repeated chromatograms have to be run at the submilligram scale; corresponding fractions may then be pooled.

It can clearly be seen from the above review of the literature that there is no standard procedure for the workup of a pluramycin antibiotic mixture.

Table 4. *Selection of TLC-Systems Suitable for Pluramycin Antibiotics*

Adsorbent	Solvent system	Compounds	R_f values	Ref.
SiO$_2$	EtOH, 28% NH$_4$OH, H$_2$O 8:1:1	Neopluramycin (2)	0.8–0.9	(*51*)
	EtOH, 14% NH$_4$OH 4:1	Kidamycin (1)	0.81	(*44*)
	EtOH, pyridine 4:1	Kidamycin (1)	0.06	(*44*)
		Triacetylkidamycin (5)	0.55	(*44*)
	Toluene, Et$_3$N 4:1	Kidamycin (1)	0.29	(*65*)
		Hedamycin (17)	0.25	(*65*)
		Rubiflavins	0.14–0.33	(*65*)
	CHCl$_3$, MeOH, 1.5 M NH$_4$OAc pH 9.5 60:40:5	PD 121,222 (14)	0.37	(*25*)
		Kidamycin (1)	0.45	(*25*)
		Neopluramycin (2)	0.85	(*25*)
	CHCl$_3$, Et$_3$N 4:1	PD 121,222 (14)	0.40	(*68*)
		Hedamycin (17)	0.64	(*65*)
		Kidamycin (1)	0.66	(*65*)
	n-BuOH, AcOH, H$_2$O 4:1:1	Neopluramycin (2)	0.25–0.3	(*51*)
SiO$_2$	CH$_2$Cl$_2$, MeOH 9:1	Triacetylhedamycin (18)	0.46	(*65*)
SiO$_2$ treated with NaHCO$_3$	CHCl$_3$, MeOH, 17% NH$_4$OH 2:1:1, lower phase	Griseorubins		(*18*)
	EtOH, 17% NH$_4$OH 4:1	Griseorubins		(*18*)
Al$_2$O$_3$ N	nBuOH, MeOH, H$_2$O 4:1:2	Griseorubins		(*18*)
Al$_2$O$_3$ paper	EtOAc sat. with H$_2$O	Kidamycin (1)	0.1	(*44*)
		Triacetylkidamycin (5)	1.0	(*44*)

4.1.2. Analytical Separations

It is by no means easy to prove that a pluramycin antibiotic on hand is indeed a single, pure compound. TLC and analytical HPLC may be used to check the purity of a pluramycin sample.

Pluramycins hardly move ($R_f = 0$–0.2) on silica gel TLC plates when simple organic eluents are used such as chloroform-methanol mixtures, acetone and the like (*51*). The solvent systems should contain a base such as ammonia, an aqueous buffer, triethylamine or pyridine. A selection of suitable TLC-systems is given in Table 4.

Analytical HPLC is of course also well suited for the investigation of the pluramycins. The procedures described above for the preparative separations can be used.

4.2. Spectroscopy

In the field of antibiotic research, where many compounds are isolated, then tested primarily for their antimicrobial activity and eventually investigated structurally, spectroscopy has a special importance. Spectra characterize a substance and often show to which class of compounds it belongs. This classification can many times be done in a straightforward manner merely from the visual aspect of the spectra. Since the overall aspect of a spectrum can be grasped much faster from an illustration than from mere values in a table, some typical spectra are shown in the Figures below.

4.2.1. UV/VIS Spectra

The overall aspect of the UV/VIS spectrum of a pluramycin is quite characteristic (cf. Fig. 2). The spectra show a prominent main absorption band around 245 nm (ε = ca. 45 000) with some shoulders, and a weaker long wavelength band at ca. 435 nm (ε = ca. 9000). The spectra do not give much information about the structure of the compound on hand, however. Two features must be mentioned: (1) the pH dependence of the spectra and (2) their dependence on the nature of the side chain at C(2).

The pH dependence of the UV/VIS spectra of the pluramycins is very obvious. Solutions of these compounds turn from orange or yellow at neutral pH to a vivid purple when made alkaline. This, of course, is typical for 1-hydroxy anthraquinone derivatives. The main absorption band is shifted from 245 to 257 nm, whereas the long wavelength

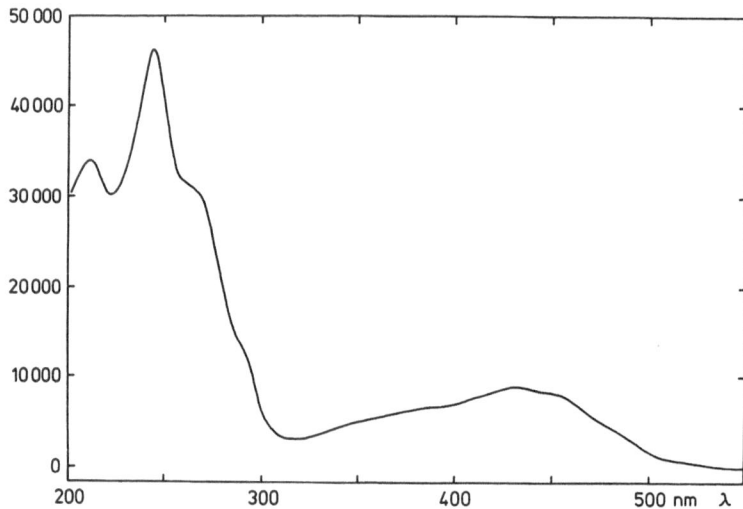

Fig. 2. UV/VIS Spectrum of hedamycin (17) in ethanol

band of the pluramycin moves from around 435 nm in neutral to ca. 550 nm in alkaline solution. Spectra taken in acidified ethanol, on the other hand, are hardly distinguishable from those measured in neutral solution. It is further noteworthy that the spectra show only a minor dependence on the solvent; measurements in ethanol, methanol or methanol/dioxane are virtually the same.

The position of the shoulder around 270 nm is sensitive to the nature of the side chain at C(2). When this side chain is not conjugated with the pyrone as *e.g.* in hedamycin (17), pluramycin A (11) and similar compounds, the shoulder appears at 263–268 nm. It is shifted towards longer wavelengths by ca. 5 or ca. 15 nm when one or two double bonds in the side chain, respectively, are conjugated with the pyrone ring. This observation was already made by FRICKE (26) in an early comparative study comprising the α- and β-indomycinones (22) and (25) and the two hydrogenated derivatives (23) and (24). These data were rather helpful in the structural elucidation of hedamycin (17) (82).

Spectral data of some representative pluramycin antibiotics are compiled in Table 5; the rubiflavinones C-1 (20) and C-2 (21) are included for comparison.

Table 5. UV/VIS Spectra of Some Pluramycin Antibiotics and Pluramycinones

Compound	λ_{max} [a]								Solvent	Ref.
Rubiflavin C-1 (7)	204 (0.73)	230 (0.93)	245 (1.00)	280s (0.59)	313s (0.40)		408 (0.23)	429 (0.23)	EtOH	(65)
Rubiflavin C-2 (8)	203 (0.69)	230 (0.98)	245 (1.00)	281s (0.63)	311s (0.45)		400 (0.23)	431 (0.23)	EtOH	(65)
Rubiflavin D (9)	203 (0.71)	214 (0.79)	244 (1.00)	268s (0.75)				434 (0.18)	EtOH	(65)
Rubiflavin A (10)	203 (0.79)		245 (1.00)	265s (0.64)	293s (0.28)			435 (0.18)	EtOH	(65)
Hedamycin (17)			244 (46200)	264s (29600)				434 (8900)	MeOH	(82)
Photohedamycin A (31)	213 (34000)		243 (48000)	263s (35000)		385 (8200)		422 (8900)	EtOH	(23)
Rubiflavin E (13)	204 (0.68)		244 (1.00)	265s (0.58)				434 (0.18)	EtOH	(65)
Kidamycin (1)	216 (37600)		243 (47000)	270s (33000)				434 (8500)	MeOH	(82)
Neopluramycin (2)	216 (0.81)		243 (1.00)	270 (0.71)				430 (0.21)	EtOH	(51)
Triacetylkidamycin (5)			238 (42900)	271 (42300)		364 (10500)			EtOH	(28)
Isokidamycin (26)			245 (44400)	272 (30800)		342s (6450)		426 (9400)	EtOH	(28)
Epoxykidamycin (6)			246 (1.00)	268s (0.67)	288s (0.32)			434 (0.17)	EtOH	(11)
Epoxykidamycin (6)			245 (1.00)	268s (0.61)	288s (0.32)			428 (0.18)	EtOH, 0.01N HCl	(11)
Epoxykidamycin (6)			257 (1.00)	283s (0.36)		335 (0.22)		555 (0.15)	EtOH, 0.01N NaOH	(11)
Kidamycin (1)	210 (130000)		258 (43500)			323 (14200)		543 (9500)	EtOH, 0.01N NaOH	(28)
Rubiflavinone C-1 (20) [b]	203 (0.75)	228 (1.00)	240s (0.89)	288 (0.62)	308s (0.48)		402 (0.27)		EtOH	(65)
Rubiflavinone C-2 (21) [b]	204 (0.54)	230 (1.00)	240s (0.90)	289 (0.67)	307s (0.54)		400 (0.29)		EtOH	(65)

[a] In parentheses: ε, or absorbance relative to the main band at ca. 245 nm; s = shoulder.
[b] Absorbances in this row relative to the band at ca. 230 nm.

4.2.2. IR Spectra

IR spectra do not give much insight in the structure of a newly isolated pluramycin antibiotic, but their overall appearance is quite characteristic for this class of compünds. The spectra of hedamycin (**17**) and kidamycin (**1**) are reproduced in Fig. 3 as examples. The carbonyl absorptions are rather typical for the different antibiotics. In hedamycin (**17**), a compound where the side chain at C(2) is not conjugated with the pyrone, the resonances of one of the quinone carbonyls and of that of the pyrone almost coincide at ca. 1650 cm^{-1}, whereas the hydrogen bonded carbonyl has its resonance at 1625 cm^{-1}. A similar band is observed around 1640–1625 cm^{-1} for kidamycin (**1**), where the C(2) side chain is conjugated with the pyrone. This conjugation shifts the pyrone C=O resonance towards lower frequencies thus broadening the whole carbonyl absorption (*82*). Acetylation of one or both sugar rings leads to the appearance of an ester carbonyl resonance at 1740 cm^{-1}; acetylation of the phenol to one at 1775 cm^{-1}. Furthermore, in this latter case one of the quinone bands is now observed at 1675 cm^{-1}, since it is no longer hydrogen bonded (*28*).

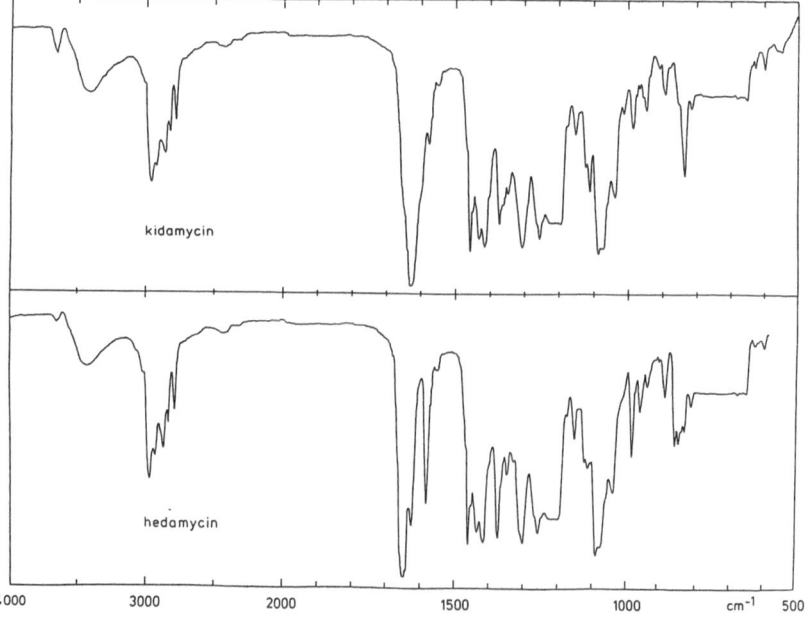

Fig. 3. IR Spectra of hedamycin (**17**) and kidamycin (**1**) in CHCl$_3$

4.2.3. ¹H-NMR Spectra

The ¹H-NMR spectra of the pluramycin antibiotics are dominated by the resonances of the aliphatic protons of the two sugars and of the side chain at C(2) (see Fig. 4). This region of the spectrum is rather crowded, and the signals heavily overlap at operating frequencies below 300 MHz. Thus, in less recent spectra, not all the chemical shifts could be determined unambiguously. The resonances of C(6′) and C(6″), of any olefinic hydrogens of the side chain and of the three protons attached to the anthra[1,2-*b*]pyran appear between 5 and 10 ppm, and the signal of the hydrogen bridged phenol is at lowest field (ca. 14 ppm). Table 6 contains the chemical shifts and coupling data for some typical pluramycin antibiotics.

The signal for H-C(9) is at a rather constant chemical shift of 8.33 ppm; the slight move observed upon epimerisation of ring F (cf. kidamycin (**1**) and isokidamycin (**26**)) does not seem to be significant as the resonance in triacetylisokidamycin (**27**) is again at 8.33 ppm (*28*) (*65*). In the photohedamycins, however, where ring E is an enol ether and thus conjugated with the anthra[1,2-*b*]pyran system, the resonance of H-C(9) is distinctly shifted upfield to 7.72 ppm.

The hydrogen atom at C(6) has its resonance at 7.95 ppm in compounds where the side chain is conjugated with the pyrone ring; the resonance is shifted slightly but significantly downfield to 8.00 ppm when there is no such conjugation. Acetylation of the phenol as in triacetylhedamycin (**18**) also influences the chemical shift of H-C(6):

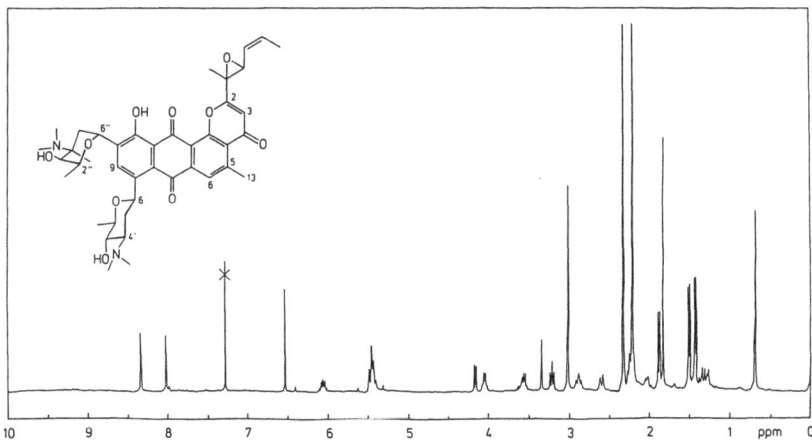

Fig. 4. 360 MHZ ¹H-NMR spectrum of rubiflavin A (**11**) in CDCl₃ ; the phenol resonance (14.1 ppm) is not shown (reprinted from (*67*) with permission)

Table 6. ^1H-NMR Spectra of Selected Pluramycin Antibiotics[a]

Position	Kidamycin (1) (86) 90 MHz	Isokidamycin (26) (65) 90 MHz	Expoxykidamycin (6) (11) 300 MHz	Rubiflavin A (10) (65) 360 MHz	PD 121,222 (14) (68) 200 MHz	Hedamycin (17) (86) 90 MHz	Triacetyl-hedamycin (18) (65) 90 MHz	Photohedamycin A (31) (23) 360 MHz
3	6.38s	6.38s	6.48s	6.52s	6.56s	6.46s	6.45s	6.46s
6	7.96d (0.6)	7.95s	8.00d	8.02s	8.02s	8.00d (0.6)	7.86d (0.5)	7.96d (1)
9	8.32s	8.40s	8.32d	8.34s	8.35s	8.33s	8.33s	7.72d (1)
11-OH	14.1 br	14.27s[b]	14.07s	14.1s	°	14.1br	–	13.8br
13	3.01s br	3.00s	3.00s	3.01s	3.02s	2.99s br	2.96d (0.5)	3.00s
15	2.00s	2.00s	1.83s	1.84s	1.71s	1.96s	1.88s	1.96s
16	7.49q (7) br	7.49q (7)	3.46q	4.16d (8)	4.86d (9.3)	3.32d (6)	3.41d (5.5)	3.33d (5)
17	2.04d (7)	2.02d	1.54d	5.43m (11/8/2)	5.43ddd (11.2/9.3/1.5)	2.89dd (4.7/2.1)	2.86dd (5.5/2.2)	2.89dd (5/2)
18	–	–	–	6.06dq (11/7)	5.71dq (11.2/6.8)	3.11qd (5/2.2)	3.12qd (5/2.2)	3.11qd (5/2)
19	–	–	–	1.89dd (7/2)	1.66dd (6.8/1.5)	1.44d (5.3)	1.43d (5)	1.44d*[d]
2'	3.56m	3.54m (8)	3.58dq	3.57dq (8.7/6.3)	3.56m	3.55m	3.65dq (9/6)	4.33dq (9/6)
3'	3.20t (9)	3.35t (9)	3.21dd	3.21t (9)	3.21t (9)	3.19t (9)	4.90t (9)	3.68t (9)
4'	2.97m[b]	2.89m	2.98m	2.89ddd (11.5/9.3/3)	2.88m	2.93br	°	3.48dd (9/2)
5'	ca. 1.2 / ca. 2.6	° / °	1.40m / 2.27m	1.32q (11.5) / ca. 2.25	° / °	ca. 1.2* / ca. 2.5*	° / °	4.98d (2)
6'	5.48m	5.45d br (9)	5.45m	5.45m (10)	5.45m	5.45m	5.37m	–
7'	1.42d (5.9)	1.38d (6)	1.45d	1.43d (6.3)	1.43d (6.4)	1.43d (5.9)	1.31d (6)	1.43d*[d]
4'N(CH$_3$)$_2$	2.33s*	2.38s*	2.41s	2.33s*	2.33s*	2.32s*	2.32s*	2.42s

Table 6 (*continued*)

Position	Kidamycin (1) (86) 90 MHz	Isokidamycin (26) (65) 90 MHz	Expoxykidamycin (6) (11) 300 MHz	Rubiflavin A (10) (65) 360 MHz	PD 121,222 (14) (68) 200 MHz	Hedamycin (17) (86) 90 MHz	Triacetyl-hedamycin (18) (65) 90 MHz	Photohedamycin A (31) (23) 360 MHz
2″	4.05q br (6)	3.85q br (7)	4.08qd	4.05q br (6.3)	4.03q br (6.5)	4.04q br	4.33qd (6.5/5)	4.09qd (6/2)
3″	3.37d (3.5)[b]	3.30s br	3.42s br	3.34s br	3.34s br	3.35s br	5.22d (5)	3.44d (2)
5″	ca. 2.1	c	2.29m	ca. 2.25	c	ca. 2.1m*	c	2.17dd (14/6)
	ca. 2.6	c	2.58dd	2.60dd (14/3)	c	ca. 2.7m*	2.65m	2.32 dd (14/6)
6″	5.48m	4.91 d br (10.5)	5.45m	5.45m (10)	5.45m	5.45m	5.37m	5.43 t (6)
7″	1.50d (6.2)	1.49d (7)	1.51d	1.51d (6.3)	1.51d (6.4)	1.51 d (5.9)	1.43d (6)	1.45d*[d]
8″	0.70s	1.21s	0.78s	0.69s	0.70s	0.71s	0.99s	0.87s
4″N(CH$_3$)$_2$	2.21s*	2.22s*	2.31s	2.22s*	2.22s*	2.22s*	2.28s*	2.26s
Acetyl groups							2.14s 2.19s 2.51s	

[a] Chemical shifts in ppm relative to TMS, in parentheses: coupling constants in Hz. Solvent: CDCl$_3$. ^{1}H-NMR spectral data of the following compounds have also been published: neopluramycin (2) (50); triacetylkidamycin (5) (28); rubiflavin C-1 (7), -C-2 (8), -D (9), -E (13), triacetylrubiflavin A (12) (65); pluramycin A (11) (33); chromoxymycin (19) (47); triacetylisokidamycin (27) (28, 65); photohedamycins B (32), -C (33), -D (34) (23).

[b] Value from (28).

[c] Chemical shift not determined (in most cases due to signal overlap).

[d] Coupling constant not determined due to signal overlap.

* Similar assignments within a column may be interchanged.

the resonance now appears at 7.86 ppm. In well resolved spectra, the benzylic coupling of H-C(6) with the methyl group at C(5) can be observed. The resonance is then split into a narrow doublet with 0.5–1 Hz; the outer lines of the expected quartet are not observed.

The chemical shift of H-C(3) is most sensitive to the nature of the side chain at C(2). With side chains conjugated with the pyrone, this resonance is at 6.38 ppm whereas in non-conjugated compounds it is at 6.45–6.56 ppm. This signal, which does not overlap with anything else, is an excellent means for the detection of mixtures. A pure compound must have a clean, single line here. Crude rubiflavin or the chromatographic fraction containing the rubiflavins C-1, -C-2, and -D were clearly recognized as mixtures as they gave multiple lines for H-C(3) and also for H-C(6). Careful examination of the spectrum of rubiflavin A (10) (Fig. 4) shows that a slight contamination by a pluramycin antibiotic with a conjugated side chain is present (probably kidamycin (1) = rubiflavin B).

The resonances of H-C(6') and H-C(6'') heavily overlap and form a multiplet at 5.45 ppm. Acetylation of the two sugars moves both signals to 5.37 ppm. In compounds where only one of the two sugars is esterified, such as neopluramycin (2) or 11,3''-diacetylkidamycin (3), the two signals can be observed separately. This is also true for isokidamycin (26), where H-C(6'') is now in an axial position and therefore shifted upfield with respect to kidamycin. The signal is observed as a broadened doublet (J = 10.5 Hz) at 4.91 ppm. The resonance of H-C(6') can now also be observed unobstructed as a broadened doublet at 5.45 ppm with a 9 Hz splitting. On the other hand, in the photohedamycins A, -B, and -C (compounds (31, 32, and 33) see Scheme 1), where there is no H-C(6'), the resonance for H-C(6'') can clearly be seen; it is a triplet with a 6 Hz coupling constant.

In the upfield region of the spectrum, the methyl signals are easily detected. The aryl methyl group at C(5) gives rise to a singlet with a rather constant chemical shift of 3.00 ppm. The splitting due the benzylic coupling with H-C(6) was observed only in a few cases. The two tall singlets corresponding to the two dimethylamino grops are also "landmarks" of the spectrum; they appear between 2.2 and 2.4 ppm and are hardly influenced by structural changes. In contrast, the resonance of the methyl group at C(4''), which in pluramycins with non acetylated ring F is at quite high field (0.7–0.8 ppm), is distinctly shifted downfield to 1.0 and 1.2 ppm upon acetylation or epimerization of ring F, respectively. An additional singlet methyl resonance corresponding to the methyl group at C(14) of the side chain appears between 1.7 and 2 ppm; its chemical shift varies according to the constitution of the side chain. Acetylated pluramycins show in the same region, of course, the signals for the acetyl methyl groups (2.15–2.2 ppm for the sugars, and around 2.5 ppm for the phenol). The three doublet methyl resonances observed correspond to the methyl groups at C(2'), C(2'') and at the end of the side chain. The methyl group of ring E absorbs at 1.43 ppm or at 1.31 ppm, depending on whether the hydroxyl group at this ring is free or acetylated; the corresponding signals for the ring F methyl group are at 1.51 and 1.43 ppm. The terminal methyl group of the side chain occurs in the range of 1.4 to 2.05 ppm; the chemical shift varies with the constitution of the side chain. The three methyl doublets just discussed often overlap, especially when a low field NMR spectrometer is used. It was, however, possible to resolve these signals in the case of hedamycin (17) at 100 MHz when the spectrum was recorded in deuteropyridine (84).

The remaining protons of the sugars and of the side chain give often rise to complex resonances due to coupling with several neighbours. Most of these resonances have been assigned unambiguously, but the most problematic in this respect are the sugar methylene protons. Their multiply-split and therefore not very tall resonances are quite often hidden under large methyl signals. Assignments of these resonances in particular, and of the whole proton NMR spectra of pluramycins in general, have lately been much facilitated by the advent

Table 7. ¹H-NMR Spectra of some Pluramycinones[a]

Position	Kidamycinone methyl ether (56) (36) 100 MHz	β-Indo-mycinone (25) (15)	Rubiflavinone C-1 (20) (65) 250 MHz	Rubiflavinone C-2 (21) (65) 250 MHz
3	6.35 s	6.52 s	6.45 s	6.42 s
6	7.91 s	8.03 s	8.05 d (0.5)	8.04 d (0.5)
8	7.88 d (8)	7.79 q	7.84 dd (7.5/2)	7.84 dd (8/2)
9	7.68 t (9)	7.65 t	7.69 t (8)	7.68 t (7.5)
10	7.44 d (8)[b]	7.30 q	7.35 dd (8/2)	7.37 dd (8/2)
11-OH	–	13.20 s	13.3 s	13.0 s
13	2.98 s	2.99 s	3.02 d (0.5)	3.02 d (0.5)
15	2.01 s	1.69 s	2.07 s br	2.08 s br
16	7.2–7.5 m	2.85 m	8.54 d br (12)	7.96 d br (11)
17	2.04 d (6)	5.4 m (11)	6.54 ddq (12/11/1)	6.59 ddq (11/15/1)
18	–	5.7 m (11)	6.10 dq (11/7)	6.38 dq (15/7)
19	–	1.62 d	2.21 dd 67/1)	1.99 d br (7)

[a] Chemical shifts in ppm relative to TMS, in parentheses: coupling constants in Hz. Solvent: CDCl₃.
[b] Value from (37).

of the newer NMR techniques (6). Thus, the three largomycin FII chromophore constituents, whose structures were recently elucidated (11, 33), were all investigated with the aid of proton-proton correlated 2D spectra; the same is true for chromoxymycin (19) (47).

The difference spectroscopy technique, where the difference of the fully coupled and selectively decoupled spectra is recorded (58), was of great help for the identification of the resonances stemming from the side chains of the different rubiflavins (65). The detection of such side chain resonances and their interpretation is of course much easier with the pluramycinones, where there are no sugar resonances to obscure the picture. Unfortunately, only very few pluramycinones have been isolated and investigated so far; due to their biological inactivity they were mostly not even looked for in the fermentation broths. In Table 7 the chemical shifts of the known pluramycinones are compiled.

Proton NMR spectroscopy was not only used for the structural elucidation of the pluramycins, but also for the investigation of their configurations and conformation (see section 4.4.).

4.2.4. ^{13}C-NMR Spectra

Carbon-13 NMR spectroscopy is an excellent tool for structure elucidation of pluramycin antibiotics. The chemical shifts are very sensitive to even minor changes in the structures. The major drawback of this method is that even with today's high field spectrometers appreciable amounts – a few milligrams – of the substances are needed for good spectra. Five or ten years back the limit was around 15–20 mg for a broadband decoupled spectrum. This demand made measurement of a ^{13}C-NMR spectrum impossible for those natural pluramycins or derivatives that were only available in minute amounts. Off-resonance decoupled spectra were only rarely reported, since for their measurement even more substance or longer measuring time is needed.

The chemical shifts of some representative compounds are compiled in Table 8. The basis for the assignments given is an extensive study of hedamycin (17), kidamycin (1) and their derivatives (86) which provided all the assignments for the anthra[1,2-b]pyran skeleton and the two sugar rings in their free and acetylated forms. The assignments were made by comparison of the spectra within this series of compounds taking into account structural variations such as differences in acetylation or in the side chain at C(2). Off-resonance and selective proton decoupling – well established techniques in carbon magnetic resonance – were further used. The recently developed advanced NMR techniques, such as 2D heterocorrelated spectroscopy or the measure-

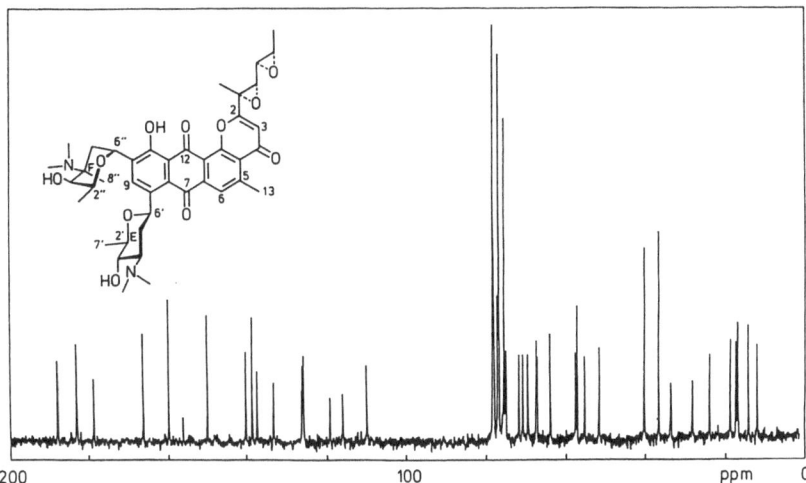

Fig. 5. 22.63 MHz ^{13}C-NMR spectrum of hedamycin (**17**) in CDCl$_3$

ment of carbon-carbon connectivities, have so far not been applied to the pluramycins due to the relativly large amount of material that is needed for these techniques.

The general aspect of the ^{13}C-NMR spectra is quite typical (see Fig. 5); they are nicely separated into two moieties, lacking resonances in the region between 78 and 108 ppm except for the spectra of the photohedamycins A, -B and -C, where the enol resonance at ca. 95 ppm is very characteristic. The sp^2-region is typical for the anthra[1,2-*b*]pyran system and, eventually, olefinic carbons from the side chain, whereas the sp^3-region contains all the sugar resonances and the bulk of the side chain signals as well as the resonance of the aromatic methyl group (C(13)) which shows a rather constant chemical shift of ca. 24 ppm.

A spectral comparison of a new pluramycin antibiotic with the known compounds of this family is of great value. The chemical shifts within the different structural elements of the pluramycins are very constant from compound to compound (see Table 8). From such a comparison it was *e.g.* quite obvious that hedamycin (**17**) had the same unacetylated sugar rings as kidamycin (**1**); the spectra of the two compounds were completely superimposable with respect to the sugar resonances (*82*).

Among the carbon atoms of the anthra[1,2-*b*]pyran system, the carbonyl groups and the other sp^2-carbon atoms bonded to oxygen

Table 8. *Carbon Chemical Shifts of*

C-Atoms	Kidamycin (1) (86)	Neoplu-ramycin (2) (50)[b]	11,3''-Diacetyl-kidamycin (3) (86)	3',3''-Diacetyl-kidamycin (4) (86)	Triacetyl-kidamycin (5) (86)	Isokida-mycin (26) (86)	Epoxy-kida-mycin (6) (11)
2	163.7	163.9	163.6	163.7	163.6	163.6	168.0
3	108.7	108.8	108.7	108.7	108.7	108.7	109.7
4	179.2	179.5	179.7	179.3	179.6	179.2	178.9
4a	125.8	125.9*	125.8	125.8*	125.8	125.9*	125.9*
5	149.6	149.6	148.2	149.6	148.1	149.4	149.9
6	125.4	125.4	124.4	125.3	124.4	125.3	125.8
6a	137.0	137.2	136.8	137.0	136.7	136.9	137.0
7	183.0	183.3	184.2	183.0	184.1	182.8	183.2
7a	125.8	126.2*	128.7	126.1*	128.7	125.6*	126.3*
8	140.0	140.6	144.1*	140.8	143.7	140.0	139.7
9	133.0	132.6	131.0	132.4	130.8	132.6	132.9
10	138.4	138.5	144.3*	138.3	144.5	139.0	137.8
11	159.7	159.3	145.4	159.4	145.4	158.7	159.7
11a	116.0	115.8	127.0	115.8	127.0	115.4	116.1
12	188.1	188.0	181.3	187.9	181.2	187.7	187.9
12a	118.9	118.9	121.0	118.8	121.0	118.6	119.1
12b[c]	155.7	155.8	154.9	155.7	154.9	155.6	156.0
13	24.0	24.1	23.8	24.0	23.8	24.0	24.3
14	127.2	127.3	127.6	127.2	127.5	127.1	57.6
15	12.1	12.1	12.0	12.1	12.0	12.0	13.8
16	134.2	134.2	133.3	134.2	133.3	134.0	62.0
17	14.9	15.0	14.7	15.0	14.7	14.9	14.1
18							
19							
1'							
2'	77.3	77.7	77.8	75.7*	75.7	77.6	77.2
3'	71.9	71.8	71.7	73.1	73.0	71.5	71.3
4'	67.4	67.5	67.6	64.8	64.8	67.6	67.8
5'	28.3	28.6	29.6	30.9	31.6	28.3	28.8
6'	75.2	75.4	75.6*	75.3*	75.5	75.6	74.8
7'	18.9	18.9	18.9	18.5	18.5	18.7	18.8
4'N(CH₃)₂	40.4	40.5	40.5	40.8	40.7	40.5	40.3
2''	67.2	69.8	70.2	69.9	70.2	72.3	67.5
3''	70.8	64.9	64.8	64.8	64.8	70.3	70.3
4''	57.4	57.6	57.9	57.7	57.9	58.7	57.6
5''	33.6	41.0	42.0	41.1	42.0	37.0	33.3
6''	69.5	76.4	75.8*	76.3*	75.7	71.0	69.4
7''	17.6	15.0	14.7	15.0	14.7	18.0	17.4
8''	12.3	13.7	13.8	13.7	13.8	10.9	13.1
4''N(CH₃)₂	36.8	39.3	39.2	39.4	39.1	36.7	36.6
Acetyl groups		170.6 21.3	170.3 168.9 21.1 (2C)	170.4 (2C) 21.2 (2C)	170.5 170.2 168.9 21.3 21.1 (2C)		

^{a} Chemical shifts in ppm relative to TMS. Solvent: $CDCl_3$. Spectra of the following (27) (86); photohedamycin C (33) (23); chromoxymycin (19) (47).
^{b} Reassigned.
^{c} In some of the references called C(1a).
*, ** Similar assignements within a column may be interchanged.

Selected Pluramycin Antibiotics[a]

Rubi-flavin A (10) (65)	Plura-mycin A (11) (33)	Diacetyl-plura-mycin A (12) (50)[b]	PD 121,222 (14) (68)	Heda-mycin (17) (86)	Photo-heda-mycin A (31) (23)	Photo-heda-mycin B (32) (23)	Photo-heda-mycin D (34) (23)
167.5	167.3	167.4	170.2	166.3	166.1	169.6	169.8
110.0	109.9	109.5	110.8	110.0	110.1	112.7	112.8
179.0	178.8	179.0	179.0	178.7	178.7	178.8	178.6
125.9*	126.0	126.1*	126.1*	125.8*	126.1	126.1	125.9
149.8	149.7	148.3	150.1	149.7	150.1	150.5	150.8
125.9	125.7	124.9	125.8*	125.9	125.9	125.6	126.7
137.4	137.2	136.7	137.2	137.3	137.1	136.9	135.6
183.3	183.2	184.1	183.1	183.1	181.2	181.2	181.3
126.4*	126.1*	128.5	125.8*	126.2*	128.8*	128.7*	128.5*
140.1	140.2	143.7	139.9	140.2	129.0*	129.1*	133.8*
133.1	134.0	130.9	133.3	133.1	136.2	136.1	131.9**
138.5	138.2	144.4	138.6	138.6	139.5	139.3	141.2
159.9	159.3	145.2	159.9	159.8	160.2	160.3	160.2
116.2	115.7	127.2	116.0	116.1	116.6	116.6	115.7
188.1	187.7	not obsd.	188.3	188.0	188.0	188.4	188.0
119.2	119.1	121.3	118.9	119.2	119.0	118.8	119.1
156.2	155.9	155.1	155.8	156.1	156.0	156.3	156.5
24.1	24.1	23.9	24.2	24.1	24.3	24.4	24.3
59.1	59.0	58.9	75.6	57.7	57.6	43.7	43.8
14.5	13.8	14.1	23.5	14.5	14.4	15.5	15.4
61.6	61.7	61.9	71.7	63.9	63.9	71.2	71.3
123.3	123.1	123.4	127.4	55.4	55.4	60.7	60.6
134.0	132.4	133.8	130.2	51.8	51.9	51.7	51.9
13.8	14.3	13.9	13.5	17.2	17.3	17.0	17.0
							196.1
77.3	77.5	75.8	77.2	77.3	76.7	76.8	131.4**
71.9	71.4	73.0	71.7	71.9	68.2*	68.3*	144.8
67.4	67.6	64.7	67.3*	67.4	66.1*	66.6*	74.8
28.4	28.7	31.6	28.4	28.3	95.2	94.4	69.9
75.2	75.1	75.4	75.0	75.2	156.7	157.4	17.7*
18.9	18.8	18.4	18.9	18.9	17.6	17.6*	
40.4	40.3	40.7	40.3	40.4	40.8	40.6	
67.2	69.8	70.2	67.4*	67.3	67.7	67.6	67.8
70.9	65.0	64.7	70.7	70.9	71.1	71.1	71.2
57.3	57.5	57.9	57.9	57.3	57.4	57.5	57.6
33.6	40.6	42.0	33.3	33.7	34.8	34.6	35.3
69.7	77.1	75.8	69.6	69.6	69.0	68.7	68.1
17.7	14.8	14.8	17.6	17.6	17.6	17.6*	17.5*
12.3	13.7	13.9	12.6	12.3	13.3	13.4	13.7
36.8	39.2	39.1	36.8	36.8	37.2	37.2	37.4
	170.5	170.5					
	21.2	170.3					
		169.1					
		21.3					
		21.2 (2C)					

compounds have also been published: triacetylhedamycin (18), triacetylisokidamycin

have their resonances at lowest field. Under standard experimental conditions the resonance of C(12b) is of rather low intensity, probably because of a long spin-lattice relaxation time; this carbon is far away from any protons which might help its relaxation. The three protonated carbon atoms, C(3), C(6), and C(9), are easily assigned from off-resonance decoupled spectra. The resonance of C(3) is at particularly high field since this carbon is the β-carbon of an enol ether like structure. Also at rather high field are the resonances of the two ring junction carbons C(11a) and C(12a); both have an oxygen function in their *ortho* positions.

Altertions in the substitution pattern of the chromophore are reflected by rather sensitive changes in the chemical shifts. Thus the aspect of the sp^2-region of the spectrum of a pluramycin antibiotic readily indicates whether the substituent at C(2) is conjugated with the pyrone or not, such as *e.g.* in kidamycin (**1**) or hedamycin (**17**), respectively. In kidamycin, the easily identifiable resonance of C(2) is at higher field (at 163.7 ppm) than in the spectrum of hedamycin (166.3 ppm), and the resonances of C(3) are at 108.7 and 110 ppm, respectively. Small influences upon the chemical shifts of C(4) and C(12b) can also be noticed.

Acetylation of the phenolic hydroxy group leads to drastic changes in the chemical shifts of a large number of carbon atoms of the anthra[1,2-*b*]pyran system. The carbonyl group C(12) and the phenol at C(11) are no longer involved in a hydrogen bond, and changes will be most prominent for these two carbon atoms. But due to the extended system of conjugation even the chemical shifts of carbon atoms which are rather remote from C(11) are influenced. Carbon atoms C(11) and C(12) have their resonances shifted massively upfield (by ca. 14 and ca. 7 ppm, respectively). On the other hand, those carbons *ortho* and *para* to the phenolic hydroxy group, *i.e.* C(10), C(11a), and C(8), move downfield upon acetylation by ca. 6, 11, and 3.5 ppm, respectively. Slight but characteristic acetylation shifts of ca. 1–2 ppm are further noted for C(6), C(9), and C(12b) (upfield), and for C(7) and C(12a) (downfield). It must be pointed out that when the phenol is acetylated C(12) seems to have an extremely long relaxation time. The signal of this carbon becomes very weak and is easily lost in the noise when the spectral parameters such as flip angle and pulse repetition time are not adjusted appropriately (*50, 86*).

The sugar resonances appear in the expected spectral regions: the carbons bearing oxygen atoms are between 57 and 78 ppm, the dimethylamino groups around 36–41 (two tall signals, each corresponding to two carbon atoms) and the C-methyl groups between 10 and 20 ppm. The signals of the two methylene carbons, C(5′) and C(5″), are readily detected in the off-resonance decoupled spectrum and by the fact that they often are somewhat broadened and therefore less tall.

The resonances of the ring E carbon atoms have rather constant chemical shifts within a series of comparable compounds. Upon acetylation of the 3' hydroxy group all these resonances – with the exception of that of C(6') – experience distinct shifts. The signal for C(3') thus moves downfield from ca. 72 to ca. 73 ppm, whereas those for C(2') and C(4') are both shifted upfield by ca. 2 and ca. 3 ppm, respectively. A large downfield shift of ca. 3 ppm is observed for C(5'); in contrast, the methyl group C(7') and the dimethylamino resonances move only slightly. These acetylation shifts are in reasonable accord with what is usually observed for an equatorial hydroxy group (*90, 97*).

Ring F was found to have a rather flexible conformation in solution (see section 4.4.2.); accordingly, the carbon chemical shifts do not show constant values within the series of comparable compounds measured, but exhibit slight variations. Acetylation of the 3'' hydroxy group leads to extreme shifts of the resonances of most carbon atoms. Only the resonance of C(4'') at ca. 57.5 ppm is hardly affected. The most drastic effects are observed for the resonances of C(5'') and C(6''), which are shifted downfield by ca. 8 and 6.5 ppm, respectively. C(3'') is shifted upfield by a similar amount, *i.e.* from ca. 70.5 to ca. 65 ppm. In contrast, the acetylation shift measured for C(2'') is only about 2.5–3 ppm downfield. It is noteworthy that even the methyl resonances which in ring E hardly are affected by acetylation of the sugar show appreciable shifts upon acetylation of ring F: The dimethylamino group moves from 36.8 to 39.3 ppm, the methyl group C(7'') from ca. 17.5 to ca. 15 ppm, and the methyl group C(8'') from ca. 12.5 to ca. 13.8 ppm. These acetylation shifts do not resemble what was reported for axial or equatorial hydroxy groups (*90, 97*) and must therefore be interpreted in terms of a drastic change of conformation of ring F upon acetylation.

Carbon-13 NMR spectroscopy is also a good tool for the detection of epimerization at C(6'') which takes place upon mild acid treatment of a pluramycin antibiotic (*28*). The epimerized ring F (as in isokidamycin (**26**)) gives a well defined set of carbon resonances which upon acetylation of the hydroxy group again give a characteristic pattern. The acetylation shifts are not very large and point to an axial hydroxyl group (*86*).

Acetylation or epimerization of ring F influence only negligibly (less than 1 ppm) the chemical shifts of the anthra[1,2-b]pyran carbon atoms as can be seen from a comparison of kidamycin (**1**) with neopluramycin (**2**) and isokidamycin (**26**) (see Table 8). On the other hand, nothing is known so far about the influence of acetylation of ring

E, since no compound has yet been detected or prepared in which only ring E is acetylated.

The side chain carbon resonances of a new pluramycin are readily detected as they do not belong to the well established sets of resonances typical for the chromophore and the two sugar rings. The assignments of the side chain resonances are, however, not always trivial. We have frequently made use of model compounds, such as *e.g.* the ester (**38**) which resembles the side chain of hedamycin. Comparisons with such compounds not only helped assignment of the resonances (*12, 13*), but also shed light on relative configurations within the side chain (see section 4.4.1.). But even then mistakes are not excluded. The choice of a suitable model compound is of great importance. Some years ago we were lead to erroneous assignments of the pluramycin A side chain carbon resonances from a comparison with 2-heptene, which proved to be an unsuitable model (*85*). The error was detected when additional specific proton decoupling experiments were carried out with the model compound (**39**) (*13*).

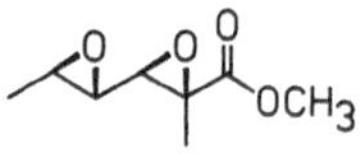

(**38**) 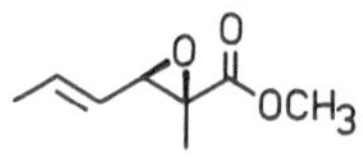   (**39**)

4.2.5. Mass Spectra

The early reports on pluramycin antibiotics seldom contained mass spectral data. The reason is that EI mass spectra, the only method routinely available at that time, are very difficult to obtain with the pluramycins due to their low stability and low volatility. SCHMITZ *et al.* reported in 1967 on the mass spectrometric determination of the molecular weight of hedamycin (**17**); these workers, however, obtained a result which was two atomic mass units too high (*79*). The error was not detected since the elemental analysis could not reveal it. Ten years later a correct molecular ion could be obtained and persilylation of hedamycin further helped, but the rather heavy derivative obtained was not easy to measure either (*82, 84*). Newer techniques, recently reviewed by HOWE and JARMAN (*40*), now make mass spectrometry a valuable tool for confirmation of molecular weights in the pluramycin series. The field desorption technique was used in 1978 to prove that hedamycin (**17**) clearly had a molecular weight of 746 (*82*). Soon after that, MACFARLANE (*56*) confirmed the molecular weights of several

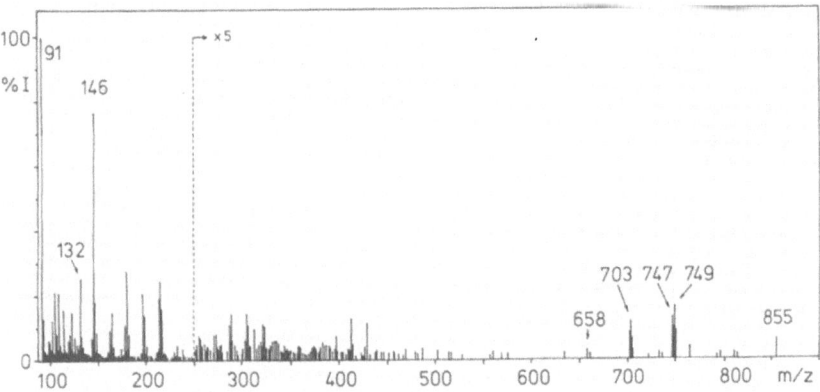

Fig. 6. FAB mass spectrum of hedamycin (**17**) (M = 746) in thioglycerol (the signal at m/z 91 is from the matrix)

antibiotics, among which hedamycin was included, using his californium-252 plasma desorption method.

Fast atom bombardment (FAB) mass spectrometry is now the most useful technique for the pluramycin antibiotics. The molecular ion, mostly in its protonated form, sometimes as an adduct with a molecule of the matrix, can usually be seen clearly. The method was thus successfully used in almost all recent structure determinations of such antibiotics. A critical point in FAB MS is the selection of a suitable matrix. The antibiotic is usually dissolved in a rather volatile solvent and this solution is then addded to the actual matrix substance. Diamylphenol was used for measurement of the spectrum of PD 121,222 (**14**) (*68*), whereas methanol/thioglycerol proved suitable for investigation of the photohedamycins (*23*) and for hedamycin (**17**) itself. In addition to the protonated molecule the protonated reduction product (at M + 3) and the protonated thioglycerol adduct (at M + 109) were observed (cf. Fig. 6). Whereas spectra in all other matrices studied yielded no prominent fragments in the high mass region of the spectrum, those obtained in thioglycerol gave signal clusters around M-43 and M-88, probably corresponding to the elimination of one and two molecules of dimethylamine, respectively, from the protonated and/or reduced molecule. In an alternate procedure, hedamycin and all the rubiflavins were first dissolved in chlorobenzene; these solutions were then added to *m*-nitrobenzyl alcohol, the actual matrix substance (*66*). The FAB spectra obtained by this method were dominated by [M + H]⁺; hardly any reasonable fragments could be made out (cf. Fig. 7). We have no explanation yet for the signal at m/z 604 in the spectrum of hedamy-

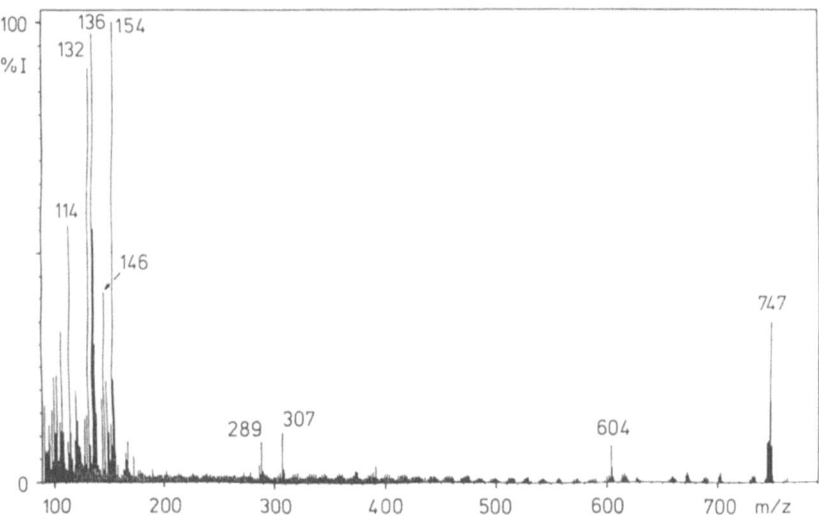

Fig. 7. FAB mass spectrum of hedamycin (17) (M = 746) in *m*-nitrobenzyl alcohol (the signals at m/z 136, 154, 289, 307 are from the matrix)

cin (17) obtained in *m*-nitrobenzyl alcohol (Fig. 7); the peak at m/z 114, on the other hand, might correspond to the protonated side chain.

GONDA, BYRNE *et al.* recently published somewhat more detailed analyses of the FAB spectra of the largomycin FII chromophore constituents obtained in a 1:1 mixture of dimethyl formamide and glycerol (*11, 33*). Again, the positive ion spectra were dominated by $[M + H]^+$, whereas the negative ion spectra yielded $M^{\bar{\cdot}}$. The elemental composition of the molecular ion could be determined by mass matching. No high mass fragments could be seen. However, the low mass end of the spectra gave some prominent signals. An ion corresponding to ring F was observed at m/z 172 (see Scheme 4); when this ring was acetylated as in pluramycin A (11) the corresponding peak occurred at m/z 214. This ring F or its acetylated form gave further rise to fragment ions at m/z 146 and 188, respectively, which often appeared as the base peak; they correspond to a formal loss of acetylene from these sugar fragments m/z 172 and 214 (*11, 33*). Similarly, ring E gave signals at m/z 158 and 132; the latter would again correspond to the elimination of C_2H_2 from the sugar.

These typical sugar fragment ions were, in part, also present in the spectra of the photohedamycins A (31), -B (32), -C (33), and -D (34) determined in thioglycerol. The signal at m/z 146, characteristic for ring F, was the base peak in all four spectra; the ion at m/z 172, corresponding to the intact ring F fragment, could, however, not be detected. Ring E in the compounds (**31**, **32**, and **33**) is an enol ether and thus lacks two hydrogen atoms compared with the native hedamycin (**17**). Nevertheless, the spectra of these three compounds showed the signal at m/z 132. This ion is apparently very stable and characteristic for the part of ring E comprising the carbon atoms 2′ through 4′ and their respective substituents. We do not think, however, that it is a secondary ion formed by fragmentation of a previously split off sugar as suggested by GONDA, BYRNE et al. (*11, 33*), since then formally C_2 would have to be eliminated from the enol ether ring E, which is not very plausible. Thus, the occurrence of m/z 132 in the spectra of the photohedamycins (**31**, **32**, and **33**) seems to indicate that the prominent sugar fragment ions at m/z 188, 146 and 132 do not arise from the elimination of acetylene from the previously split off sugar, but rather are formed by direct fission of the parent compound and hydrogen migration as outlined in Scheme 4. In the spectrum of photohedamycin D (**34**), as expected, the signal at m/z 132 is absent, since what is left over of the former ring E is no longer able to form this ion.

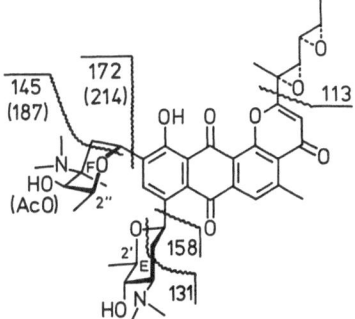

Scheme 4. Mass spectral fragmentation of pluramycin antibiotics

4.3. Chemical Reactivity

4.3.1. General Chemical Characterization

Pluramycin antibiotics show the typical reactions of 1-hydroxyanthraquinones. Neutral or weakly acidic solutions of these compounds are yellow, whereas alkaline solutions are purple. Reduction with hydrogen (*79, 84*), zink or dithionite (*18*) yields products which are readily reoxidized in the atmosphere. The hydroquinone form could be trapped

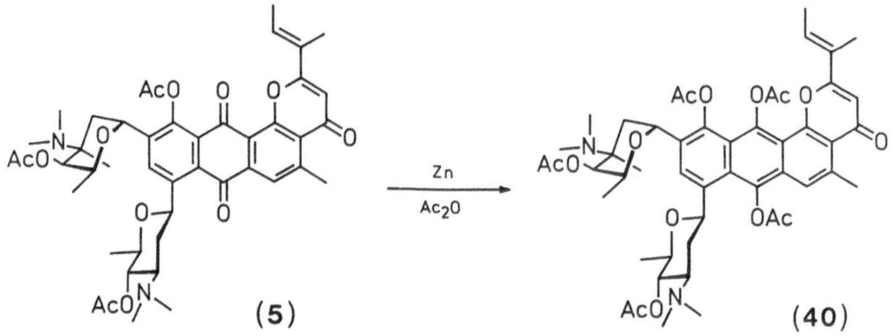

Scheme 5. Reductive acetylation of kidamycin

as the ester (40) (see Scheme 5) when triacetylkidamycin (5) was reduced in acetic anhydride with zinc (28) (cf. also (18)).

The formation of colored complexes with many metal ions was reported for hedamycin (17) (84) and other compounds (51): reddish brown with Fe^{3+}, purple with Mg^{2+}, violet with Ba^{2+} or Co^{2+}, dull red with Cu^{2+}, brownish violet to blue with Ni^{2+}. Surprisingly, for pluramycin A (11) (57) and the rubiflavin mixture (1) negative results with ferric chloride were reported; these should be considered as erroneous observations.

The pluramycins prove to be chemically rather labile. They readily decompose in solution and upon irradiation with UV- and daylight; this makes separation and purification as well as degradation and derivatization very difficult.

4.3.2. Degradation, Decomposition

The pluramycin antibiotics seem to be quite stable in solid form when kept in the dark. However, they decompose readily in solution, especially when exposed to light.

Decomposition also occurs at higher temperature. The biological activity of the largomycin FII chromophore (components so far identified: pluramycin A (11), rubiflavin A (10), and epoxykidamycin (6)) dropped to 40% after 2 h treatment at 60° C in the dark (33). Heating pluramycin A in 0.1 M ammonium acetate solution primarily hydrolyzed the ester function, converting the substance to rubiflavin A before further decomposition took place (33). Furthermore, none of the pluramycin antibiotics so far obtained in the crystalline state showed a clean melting point; the compounds decompose before melting properly (28, 57, 79).

It is noteworthy, that heat deactivation as well as photodeactivation are reduced in the presence of proteins. Thus, a stabilizing effect of serum on pluramycin A (11) was described (57), and with largomycin FII the stability of the native protein antibiotic was clearly greater than that of the protein-free chromophore (33).

Controlled degradation of pluramyicin antibiotics is difficult; only a few useful reactions have been described. Since the compounds do not contain glycosidic linkages, it is not possible to achieve hydrolysis into sugars and aglycones by mild acid treatment. When kidamycin (1) is refluxed with p-toluenesulfonic acid in chloroform, epimerization at C(6″) takes place and a well defined product, isokidamycin (26), is obtained (see Scheme 6) (28). Treatment of the indomycins with 50% KOH at elevated temperature liberated dimethylamine (80). A milder method of deamination consists of oxidizing the pluramycin antibiotic (*e.g.* triacetylkidamycin (5)) with perbenzoic acid and subjecting the resulting *N*-oxide to a Cope elimination by refluxing it in benzene; the bis[des(dimethylamino)] derivative (41) is obtained (see Scheme 7) (28).

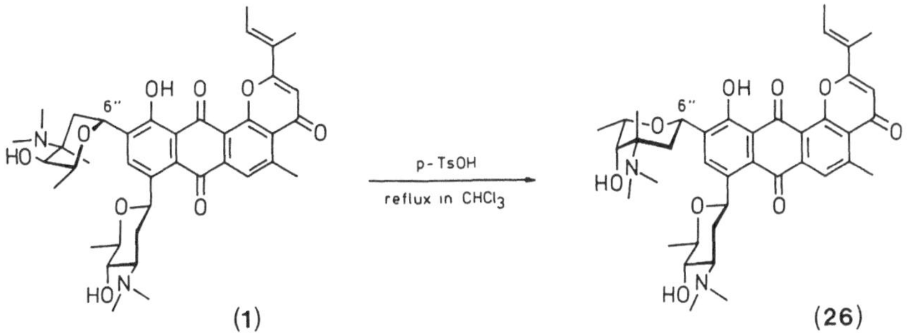

Scheme 6. Epimerization of kidamycin

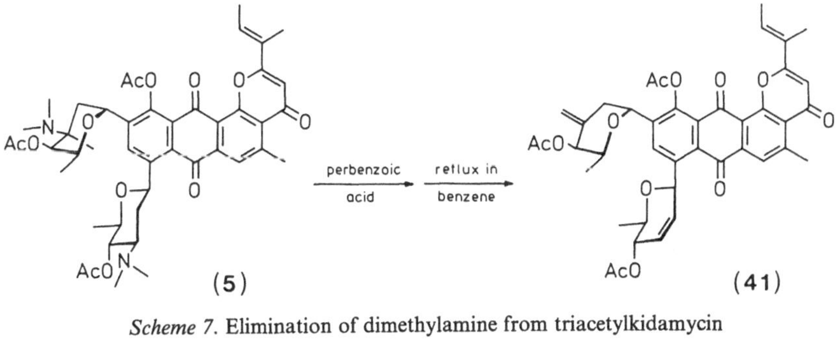

Scheme 7. Elimination of dimethylamine from triacetylkidamycin

Prompted by reports of biochemists that hedamycin (17) rapidly lost its biological activity in solution and by our own observations that hedamycin degraded to ultimately water soluble, polar products when allowed to stand in solution in the daylight, a study of this photodegradation was undertaken. Hedamycin (17) was irradiated in different solvents in the presence or in the absence of oxygen (23). The color of the solutions which at the beginning was a clear yellow changed to orange-brown during irradiation. The products obtained were separated by HPLC and their structures determined mainly by NMR spectroscopy. When hedamycin (17) was irradiated in the presence of oxygen, only one product could be isolated: photohedamycin A (31) (see Scheme 2). This compound was easily identified as an enol ether. A comparison of its NMR spectra with those of hedamycin (17) showed that the constitutions of the two compounds were the same with the exception of ring E. Proton NMR spectrometry revealed an olefinic proton resonance at 4.98 ppm in ring E of (31). Two new carbon resonances for this partial structure appeared at 156.7 and 95.2 ppm clearly indicating an enol ether structure. We assume this product to be formed in the following way: photoreduction of the anthraquinone part of hedamycin to the hydroquinone level in analogy to the described photoreduction of anthraquinones (34, 102) is probably the first step. The well-positioned ring E may thereby act as an intramolecular hydrogen donor and lose two hydrogen atoms. The resulting hydroquinone is then reoxidized to the quinone level by the oxygen present in the solution or by the atmosphere during work-up. This formation of a ring E enol ether seems to be a general reaction of pluramycins, since kidamycin (1) showed the same behaviour yielding photokidamycin A (42).

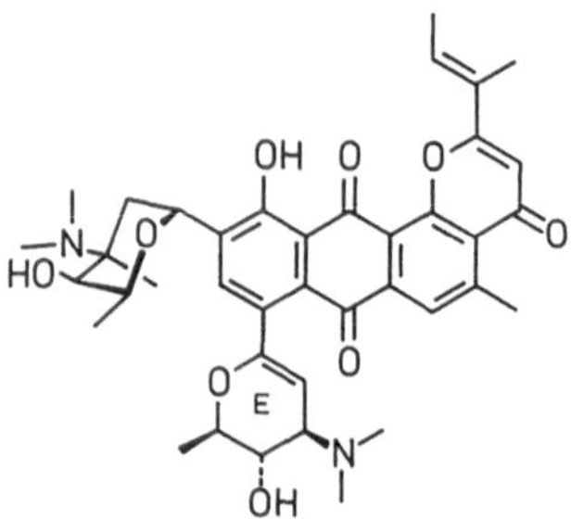

photokidamycin A   (42)

When hedamycin (17) is irradiated in the absence of oxygen, additional photoproducts besides (31) are detected (see Scheme 2): the two diastereoisomeric photohedamycins B (**32**) and -C (**33**) and photohedamycin D (**34**). In the degradation products (**32**) and (**33**) ring E is again the enol ether already described. However, the inner epoxide of the side chain at C(2) was opened reductively. These two products seem to be degradation products of (**31**). Their formation depends on the presence of a hydrogen donor; similar products could not be found in a model study where the chromone (**43**) was photolyzed under similar conditions but with no hydrogen donor present (*22*). On the other hand, it has been reported that when epoxy ketones are irradiated in the presence of tri-*n*-butyltin hydride, reductive opening of the oxirane ring occurs (*48*). Photohedamycin D (**34**) then shows, that upon prolonged irradiation, the enol ether ring is further attacked and degraded to an acyclic unsaturated ketone. The mechanism of formation of this new side chain is not clear. Formally, it results from hydrolysis of the enol ether ring in a compound such as (**32**) or (**33**) and subsequent elimination of dimethylamine.

(**43**)

The few compounds isolated in our studies do certainly not exhaustively describe the photochemical behaviour of pluramycin antibiotics. When dealing with such antibiotics in the laboratory, very often degradation to rather polar products is observed. We were not able to identify these products as they tend to decompose further during work-up. The same mixture of degradation products gave different HPLC peak patterns when allowed to stand for increasing periods of time before chromatography; less polar substances disappeared and more polar ones were formed. The few photoproducts so far identified cannot account for the rapid and complete loss of biological activity observed for solutions of hedamycin, since (**31**) was found to be only about 15 times less cytotoxic than hedamycin (**17**) itself (*24*); further degradation must be involved.

The photolability of the largomycin FII chromophore, a mixture containing pluramycin A (**11**), rubiflavin A (**10**), and epoxykidamycin (**6**) among other components, was studied by GONDA *et al.* (*33*); no attempt to isolate the degradation products was, however, made. Total loss of biological activity after 1–1.5 hours of irradiation was observed along with a color change of the solution from yellow to orange.

4.3.3. Derivatization

Derivatization was extensively studied by FURUKAWA *et al.* in the case of kidamycin (**1**) (*28*). Peracetylation could be achieved with acetic anhydride in pyridine; the phenol as well as the two sugar hydroxy groups were esterified by this procedure, the product being the triacetate (**5**). Only the sugar OH-groups were attacked when the acetylation was carried out in acetic anhydride in the presence of sodium acetate and (**4**) was isolated. Partial hydrolysis of triacetylkidamycin (**5**) to the 11,3″-diacetate (**3**) was observed upon refluxing (**5**) in methanol for 30 min. Whereas peracetylation and the selective esterification of the sugar worked equally well with hedamycin (**17**) (*65, 105*), the partial ester hydrolysis mentioned above gave only degradation products when carried out with triacetylhedamycin (**18**). The nature (and chemical lability or inertness) of the C(2) side chain thus plays an important role in derivatizing reactions.

In order to obtain derivatives suitable for X-ray analysis, the synthesis of many heavy atom derivatives was studied. Esterification of isokidamycin (**26**) with *m*-bromobenzoyl or *m*-iodobenzoyl chloride gave

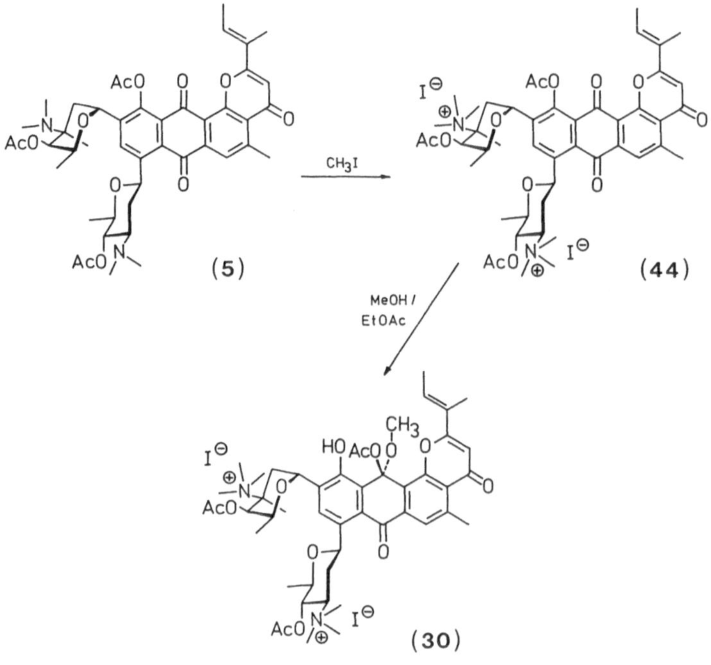

Scheme 8. Quaternization of triacetylkidamycin

the well defined compounds (**28**) and (**29**), respectively (see Scheme 1) (*28, 29*); an analogous bis-*m*-iodobenzoate of hedamycin could also be prepared (*65*) – it was, however, not obtained in as crystal form suitable for X-ray analysis. Quaternization of the dimethylamino groups was also undertaken: When triacetylkidamycin (**5**) was dissolved in methyl iodide, crystals of the bis methiodide (**44**) separated from the solution overnight (*28*). The most spectacular derivative (**30**) was, however, obtained unexpectedly when this bis-methiodide was recrystallized from methanol/ethyl acetate (*30*). The compound (**30**) is formally derived from (**44**) by the addition of methanol to the quinone and subsequent acetyl migration (see Scheme 8).

4.4. Stereochemistry

4.4.1. Configuration

The absolute configuration was determined only for kidamycin (**1**) and hedamycin (**17**) and found to be the same. The assumption that all other pluramycins probably also share this same absolute configuration seems reasonable. X-ray crystallography (using the anomalous dispersion procedure) was used to determine the absolute configuration of the quaternized triacetylkidamycin methanolysis product (**30**) (*30*) and of the isokidamycin derivatives (**28**) and (**29**) (*29*). The absolute configuration of hedamycin (**17**) was related to that of the kidamycin derivatives by comaprison of the CD curves of hedamycin and kidamycin (*105*).

The relative configurations within the sugars can be determined by NMR spectroscopy. Quite typical ^{1}H$-^1$H coupling patterns are obtained in ring E which allow a reasonable assignment of the configurations (*84*). In contrast, proton-proton coupling in ring F does not help much, since the sequence of coupling protons is interrupted at C(4″), and since the conformation of ring F in solution is a rather flexible one (see below). Carbon-13 NMR spectroscopy which is very sensitive to steric changes in molecules is very well suited to relate the sugar configurations of a new pluramycin antibiotic to those of a compound whose configurations have been determined unambiguously, *e.g.* by an X-ray structure determination (*82*).

Since the pluramycins have different side chains at C(2), configurational data obtained for the side chain of one compound cannot be easily transferred to a different compound by simple spectroscopic comparisons. Thus, a special effort is necessary to establish the relative configurations in this particular structural fragment. Assignments have been made mostly on the basis of NMR spectral data using well established procedures. Double bond configurations can be determined from the size of the proton-proton coupling constants or by observation of γ-effects on the carbon chemical shifts. NOE experiments and difference spectroscopy were also used for this purpose. The size of allylic couplings should, however, only be used with utmost care for the assignment of double bond configurations in acyclic systems (85). The substitution pattern and thus the configurations of epoxides in the C(2) side chain have been investigated mostly by ^{1}H- and ^{13}C-NMR spectral comparisons with esters such as (38) or (39) which served as suitable model compounds (13, 85). Whereas the configurations of double bonds and epoxides are more or less easily tracked down due to the rigidity of these systems, the configurations of those carbon atoms which are not involved in either of these systems are very difficult to determine: It is almost impossible without derivatization and has therefore rarely been done.

Another difficult task is the correlation of the relative configurations in the side chain with those in the sugar moieties; since these two structural elements are separated from each other by the planar anthra[1,2-b]pyran system and since degradation of pluramycins is hardly possible, the only solution to this problem seems to be X-ray structure determination.

4.4.2. Conformations

The X-ray structures of the kidamycin derivative (30) (see Fig. 8) (30), of the isokidamycin derivatives (28) and (29) (29), and of hedamycin (17) (see Fig. 9) (105) – besides confirming the constitutions and configurations – yielded information about the conformations of these compounds in the crystalline state. All structures agree inasmuch as the anthra[1,2-b]pyran part is more or less flat and thus conformationally uninteresting. Much more interesting are, of course, the conformations of the sugar rings. Ring E always adopts the chair conformation in the crystal with all the substituents in equatorial positions.

For ring F, the structure of the kidamycin derivative (30) shows a twisted boat conformation where the aryl and trimethylammonium substituents are in equatorial positions, whereas the hydroxy and C(4″)-methyl groups are axially oriented (see Fig. 8). It must be pointed out, however, that (30) is not an anthra[1,2-b]pyran derivative and that its sugars are acetylated and derivatized to the quaternary ammonium iodides. Thus the conformation found is characteristic for this specific compound but not necessarily also for the parent kidamycin (1).

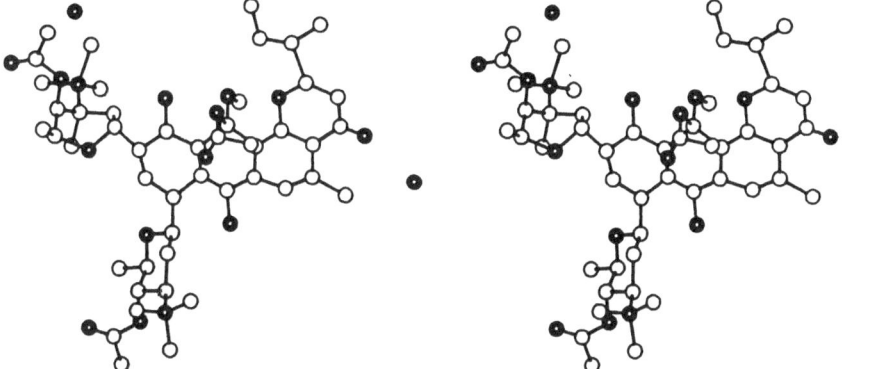

Fig. 8. Stereoscopic view of the kidamycin derivative (**30**) (Redrawn using the atomic coordinates from (*29a*))

Ever since the publication of the paper by FURUKAWA et al. (*30*), formulae of pluramycin antibiotics were drawn with ring F in a boat conformation. The X-ray structure determination of hedamycin (**17**), however, was carried out with the underivatized antibiotic using direct methods. Surprisingly, ring F in this case was clearly in a chair conformation. The aryl substituent was now in an axial position, but in order to relieve strain the lobe of the chair carrying this aryl substituent was flattened out (see Fig. 9). According to these findings, we draw now

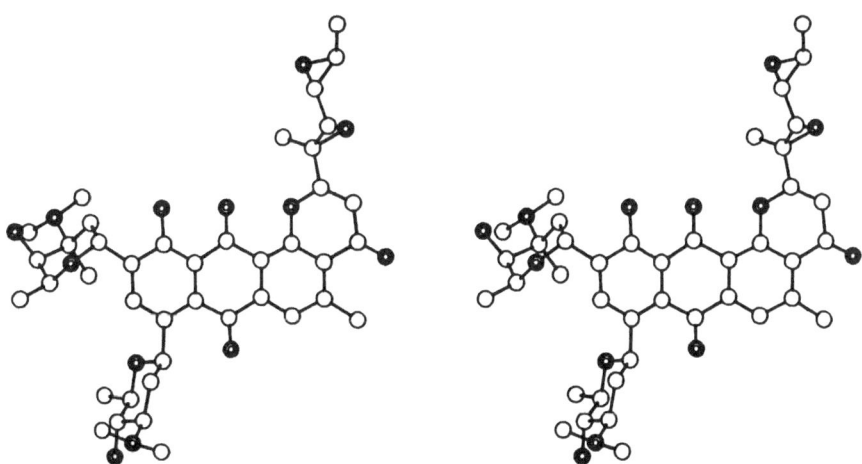

Fig. 9. Stereoscopic view of hedamycin (**17**)

the structures of pluramycins with ring F in a chair conformation. The particular substitution pattern of ring F seems to induce considerable strain. This can be relieved when ring F either adopts a twist conformation or by angle deformation or ultimately by epimerization at C(6''). Treatment of kidamycin (1) with p-toluenesulfonic acid (see Scheme 6) provokes this epimerization at C(6'') leading to isokidamycin (26) (28). In this compound, ring F now adopts a chair conformation with the aryl substituent oriented equatorially (29).

The X-ray crystallographic studies mentioned, further yielded information about the orientation of the C(2) side chain with respect to the pyrone ring A, i.e. about the torsion angle C(3)−C(2)−C(14)−C(16). This torsion angle is −62° in hedamycin (17) (21) and ca. 180° in the kidamycin and isokidamycin derivatives (28, 29, and 30). In solution, information about this torsion angle was obtained in a few cases from NOE experiments. Irradiation of the methyl protons at C(15) lead to an incerased integral for the resonance of H-C(3) in pluramycin A (11) (50) and the rubiflavinones C-1 (20) and -C-2 (21) (65), indicating a spatial proximity of the side chain methyl group and the proton at C(3).

The conformations of hedamycin and kidamycin in solution were studied using NMR spectroscopy (86). Ring E adopts the same chair conformation in solution that was found in the crystal as shown by the $^1H-^1H$ coupling constants measured. Furthermore, the ^{13}C-NMR shifts observed for the ring E carbon atoms upon acetylation of the C(3') hydroxy group of kidamycin (1) and of hedamycin (17) were in reasonable agreement with the acetylation shifts determined for trans-4-t-butylcyclohexanol (45) (90) and D-glucose (97), two compounds with equatorial hydroxy groups. This further corroborates that ring E is in solution in the same chair conformation as in the crystal.

(45)   (46)

Ring F in isokidamycin (26) was found to adopt a chair conformation in the crystal with the aryl and dimethylamino substituents in equatorial positions and the OH group axially oriented. Thus, the ^{13}C-NMR acetylation shifts of (26) were now compared with those measured for cis-4-t-butylcyclohexanol (46), a compound with an axial hydroxy group. The reasonable agreement between the acetylation shifts for the two compounds pointed to the fact, that also ring F of isokidamycin is in solution in the same chair conformation as in the crystal.

The crystal conformation of ring F of hedamycin and kidamycin was shown to depend strongly on the substitution pattern. In hedamycin (17), it was a flattened chair, whereas in the kidamycin derivative (30), with quaternized amino groups and acetylated hydroxy functions, it was a twisted boat form. We were therefore not surprised to find this ring to be rather flexible in solution too. There was no way to correlate the ^{13}C-NMR acetylation shifts with either *trans*-4-*t*-butylcyclohexanol (45) with an equatorial OH group or with the corresponding *cis*-isomer (46) with an axial hydroxy group. Thus, we assume that a drastic change of conformation must take place upon acetylation of the ring F of pluramycin antibiotics (86) and that this ring is conformationally rather flexible in solution.

5. Total Synthesis

5.1. Pluramycinones

Only one synthetic effort towards a pluramycinone derivative (*viz.* kidamycinone methyl ether (56)) has come to our attention (*36, 37*). HAUSER and coworkers used a sequence of progressive, regioselective annelation reactions. They started with 7-methoxy-3-phenylsulfonyl-1(3*H*)-isobenzofuranone (47) (see Scheme 9), whose anion was condensed with methyl crotonate. Methylation then gave the naphthalene derivative (48). This was brominated at the *C*-methyl group, whereupon the bromine was substituted with thiophenoxide. The resulting sulfide was oxidized to the sulfoxide (49). The anion of this intermediate was then brought to reaction with 3-pentene-2-one and subsequent thermal elimination of benzenesulfenic acid gave the anthrahydroquinone ether (50). This compound possesses the structural elements of 2-hydroxyacetophenone which is a good starting material for the synthesis of benzopyrones (*20*). The construction of the pyrone part of the target molecule, however, proved to be more difficult than was expected from model studies with 2-hydroxyacetophenone and tigloyl chloride (*37*). In a "Baker-Venkataraman" like reaction sequence (*4*), (50) was acylated with tigloyl chloride. Transfer of the tigloyl group to the acetyl function could, however, only be achieved after oxidation to the anthraquinone level (51). The final cyclization of the diketone (52) did not yield the target molecule (56) but rather (53). In a second attempt, the dianion of compound (50) was condensed with tiglaldehyde yielding

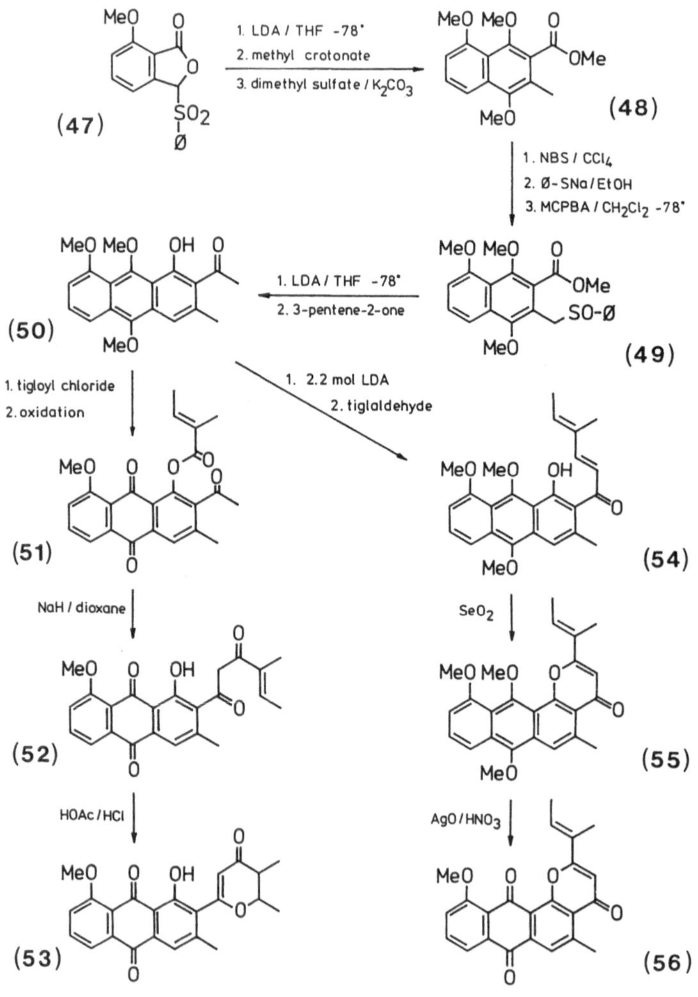

Scheme 9. Synthesis of kidamycinone methyl ether

the dienone (54). After a number of unsuccessful trials this compound
could finally be cyclized and dehydrogenated to the tetracyclic intermediate (55) using selenium dioxide in *t*-amyl alcohol. Oxidation of (55)
with silver oxide in dilute nitric acid then gave the desired kidamycinone
methyl ether (56). Koo's procedure (*52*), which involves condensation
of a 2-hydroxyacetophenone with esters followed by an acid catalyzed
cyclization of the 1,3-diketones obtained, was apparently not tried by
HAUSER *et al*. This procedure was somewhat later used successfully

in our laboratory for the synthesis of a variety of benzopyrone derivatives related to the pluramycins (83).

5.2. Sugar Moieties

Total syntheses of derivatives of the two sugars involved in the pluramycin family, angolosamine (57) and N,N-dimethylvancosamine (58), have been published. These reactions were, however, not carried

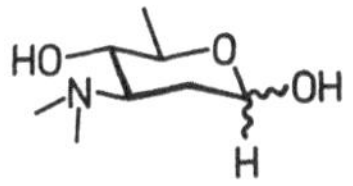

(57) 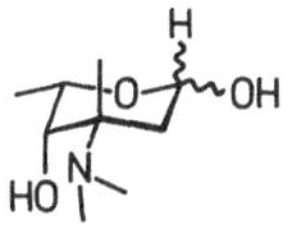   (58)

out with the pluramycins in view. Thus, only leading references to this type of work without any further comment are given here; details can be found in the recent review by HAUSER and ELLENBERGER (35 a). Angolosamine hydrochloride has been synthesized (3), whereas N,N-dimethylvancosamine itself has not yet been prepared, but syntheses of N-benzoylvancosamine (27) and of the methyl glycoside of N-acetylvancosamine (91) can be found in the literature.

5.3. Pluramycin Antibiotics

Neither a total synthesis nor any attempt to synthesize a pluramycin antibiotic from a natural or synthetic aglycone and the sugars using C-glycosylation reaction have so far been published.

6. Biosynthesis

The biosynthesis of the pluramycins has so far hardly been investigated experimentally. FRICKE (26) as well as FURUKAWA (28) suggested, however, that the 23 carbon atoms (or 21, for compounds with C_4 side chains) of the anthra[1,2-b]pyran part and of the side chain at C(2) originated from a polyketide (see Scheme 10). The methyl group in the side chain is thought to come from the C_1-pool whereas either C(13) or C(19) must arise from a decarboxylation of the terminal

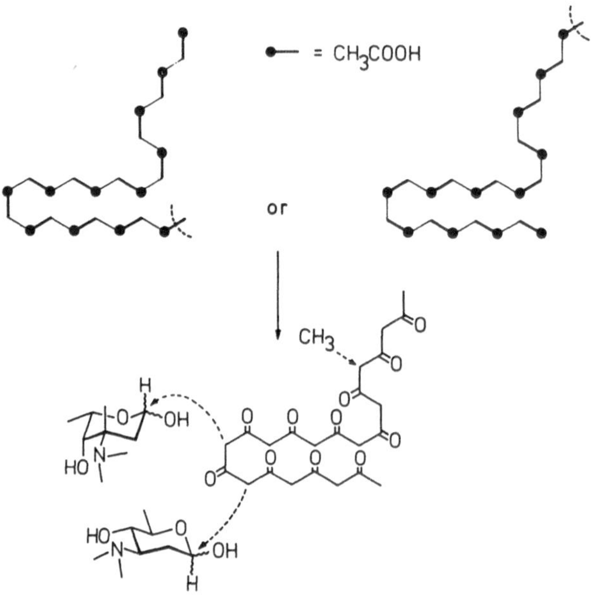

Scheme 10. Hypothetical biogenetic pathway to the pluramycins

acetate unit in the polyketide. The sugar moieties might be attached
to the pluramycinone skeleton as preformed units at some later stage.

The overall picture drawn by Fricke and Furukawa is most certainly correct, as it is well known that microbial anthraquinones may
be formed along the polyketide pathway (*94, 95*). However, it is by
no means certain in which direction the polyketide chain is built up.
Either C(13), as suggested by Fricke, or C(19) might be the starting
point. The side chain methyl group, C(15), might – instead of coming
from the C_1-pool – as well originate from the incorporation of a propio-
nate unit into the growing polyketide (*26*), as has been observed in
macrolide antibiotics (*74*).

The different types of side chains observed in the pluramycin family
all represent different levels of oxidation of the same biogenetic precursor. It is noteworthy that with the detection of the epoxydized kidamycin derivative (**6**) as one of the largomycin FII chromophore constituents (*11*), the first compound with an oxygenated four carbon side
chain has been added to the family.

Compounds with four- and six-carbon side chains are produced
side by side by certain microorganisms. The strain of *Streptomyces*
SC 3728, which produces the rubiflavin mixture (*67*), and the

S. pluricolorescens strain, from which Largomycin FII was obtained (*11, 33*), are two examples.

So far, only one experimental approach towards the biosynthesis of a pluramycin antibiotic has come to our attention. Carbon-13 labelled sodium acetate was fed to a culture of a hedamycin-producing strain of *Streptomyces griseoruber*. However, the strain used apparently had mutated and did no longer produce appreciable amounts of the antibiotic. The product that could be isolated from the feeding experiment showed about a twofold enrichment at the labelled sites and clearly proved the expected polyketide origin of hedamycin (**17**) (*81*).

7. Biological and Biochemical Behaviour

7.1 General Remarks

The pluramycin antibiotics have interesting biological features. They are effective against a large number of microorganisms, they are antitumor active and they show a distinct interaction with DNA. This latter feature is perhaps the basis for all their biological activities observed as well as for their rather high general toxicity. The therapeutic index is rather low: 5–10 for pluramycin A (**11**), but was increased to about 160 when pluramycin A was complexed with human serum albumin (*73*). Pluramycin antibiotics have no medicinal use today. This is certainly due to their low therapeutic index and high general toxicity. In addition, the chemical instability of the pluramycins makes their handling very difficult. Biochemical investigations are hampered by the omnipresent uncertainty whether the observed effects are really due to the pluramycin antibiotic under investigation or to some highly active but not detected degradation product or to homologous compounds.

The first biochemical investigations of a newly isolated pluramycin antibiotic were usually undertaken in order to establish that the compound on hand – and whose chemical structure was not yet known – was indeed a novel antibiotic. Thus in order to get an overview of the spectrum of its biological activities the action against bacteria, fungi, cell cultures and experimental tumors was determined.

A second, often reported approach was to include one or two pluramycins in a comparison study involving a single or a few test organisms and a broad selection of antibiotics of many different kinds (see *e.g.* (*5, 77, 87*)). Usually such studies did not shed much light on the biological activity or the mode of action of the pluramycin antibiotics involved.

The most interesting biochemical aspect of the pluramycins, their interaction with DNA, has hardly been investigated systematically. A lot of papers have appeared to the subject; however, they were published at a time when the chemical constitutions of these substances were not yet known. Thus, some of the results and interpretations put forward in those articles have to be taken with caution. Quantitative differences in the activities of hedamycin (17) and rubiflavin were *e.g.* said to be due to the rubiflavin molecule being roughly half the size of hedamycin (*100*). However, this cannot be the case, since we recently showed that 'rubiflavin' is actually a mixture containing several pluramycin antibiotics (*67*).

Investigations of structure-activity relationships became only possible after structure elucidation of some of the pluramycin antibiotics. Very little has been done so far. A comparison of hedamycin (17) and photohedamycin A (31) on the one hand and of kidamycin (1) and isokidamycin (26) on the other hand using a simple test with HeLa cells gave the first insight in the role the two sugars play. Photohedamycin A (31) was less active than hedamycin (17) by a factor of ca. 15, whereas the activity ratio of kidamycin (1) to isokidamycin (26) was ca. 30 (*24*). Acetylation of the sugars also influences the biological activity as was shown by the lower toxicity of triacetylkidamycin (5) as compared with kidamycin (1) (*45*). These results clearly point out that the precise geometry of the sugars is vital to biological activity. It must be remembered that with the anthra[1,2-*b*]pyrans lacking sugar moieties (the 'aglycones' or 'pluramycinones') no biological activity whatsoever has been detected so far. Nothing is known yet about the role of the C(2) side chain. Some of these, such as the diepoxide side chain of hedamycin (17), might act as alkylating agents. However, so far no one has ever tried to isolate *e.g.* alkylated bases after incubation of DNA with a pluramycin antibiotic. The biological activity depends on the nature of this side chain. Kidamycin (1), with the simple olefinic substituent proved to be less cytotoxic than hedamycin (17) by about two orders of magnitude in our simple HeLa cell test (*24*); kidamycin is also less toxic for mice and less active in antimicrobial tests (*8, 44*).

Besides a few conference abstracts only a single paper has dealt with the pharmacokinetics of a pluramycin antibiotic. In 1973, Umezawa *et al.* reported (*96*) on the fate of triacetylkidamycin (5) given intravenously to mice. The compound disappeared from the blood within two hours and was found thereafter in various organs, such as liver, lung and kidney.

In the following chapters the focus will be on pluramycin A (11), hedamycin (17), and kidamycin (1) and its triacetate (5), *i.e.* on antibiotics whose molecular structures have been determined. Where appropriate, data from other pluramycin antibiotics will also be included.

7.2. Toxicity

Studies with mice show that all pluramycins have high acute toxicity. Hedamycin (17) is clearly the most potent compound, having an LD_{50} of 300 µg/kg when given intraperitoneally (8). For pluramycin A (11) the LD_{50} value is around 10 mg/kg (71, 88), for kidamycin (1) around 18 mg/kg (44, 46). Acetylation seems to lower the acute toxicity as shown by the LD_{50} of 50 mg/kg determined for triacetalkidamycin (5) (45), whereas salt formation increases the toxicity. Pluramycin A tartrate and ascorbinate are more toxic than the free base (71, 88). This might be due either to better absorption of the antibiotic or perhaps to a toxic decomposition product; pluramycin A salts proved to be less stable in solution than the free base (71, 88).

Pluramycin A (11) is to our knowledge the only compound that was tested for chronic toxicity (88). When 600 µg/kg of this compound were given i.p. daily to mice for 30 days, all animals survived and showed only a 3% loss of weight. Daily administrations of 1250 µg/kg for 30 days killed $^1/_5$ of the animals and led to a slightly increased loss of weight. Injections of 2500 µg/kg per day were lethal. Somewhat different values were observed when pluramycin A ascorbinate was used (71).

7.3. Activity Against Bacteria, Yeasts and Other Microorganisms

Pluramycin antibiotics have been found to be active against a wide variety of microorganisms. As an example, the antimicrobial spectrum of hedamycin (17) is given in Table 9. The minimal inhibitory concentrations for this compound are in the range of 0.003–25 µg/ml (8). The activities of the rubiflavin mixture (1.4–37.5 µg/ml (1)) and of pluramycin A (11) (6–100 µg/ml (88)) are somewhat lower, whereas kidamycin (1) and its triacetyl derivative (5) are significantly less active (0.19–1.56 mg/ml and 0.78–6.5 mg/ml, respectively (45)). It is noteworthy that the recently detected epoxidized kidamycin (6) is again highly active (11); the minimal inhibitory concentrations are in the range of 0.005–50 µg/ml and are thus comparable to those of hedamycin (17).

It should be pointed out here that pluramycin A (11) was also able to induce morphological changes in bacteria. When a growing culture of E. coli was exposed to the antibiotic, filament formation

Table 9. *Antimicrobial Spectrum of Hedamycin* (**17**)

Test organism	Minimal inhibitory concentration µg/ml	Test organism	Minimal inhibitory concentration µg/ml
Anaerobic bacteria		*Salmonella paratyphi* A	1.25
Peptococcus prevotii ATCC 9321	0.1	*S. paratyphi* B	1.25
Clostridium chauvoei ATCC 10092	3.2	*S. typhosa*	0.31
		Shigella dysenteriae	1.25
Aerobic and facultative bacteria		*S. sonnei*	0.31
Aerobacter aerogenes	3.12	*Staphylococcus aureus* Smith	0.0062
Alcaligenes faecalis ATCC 8750	0.31	*S. aureus* 1633-2	0.031
Bacillus subtilis ATCC 6633	0.125	*S. aureus* 52-75	0.016
Corynebacterium xerosis	0.031	*S. pyogenes* Digonnet	0.003
Diplococcus pneumoniae	0.0031		
Escherichia coli 01495	1.6	Yeasts	
E. coli ATCC 8739	3.1	*Candida albicans*	25.0
Klebsiella pneumoniae	1.6	*Kloeckera brevis* ATCC 9774	0.8
Mycobacterium phlei	0.031	*Saccharomyces cerevisiae*	3.2
M. ranae	0.125		
M. smegmatis 607	0.25	Protozoa	
Proteus morganii	6.25	*Crithidia fasciculata*	12.5
P. rettgeri	3.12	*Ochromonas malhamensis*	1.6
P. vulgaris ATCC 9930	6.25	*Tetrahymena pyriformis*	0.8
Pseudomonas aeruginosa	6.25		

Data from (*8*).

of the cells was observed along with cell destruction; they disintegrated with arc-like curvature (*60*).

7.4. Interaction with Phages and Lysogenic Bacteria

Hedamycin (**17**) and pluramycin A (**11**) were both found to induce lambda phage production in lysogenic *E. coli* W 1709 (*77*). The authors found that this property was quite useful in screening for antineoplastic agents. The minimal inducing concentrations were given as 1 and 0.5 µg/ml for hedamycin and pluramycin A, respectively. With a different test, BRADNER *et al.* (*8*) found that induction was effected already with a hedamycin concentration as low as 0.0125 µg/ml. At higher concentrations hedamycin acts as an antiphage agent. A content of 0.25 µg/ml inactivated 99% of the free lambda phage (*38*).

7.5. Cytotoxicity

The antitumor and cytotoxic activities are among the most interesting biological properties of the pluramycin antibiotics. These compounds were found to be active against a wide variety of tumors and cell cultures.

Hedamycin (17) inhibited the growth of HeLa cell cultures and had an ID_{50} of 0.00013 µg/ml. The growth of a transplanted tumor, Walker 256, was significantly inhibited, as was that of adenocarcinoma of the duodenum in hamsters. However, this compound had no effect on adenocarcinoma 755, myeloid leukemia C-1498 or lymphatic leukemia L-1210 *in vivo* (8).

Kidamycin (1) had a significant life-prolongation effect on mice bearing Ehrlich ascites tumors at single i.p. doses from just below the LD_{50} (18 mg/kg) to $^1/_{16}$ of the LD_{50}. The compound also showed some effect against leukemia L-1210 when given i.p. in eight daily doses of $^1/_{10}$ of the LD_{50}. Slight effects were noted on Sarcoma-180 (solid type), NF-sarcoma and Yoshida sarcoma (46).

Pluramycin A (11) is the most thoroughly investigated compound of this family. It was included in a large number of comparative studies involving a wide variety of substances, tumors, cell cultures or microorganisms. In various experiments, pluramycin A proved to be active against Ehrlich ascites as well as Ehrlich subcutaneous solid type of carcinoma (71, 88). It is noteworthy that in these experiments the free base was more active than the ascorbic acid salt (71). Pluramycin A (11) was also effective against mouse sarcoma 180 and Yoshida rat sarcoma (71). Growth inhibition was noted for cultures of human KB cells (ID_{50} : 0.003 µg/ml) (87), and HeLa cells (59, 61). The inhibitory effect on KB cells, Ehrlich ascite cells, and mouse leukemia L-1210 cells was further monitored by incorporation of radioactive nucleotides (75). Pluramycin A tartrate was found to form a complex with a high molecular weight fraction of human serum. This complex was isolated by chromatography on Sephadex G-200; its minimal effective dose active against Ehrlich carcinoma was 8 to 16 time less that of pluramycin A alone (73).

7.6. Interaction with Nucleic Acids and Mode of Action

It is well known that DNA is a sensitive target for many antitumor antibiotics and plays an important role in the mechanism of action of such compounds (54). In many experiments the pluramycin antibiotics were found to strongly interact with nucleic acids. The most evident effect of these compounds is that they raise T_m of DNA. For calf thymus DNA elevation of T_m of some 30 degrees was reported with hedamycin (17) (42) or neopluramycin (2) (93); elevations of 20° and 11° were observed for salmon sperm DNA with pluramycin A (11) (89) and the rubiflavin mixture (101), respectively. A similar behaviour was also displayed by the polynucleotide $d(T-A)_n$; hedamycin (17) raised its T_m by 20° (42).

The antibiotics seem to form rather strong complexes with DNA. JERNIGAN et al. (41) suggested that three types of interactions were present between DNA and hedamycin (17). Type III was described as a loose aggregation at the surface of the DNA; it is easily broken up by dialysis in low ionic strength salt solutions. Binding type II is stronger as indicated by the 1 M NaCl necessary for removal of the bound antibiotic by dialysis. This type of binding might involve intercalation. The strongest interaction was called type I. It was not possible to remove the antibiotic from this interaction by dialysis, even at high salt concentrations. This binding type was thought to involve covalent bonds between hedamycin and DNA. The authors suggested that the antibiotic is first bound through the interaction of type II and that then a relatively slow (6–10 h at 25°) alkylation of the DNA bases by the hedamycin epoxides might take place. However, nobody has so far isolated base alkylation products from such complexes. There seems to be an optimal ratio of 1 hedamycin molecule per 5–10 nucleotides for these interactions. A similar ratio was derived by JOEL and GOLDBERG from measurements of the change in the UV spectrum of hedamycin when interacting with DNA (42).

This strong complex formation results in stabilization of the double helix as shown by the increased T_m and inhibition of the enzymatic degradation of DNA by snake venom phosphodiesterase (100). It also seems to be the reason for the inhibition of the DNA synthesis and subsequently of the RNA and protein syntheses observed with E. coli and HeLa cell cultures when treated with pluramycin A (11) (69).

NAGAI et al. (70) further showed that the inhibition of DNA synthesis due to pluramycin A (11) in a cell-free system of E. coli could be reversed by the addition of more template DNA. The same was true for the pluramycin A induced inhibition of Mycoplasma mycoides. These findings suggest that pluramycin A can be 'inactivated' by the addition of sufficiently DNA.

An interesting observation was made by WHITE and DEARMAN (99). When the rubiflavin mixture was added to a suspension of E. coli containing a source of electrons such as ethanol, biological reduction of the rubiflavin occurred. An EPR signal could be measured and was assigned to the semiquinone radical. The assigment was only tentative since the signal observed did not show the fine structure exhibited by the corresponding ERP signal of chemically reduced rubiflavin. This lack of fine structure (splitting) could, however, also be taken as an evidence for an association of the radical with a macromolecule – presumably DNA.

It is noteworthy that carminic acid (**59**), an anthraquinone *C*-glycoside and thus resembling somewhat the pluramycin antibiotics, was shown by LOWN *et al.* (*55*) to effect single strand scission of covalently closed circular PM-2 DNA when the compound was reduced with sodium borohydride to the hydroquinone stage. EPR experiments indicated that the reoxidation of carminic acid proceeds *via* the semiquinone radical and ultimately leads to the liberation of a free hydroxyl radical, thought to be the active reagent in the strand breaking process. Strand breaking was indeed detected when covalently closed circular PM-2 DNA treated with hedamycin (**17**) was centrifuged in alkaline sucrose gradients (*64*).

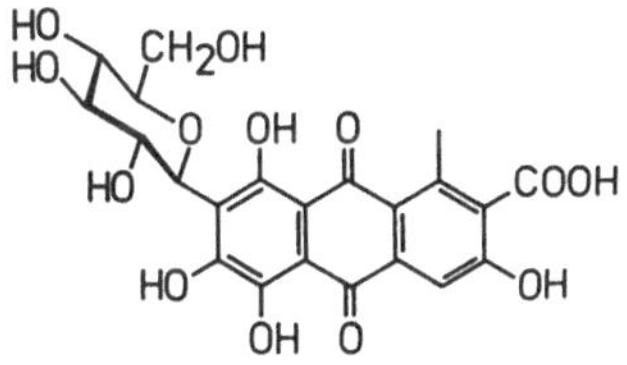

(**59**)

No systematic study of the DNA-pluramycin antibiotic interaction has been undertaken so far. However, if one assumes a similar behaviour for the different members of this family of antibiotics, the many single results described above might lead to the following speculative picture for the mode of action of the pluramycin antibiotics:

The antibiotic loosely interacts in a first step with DNA and then, in a second one, intercalates. At these stages the two sugar rings must play an important role as shown by our structure-activity studies and the inactivity of the aglycones. The intercalation stabilizes the double helix thus restricting the unwinding necessary for DNA and RNA synthesis, which are consequently blocked or slowed down. This also brings the protein biosynthesis to an end, but only at some later moment, when there is no more messenger RNA in the cytoplasm. The next phase is alkylation of nucleobases which will further stabilize the double helix and also make it succeptible to strand breaking reagents. Strand breaking or even crosslinking might also be caused by radicals formed during the reoxidation of the pluramycin antibiotics that were previously reduced in some way to the hydroquinone stage. A lot of experiments remain to be done, of course, to corroborate, correct or disprove this hypothetical pathway.

Acknowledgement

Research in the author's laboratory was accomplished thanks to the skillful collaboration of Dr. M. Ceroni, Dr. A. Fredenhagen and Dr. H. Nadig. Financial support came from the Swiss National Science Foundation and the Ciba-Stiftung, Basel. In addition, Dr. H. Nadig is thanked for critically reading this manuscript.

Note added in proof:

Very recently, a new compound related to the pluramycin antibiotics was detected by Itoh and coworkers[1]: the antibiotic SF-2330 (**60**). This compound is interesting in many respects. It is the first pluramycinone to be biologically active and also the first "aglycone" with epoxides in the side chain. Furthermore, its side chain has never been observed so far in pluramycin type antibiotics: it is a four carbon side chain with a diepoxide structure. SF-2330 is therefore also a new addition to the list of natural products with open chain 1,2:3,4 diepoxides.

The antibiotic SF-2330 is a metabolite of *Streptomyces sp.* SF-2330 and was extracted from the fermentation broth and the mycelium with ethyl acetate and aqueous acetone, respectively. Purification was achieved by repeated chromatography on silica gel columns. From 35 l of fermentation broth 115 mg of crude SF-2330 were obtained, which after recrystallization from chloroform gave 75 mg of orange needles. The elemental composition was determined as $C_{22}H_{14}O_7$ (molecular weight 370). The constitution of SF-2330 (**60**) was elucidated with spec-

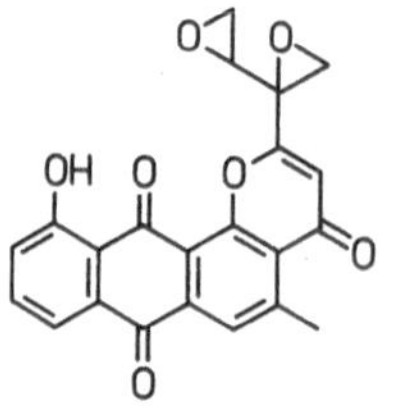

(**60**)

[1] Itoh J., T. Shomura, T. Tsuyuki, J. Yoshida, M. Ito, M. Sezaki, and M. Kojima: Studies on a New Antibiotic SF-2330. I. Taxonomy, Isolation and Characterization. J. Antibiotics **39**, 773 (1986). Itoh, J., T. Tsuyuki, K. Fujita, and M. Sezaki: Studies on a New Antibiotic SF-2330. II. The Structural Elucidation. J. Antibiotics **39**, 780 (1986).

troscopic methods (MS, ^{1}H- and ^{13}C-NMR); the configurations in the diepoxide fragment, however, remain to be determined.

The antibiotic SF-2330 is the first pluramycin aglycone that was found to be biologically active. Its acute toxicity (LD_{50}: 18.8 mg/kg mouse i.p.) is lower than that of hedamycin (17) or pluramycin A (11). The compound (60) was active against Gram-positive but not Gram-negative bacteria. The minimal inhibitory concentrations determined e.g. for *Staphylococcus aureus* (1.56 µg/ml) and *Bacillus subtilis* ATCC 6633 (0.78 µg/ml) were almost identical with the corresponding values obtained with hedamycin (17) and epoxykidamycin (6).

The newly detected antibiotic SF-2330 (60) is of great interest for the study of structure-activity relationships in the pluramycin/pluramycinone family of natural products. Since this compound lacks the sugars, the investigation of its behaviour towards DNA should tell whether rings E and F of the pluramycins are mandatory for the strong interaction with DNA of these compounds. On the other hand, the properties of SF-2330 suggest that much of the antimicrobial activity of the pluramycins with epoxidized side chains such as epoxykidamycin (6), rubiflavin A (10), pluramycin A (11), and hedamycin (17) must be due to the epoxides and be independent of the presence of the sugars. Since the feasibility of the total synthesis of a pluramycinone has been demonstrated, SF-2330 might eventually become available in quantities that would allow quite detailed studies of its mode of action.

References

1. Aszalos, A., M. Jelinek, and B. Berk: Rubiflavin, a toxic antitumor antibiotic. Antimicrobial Agents and Chemotherapy 1964, 68 (1965).
2. Aszalos, A., R.S. Robison, N.V. Kraemer, J.A. Henshaw, and M.S. Giannini: Tumimycin, seine Salze, Verfahren zu seiner Herstellung und seine Verwendung als Antibioticum. Deutsche Offenlegungsschrift 2139261 (17. 2. 1972).
3. Baer, H.H., and F.F.Z. Georges: The Synthesis of D-Angolosamine. Can. J. Chem. 55, 1100 (1977).
4. Baker, W.: Molecular Rearrangement of Some o-Acyloxyacetophenones and the Mechanism of the Production of 3-Acylchromones. J. Chem. Soc. (London) 1933, 1381.
5. Bartus, H.R., C.K. Mirabelli, J.I. Auerbach, A.R. Shatzman, D.P. Taylor, R.K. Johnson, M. Rosenberg, and S.T. Crooke: Improved Genetically Modified *Escherichia Coli* Strain for Prescreening Antineoplastic Agents. Antimicrobial Agents and Chemotherapy 1984, 622.
6. Benn, R. and H. Günther: Moderne Pulsfolgen in der hochauflösenden NMR-Spektroskopie. Angew. Chem. 95, 381 (1983).
7. Bérdy, J.: Recent Developments of Antibiotic Research and Classification of Antibiotics According to Chemical Structure. Adv. Appl. Microbiol. 18, 309 (1974).
8. Bradner, W.T., B. Heinemann, and A. Gourevitch: Hedamycin, a New Antitumor

Antibiotic. II. Biological Properties. Antimicrobial Agents and Chemotherapy **1966**, 613 (1967).

9. BROCKMANN, H.: Indomycine und Indomycinone. Angew. Chem. **80**, 493 (1968).

10. BRUFANI, M., and W. KELLER-SCHIERLEIN: Stoffwechselprodukte von Mikroorganismen. 54. Mitteilung. Über die Zuckerbausteine des Angolamycins: L-Mycarose, D-Mycinose und D-Angolosamin. Helv. Chim. Acta **49**, 1962 (1966).

11. BYRNE, K.M., S.K. GONDA, and B.D. HILTON: Largomycin FII Chromophore Component 4, a New Pluramycin Antibiotic. J. Antibiotics **38**, 1040 (1985).

12. CERONI, M., and U. SÉQUIN: The Structure of the Antibiotic Hedamycin. IV. Relative Configurations in the Diepoxide Side Chain. Tetrahedron Letters **1979**, 3703.

13. CERONI, M., and U. SÉQUIN: Determination of the Relative Configurations in the Side Chains of the Antibiotics Hedamycin and Pluramycin A; Synthesis and NMR Data of Suitable Model Compounds. Helv. Chim. Acta **65**, 302 (1982).

14. CORBAZ, R., L. ETTLINGER, E. GÄUMANN, W. KELLER-SCHIERLEIN, L. NEIPP, V. PRELOG, P. REUSSER, and H. ZÄHNER: Stoffwechselprodukte von Actinomyceten. 2. Mitteilung. Angolamycin. Helv. Chim. Acta **38**, 1202 (1955).

15. DAHM, K.H.: Personal communication.

16. DAHM, K.H.: Unpublished results.

17. DORNBERGER, K., U. BERGER, W. GUTSCHE, W. JUNGSTAND, K. WOHLRABE, A. HÄRTL, and H. KNÖLL: Griseorubins, a New Family of Antibiotics with Antimicrobial and Antitumor Activity. II. Biological Properties and Antitumor Activity of the Antibiotic Complex Griseorubin. J. Antibiotics **33**, 9 (1980).

18. DORNBERGER, K., U. BERGER, and H. KNÖLL: Griseorubins, a New Family of Antibiotics with Antimicrobial and Antitumor Activity. I. Taxonomy of the Producing Strain, Fermentation, Isolation and Chemical Characterization. J. Antibiotics **33**, 1 (1980).

19. ECKARDT, K.: Quinones and Other Carbocyclic Antitumor Antibiotics. In: Antitumor Compounds of Natural Origin: Chemistry and Biochemistry (ASZALOS, A., ed.), Vol II, p. 27. Boca Raton, FA: CRC Press, Inc. 1981.

20. ELLIS, G.P.: General Methods of Preparing Chromones. In: The Chemistry of Heterocyclic Compounds, Vol. 31: Chromenes, chromanones and chromones (ELLIS, G.P., ed.), p. 495. New York: John Wiley & Sons 1977.

21. FREDENHAGEN, A., W. RITTER, U. SÉQUIN, and M. ZEHNDER: Solid State Conformation of a Bioxirane Related to Hedamycin. Chimia **35**, 334 (1981).

22. FREDENHAGEN, A., and U. SÉQUIN: Phototransformations of Some 2-Substituted 4*H*-Chromen-4-ones (4-chromones) Related to the Antitumor Antibiotic Hedamycin. Helv. Chim. Acta **66**, 586 (1983).

23. FREDENHAGEN, A., and U. SÉQUIN: The Structures of Some Products from the Photodegradation of the Pluramycin Antibiotics Hedamycin and Kidamycin. Helv. Chim. Acta **68**, 391 (1985).

24. FREDENHAGEN, A., and U. SÉQUIN: The Photodeactivation of Hedamycin, an Antitumor Antibiotic of the Pluramycin Type. J. Antibiotics **38**, 236 (1985).

25. FRENCH, J.C.: Personal communication.

26. FRICKE, I.: Zur Kenntnis der Indomycine, einer neuen Klasse von antibiotisch und cytostatisch wirksamer Naturstoffe. Dissertation. Göttingen 1973.

27. FRONZA, G., C. FUGANTI, P. GRASSELI, and G. PEDROCCHI-FANTONI: Synthesis of the Four Configurational Isomers of *N*-Benzoyl-2,3,6-trideoxy-3-*C*-methyl-3-amino-L-hexose from the (2*S*,3*R*)-Diol obtained from α-Methylcinnamaldehyde by Fermentation with Bakers' yeast. J. Carbohyd. Chem. **2**, 225 (1983).

28. FURUKAWA, M., I. HAYAKAWA, G. OHTA, and Y. IITAKA: Structure and Chemistry of Kidamycin. Tetrahedron **31**, 2989 (1975).

29. FURUKAWA, M., and Y. IITAKA: Structure of Kidamycin: X-ray Analysis of Isokidamycin Derivatives. Tetrahedron Letters **1974**, 3287.

29a. FURUKAWA, M., and Y. IITAKA: Structures of Kidamycin Derivatives: Triacetylmethoxykidamycin Bis(trimethylammonium) Iodide and Isokidamycin Bis(m-bromobenzoate). Acta Cryst. **B 36**, 2270 (1980).

30. FURUKAWA, M., A. ITAI, and Y. IITAKA: Crystallographic Studies of an Anthraquinone Derivative Obtained from Kidamycin. Tetrahedron Letters **1973**, 1065.

31. GERWICK, W.H., and W. FENICAL: Spatane Diterpenoids from the Tropical Marine Algae Spatoglossum Schmittii and Spatoglossum Howleii (Dictyotaceae). J. Organ. Chem. (USA) **48**, 3325 (1982).

32. GERWICK, W.H., W. FENICAL, D. VAN ENGEN, and J. CLARDY: Isolation and Structure of Spatol, a Potent Inhibitor of Cell Replication from the Brown Seaweed Spatoglossum Schmittii. J. Amer. Chem. Soc. **102**, 7991 (1980).

33. GONDA, S.K., K.M. BYRNE, P.K. HERBER, Y. TONDEUR, D. LIBERATO, and B.D. HILTON: Structure and Properties of Major Largomycin FII Chromophore Components. J. Antibiotics **37**, 1344 (1984).

34. HAMANOUE, K., K. YOKOYAMA, T. MIYAKE, T. KASUYA, T. NAKAYAMA, and H. TERANISHI: Photochemical Reactions of Chloroanthraquinones. Chemistry Letters **1982**, 1967.

35. HATA, T., T. HOSHINO, A. MATSUMAE, S. NOMURA, Y. SANO, and Y. YAJIMA: Iyomycin, a New Antitumor Antibiotic. Intern. Congr. Chemotherapy, Proc., 3rd, Stuttgart, 1963, **2**, 1032 (1964). Cf. Chem. Abstr. **65**, 6127 g (1966).

35a. HAUSER, F.M., and S.R. ELLENBERGER: Syntheses of 2,3,6-Trideoxy-3-amino- and 2,3,6-Trideoxy-3-nitrohexoses. Chem. Rev. **86**, 35 (1986).

36. HAUSER, F.M., and R.P. RHEE: 4H-Anthra[1,2-b]pyran Antibiotics. Total Synthesis of the Methyl Ether of Kidamycinone. J. Amer. Chem. Soc. **101**, 1628 (1979).

37. HAUSER, F.M., and R.P. RHEE: Anthra[1,2-b]pyran Antibiotics: Total Synthesis of O-Methylkidamycinone. J. Organ. Chem. (USA) **45**, 3061 (1980).

38. HEINEMANN, B., and A.J. HOWARD: Antiphage Properties of Compounds Possessing Both Antitumor and Inducing Activities. Antimicrobial Agents and Chemotherapy **1964**, 126 (1965).

39. HORI, Y., M. HINO, Y. KAWAI, S. KIYOTO, H. TERANO, M. KOHSAKA, H. AOKI, M. HASHIMOTO, and H. IMANAKA: A New Antibiotic, Chromoxymycin. II. Production, Isolation, Characterization and Antitumor Activity. J. Antibiotics **39**, 12 (1986).

40. HOWE, I., and M. JARMAN: New Techniques for the Mass Spectrometry of Natural Products. Fortschr. Chem. organ. Naturstoffe **47**, 107 (1985).

41. JERNIGAN, H.M., J.L. IRVIN, and J.R. WHITE: Binding of Hedamycin to Deoxyribonucleic Acid and Chromatin of Testis and Liver. Biochemistry **17**, 4232 (1978).

42. JOEL, P.B., and I.H. GOLDBERG: The Inhibition of RNA and DNA Polymerases by Hedamycin. Biochim. Biophys. Acta **224**, 361 (1970).

43. JOHNSON, A.W., R.M. SMITH, and R.D. GUTHRIE: Vancosamine: The Structure and Configuration of a Novel Amino-sugar from Vancomycin. J.C.S. Perkin I **1972**, 2153.

44. KANDA, N.: A New Antitumor Antibiotic, Kidamycin. I. Isolation, Purification and Properties of Kidamycin. J. Antibiotics **24**, 599 (1971).

45. KANDA, N.: A New Antitumor Antibiotic, Kidamycin. III. Preparation and Properties of Acetylkidamycin. J. Antibiotics **25**, 557 (1972).

46. KANDA, N., M. KONO, and K. ASANO: A New Antitumor Antibiotic, Kidamycin. II. Experimental Treatment of Cancer with Kidamycin. J. Antibiotics **25**, 553 (1972).

47. KAWAI, Y., K. FURIHATA, H. SETO, and N. OTAKE: The Structure of a New Antibiotic, Chromoxymycin. Tetrahedron Letters **1985**, 3273 (1985).

48. KELLER, P., G. EGGART, H. WEHRLI, K. SCHAFFNER, and O. JEGER: Photochemische Reaktionen. 41. Zur bimolekularen Photoreduktion cyclischer Ketone, cyclischer β-Hydroxyketone und cyclischer α, β-Epoxyketone. Helv. chim. Acta **50**, 2259 (1967).

49. KINUMAKI, A., and M. SUZUKI: Proposed Structure of Angolamycin (Shincomycin A) by Mass Spectroscopy. J. Antibiotics 25, 480 (1972).

50. KONDO, S., M. MIYAMOTO, H. NAGANAWA, T. TAKEUCHI, and H. UMEZAWA: Structures of Pluramycin A and Neopluramycin. J. Antibiotics 30, 1143 (1977).

51. KONDO, S., T. WAKASHIRO, M. HAMADA, K. MAEDA, T. TAKEUCHI, and H. UMEZAWA: Isolation and Characterization of a New Antibiotic, Neoluramycin. J. Antibiotics 23, 354 (1970).

52. KOO, J.: Synthesis in the Chromone Series. 5,8-Dimethoxy-2-substituted Chromones and Nitrogen Analogs. J. Organ. Chem. (USA) 26, 2440 (1961).

53. KUDINOVA, M.K., G.A. BABENKO, R.S. UKHOLINA, T.S. MAKSIMOVA, N.P. NECHAEVA, L.P. TEREKHOVA, and O.K. ROSSOLIMO: Antitumor Antibiotic 4418 Similar to Antibiotics of the Pluramycin-Iomycin Group. Antibiotiki 1968, 201.

54. LOWN, J.W.: Newer Approaches to the Study of the Mechanism of Action of Antitumor Antibiotics. Acc. Chem. Res. 15, 381 (1982).

55. LOWN, J.W., H.-H. CHEN, S.-K. SIM, and J.A. PLAMBECK: Reactions of the Antitumor Agent Carminic Acid and Its Derivatives with DNA. Bioorganic Chem. 8, 17 (1979).

56. MACFARLANE, R.D.: Californium-252 Plasma Desorption Mass Spectrometry (PDMS) of Antibiotic Molecules. NBS Spec. Publ. (U.S.), 519 (Trace Org. Anal.: New Front. Anal. Chem.), 673 (1979). Cf. Chem. Abstr. 91, 139904 v (1979).

57. MAEDA, K., T. TAKEUCHI, K. NITTA, K. YAGISHITA, R. UTAHARA, T. OSATO, M. UEDA, S. KONDO, Y. OKAMI, and H. UMEZAWA: A New Antitumor Substance, Pluramycin. J. Antibiotics, Ser. A 9, 75 (1956).

58. MASSIOT, G., S.K. KAN, P. GONORD, and C. DURET: The Fourier Transform Difference Spectra method. An Application to Structural Elucidation of Andranginine, a Novel Indole Alkaloid. J. Amer. Chem. Soc. 97, 3277 (1975).

59. MATSUBARA, H.: Cytotoxic Characteristics of Virus Antitumor Antibiotics in HeLa Cell Cultures and a Proposed Anticellogram for the Differentiation of These Substances. J. Antibiotics, Ser. B 13, 262 (1960).

60. MATSUMAE, A., and T. HATA: Morphological Change of Bacteria Induced by Chemotherapeutic Agents. I. A Classification of Chemotherapeutic Agents Based on Their Activities to Induce Morphological Changes of Escherichia Coli Strain B. J. Antibiotics, Ser. A 17, 164 (1964).

61. MATSUMOTO, K.: Cytological Effects of Various Antitumor Substances on HeLa Cells and Synergistic Effects. J. Antibiotics, Ser. B 14, 1 (1961).

62. MCCORMICK, M.H., W.M. STARK, G.E. PITTENGER, R.C. PITTENGER, and J.M. MCGUIRE: Vancomycin, a New Antibiotic. I. Chemical and Biologic Properties. Antibiotics Ann. 1955/56, 606.

63. MONDON, A., B. EPE, U. OELBERMANN, V. SINNWELL, and G. REMBERG: Zur Kenntnis der Bitterstoffe aus Cneoraceen – XVII. Tetrahedron Letters 23, 4015 (1982).

64. MONG, S., J.E. STRONG, and S.T. CROOKE: Interaction of Covalently Closed Circular PM-2 DNA and Hedamycin. Biochem. Biophys. Res. Commun. 88, 237 (1979).

65. NADIG, H.: Über Antibiotika des Pluramycin-Typs. Isolierung und Strukturaufklärung einiger Rubiflavin-Komponenten. Struktur von Hedamycin. Dissertation. Basel 1982.

66. NADIG, H., and U. SÉQUIN, unpublished.

67. NADIG, H., and U. SÉQUIN: A Structural Investigation of the Antibiotic Rubiflavin. Helv. Chim. Acta 63, 2446 (1980).

68. NADIG, H., U. SÉQUIN, R.H. BUNGE, T.R. HURLEY, D.B. MURPHEY, and J. C. FRENCH: Isolation and Structure of a New Antibiotic Related to Rubiflavin A. Helv. Chim. Acta 68, 953 (195).

69. NAGAI, K., N. TANAKA, and H. UMEZAWA: Inhibition of Nucleic Acid Biosynthesis in Cell-free Systems of Escherichia Coli B by Pluramycin. J. Biochemistry (Tokyo) 67, 655 (1970).

70. NAGAI, K., H. YAMAKI, N. TANAKA, and H. UMEZAWA: Inhibition by Pluramycin A of Nucleic Acid Biosynthesis. J. Biochemistry (Tokyo) **62**, 321 (1967).
71. NISHIBORI, A.: Antitumor Effect of Pluramycin A on Experimental Animal Tumors. J. Antibiotics, Ser. A **10**, 213 (1957).
72. OGAWARA, H., K. MAEDA, K. NITTA, Y. OKAMI, T. TAKEUCHI, and H. UMEZAWA: An Antibiotic, Plurallin, Consisting of a Pluramycin-like Prosthetic Group and a Glycoprotein. J. Antibiotics, Ser. A **19**, 1 (1966).
73. OGAWARA, H., K. MAEDA, and H. UMEZAWA: Pluramycin Complex with Human Serum Albumin and the Antitumor Activity. J. Antibiotics, Ser. A **19**, 141 (1966).
74. OMURA, S., A. NAKAGAWA, H. TAKESHIMA, J. MIYAZAWA, C. KITAO, F. PIRIOU, and G. LUKACS: A ^{13}C Nuclear Magnetic Resonance Study of the Biosynthesis of the 16-Membered Macrolide Antibiotic Tylosin. Tetrahedron Letters **1975**, 4503.
75. OTAKE, N., and T. SASAKI: A Screening Procedure on the Inhibitors of Substrate-Incorporation in Tumor Cells. Methodological Survey and Its Evaluation. Agric. Biol. Chem. **41**, 1039 (1977).
76. PARKER, W.L., M.L. RATHNUM, V. SEINER, W.H. TREJO, P.A. PRINCIPE, and R.B. SYKES: Cepacin A and Cepacin B, Two New Antibiotics Produced by *Pseudomonas Cepacia*. J. Antibiotics **37**, 431 (1984).
77. PRICE, K.E., R.E. BUCK, and J. LEIN: Incidence of Antineoplastic Activity Among Antibiotics Found to be Inducers of Lysogenic Bacteria. Antimicrobial Agents and Chemotherapy **1964**, 505 (1965).
78. SANO, Y., N. KANDA, and T. HATA: Iyomycin, a New Antitumor Antibiotic from *Streptomyces*. III. Isolation and Properties of Iyomycin B. J. Antibiotics, Ser. A **17**, 117 (1964).
79. SCHMITZ, H., K.E. CROOK, JR., and J.A. BUSH: Hedamycin, a New Antitumor Antibiotic. I. Production, Isolation, and Characterization. Antimicrobial Agents and Chemotherapy **1966**, 606 (1967).
80. SCHNELL, J.: Indomycine, eine Gruppe neuer Cytostatika aus Streptomyceten. Dissertation. Göttingen 1963.
81. SCOTT, A.I.: Personal communication.
82. SÉQUIN, U.: The Structure of the Antibiotic Hedamycin. II. Comparison of Hedamycin and Kidamycin. Tetrahedron **34**, 761 (1978).
83. SÉQUIN, U.: ^{13}C-NMR. Spectral Differences Between Corresponding Methyl Esters, Phenyl Esters and 2-Substituted Chromones. Helv. Chim. Acta **64**, 2654 (1981).
84. SÉQUIN, U., C.T. BEDFORD, S.K. CHUNG, and A.I. SCOTT: The Structure of the Antibiotic Hedamycin. I. Chemical, Physical and Spectral Properties. Helv. Chim. Acta **60**, 896 (1977).
85. SÉQUIN, U., and M. CERONI: Concerning the Configuration of the Side Chain in the Antibiotic Pluramycin A. Helv. Chim. Acta **61**, 2241 (1978).
86. SÉQUIN, U., and M. FURUKAWA: The Structure of the Antibiotic Hedamycin. III. ^{13}C-NMR Spectra of Hedamycin and Kidamycin. Tetrahedron **34**, 3623 (1978).
87. SMITH, C.G., W.L. LUMMIS, and J.E. GRADY: An Improved Tissue Culture Assay. II. Cytotoxicity Studies with Antibiotics, Chemicals, and Solvents. Cancer Res. **19**, 847 (1959).
88. TAKEUCHI, T., T. HIKIJI, K. NITTA, and H. UMEZAWA: Effect of Pluramycin A on Ehrlich Carcinoma of Mice. J. Antibiotics, Ser A. **10**, 143 (1957).
89. TANAKA, N., K. NAGAI, H. YAMAGUCHI, and H. UMEZAWA: Inhibition of RNA and DNA Polymerase Reactions by Pluramycin A. Biochem. Biophys. Res. Commun. **21**, 328 (1965).
90. TERUI, Y., K. TORI, and N. TSUJI: Esterification Shifts in Carbon-13 NMR Spectra of Alcohols. Tetrahedron Letters **1976**, 621.
91. THANG, T.T., F. WINTERNITZ, A. OLESKER, A. LAGRANGE, and G. LUKACS: Synthesis

of a Derivative of Vancosamine, a Component of the Glycopeptide Antibiotic Vancomycin. J.C.S. Chem. Comm. **1979**, 153.

92. Toth-Sarudy, E., I. Gado, J. Gyimesi, M. Halasz, I. Horvath, K. Magyar, L. Alfoldi, J. Bérdy, and B. Doczi: Griseofagins. Hung. Pat. 157600, 13. Juli 1970. Cf. Chem. Abstr. **73**, 119214v (1970).

93. Tsukada, I., M. Hamada, and H. Umezawa: Neopluramycin, an Inhibitor of Nucleic Acid Synthesis. J. Antibiotics **24**, 189 (1971).

94. Turner, W.B.: Fungal Metabolites, p. 74ff. London, New York: Academic Press 1971.

95. Turner, W.B., and D.C. Aldridge: Fungal Metabolites II, p. 55ff. London, New York: Academic Press 1983.

96. Umezawa, I., Komiyama, H. Takeshima, T. Hata, M. Kono, and N. Kanda: A New Antitumor Antibiotic, Kidamycin. IV. Pharmacokinetics of Acetylkidamycin. J. Antibiotics **26**, 669 (1973).

97. Vignon, M.R., and P.J.A. Vottero: RMN ^{13}C: Sur l'utilisation des esters pour l'attribution des carbones des molécules glucidiques. Tetrahedron Letters **1976**, 2445.

98. Weringa, W.D., D.H. Williams, J. Feeney, J.P. Brown, and R.W. King: The Structure of an Amino-sugar from the Antibiotic Vancomycin. J.C.S. Perkin I **1972**, 443.

99. White, J.R., and H.H. Dearman: Generation of Free Radicals from Phenazine Methosulfate, Streptronigrin, and Rubiflavin in Bacterial Suspensions. Proc. Natl. Acad. Sci. U.S. **54**, 887 (1965).

100. White, H.L., and J.R. White: Hedamycin and Rubiflavin Complexes with Deoxyribonucleic Acid and Other Polynucleotides. Biochemistry **8**, 1030 (1969).

101. White, H.L., and J.R. White: Binding of Rubiflavin to Deoxyribonucleic Acid in Relation to Antibacterial Action. Antimicrobial Agents and Chemotherapy **1966**, 227 (1967).

102. Wilkinson, F.: Transfer of Triplet-State Energy and the Chemistry of Excited States. J. Phys. Chem. **66**, 2569 (1962).

103. Yamaguchi, T., T. Furumai, M. Sato, T. Okuda, and N. Ishida: Studies on a New Antitumor Antibiotic, Largomycin. I. Taxonomy of the Largomycin-Producing Strain and Production of the Antibiotic. J. Antibiotics **23**, 369 (1970).

104. Yoshida, T., and K. Katagiri: Anthracidins A and B, New Antibiotics. Antimicrobial Agents and Chemotherapy **1963**, 63 (1964).

105. Zehnder, M., U. Séquin, and H. Nadig: The Structure of the Antibiotic Hedamycin. V. Crystal Structure and Absolute Configuration. Helv. Chim. Acta **62**, 2525 (1979).

(*Received May 16, 1986*)

Cyclosporine and Analogues – Isolation and Synthesis – Mechanism of Action and Structural Requirements for Pharmacological Activity

By R.M. WENGER, Preclinical Research, Sandoz Ltd., Basel, Switzerland

With 14 Figures

Contents

1. Introduction

1.1. History and Summary

Cyclosporine (**1**), originally named "Cyclosporin A" (for nomenclature see section 1.2), is the main component of a new family of cyclic peptides each comprising 11 amino acids. These peptides are produced as secondary fungal metabolites by *Cylindrocarpon lucidum Booth* and *Tolypocladium inflatum Gams*. Both strains of fungi imperfecti were isolated from soil samples collected in Wisconsin (USA) and Hardanger Vidda (Norway). The isolation, the culture conditions and the taxonomy of these fungi are reported by Dreyfuss *et al.* (*1*).

Cyclosporine (**1**), initially isolated for its antifungal activities – extracts of cultures exhibit *in vitro* a narrow spectrum of antifungal activities but exert only marginal effects *in vivo* – was characterized by Borel *et al.* (*2–6*) as an immunosuppressant which acted mainly on T-lymphocytes, showed antiinflammatory effects and was practically devoid of toxicity. It is unique among the presently available immunosuppressive drugs in that it reversibly inhibits only some classes of lymphocytes and does not affect hemopoietic tissues (bone marrow). More recently its spectrum of activity has been shown to include a number of antiparasitic actions. Bueding *et al.* (*7*) found that the substance had antischistosomal activity and Thommen (*8*) was the first to demonstrate its antimalarial action. Its effect on other parasitic models is currently under investigation.

Cyclosporine has been reported to be effective in preventing allograft rejections for a variety of organs in a number of different species [for reviews see (9) and (10)]. The first transplantations in humans performed with the aid of cyclosporine were reported by CALNE et al. (11) for kidney, and by POWLES et al. (12) for bone marrow. Since then the use of cyclosporine has been extended to include liver, pancreas, lung and corneal transplantations and the availability of the drug has led to a revival in heart transplantation. The recent proceedings of the First International Congress on Cyclosporine held in Houston, Texas, in May 1983 contains an impressive wealth of information on pharmacological and clinical actions of the drug (13). Cyclosporine was introduced on the market in 1983 for the prevention of organ rejection under the trade name SANDIMMUN(E)®. The extension of the use of Sandimmune to other clinical indications such as autoimmune diseases has recently been summarized (14).

In the present review only some aspects of cyclosporine will be presented. In the first section on the chemistry of cyclosporine and natural metabolites the main emphasis will be on describing the chemical and physicochemical work caried out in the course of their structure elucidation. A brief overview of the progress achieved during recent years in cyclosporine chemistry will be given. In the second section the fine specificity of monoclonal antibodies of cyclosporine will be briefly discussed to tentatively support the cyclosporine conformation obtained in the solid state as a model for our structure-activity relationships study. In the third section the biosynthesis of cyclosporine will be discussed. In the fourth section the synthesis of cyclosporine and some of its analogues will be described. In the fifth section the results obtained from specifically modified synthetic cyclosporines and some naturally occurring variants are interpreted in terms of the structural requirements for immunosuppressive activity. Although incomplete, these results are beginning to define a large area on the surface of the cyclosporine molecule, which must be involved in interactions between this drug and its receptor on or in the lymphocyte. In the last section some of the effects of cyclosporine in lymphoid cell culture systems are presented in terms of its possible mechanism of action.

1.2. Nomenclature

The name cyclosporin A was used in the scientific literature for many years. For practical reasons, it was then proposed to define "cyclosporin" as the cyclic undecapeptide with the structure of "cyclosporin A" (15). When the USAN name "cyclosporine" was accepted in the USA, this name was adopted for the basic structure (13, 16–18).

The other fungal metabolites of the same type, "cyclosporins B, C, D, E and G" have then, according to peptide nomenclature, (19) the following names: [Ala^2]cyclosporine (for cyclosporin B), [Thr^2]cyclosporine (for cyclosporin C), [Val^2]cyclosporine (for cyclosporin D), [Val^{11}]cyclosporine (for cyclosporin E) and [Nva^2]cyclosporine (for cyclosporin G). Meanwhile the following other generic names have been accepted in other countries: England, "cyclosporin" (BAN name); France,"ciclosporine", Spain "ciclosporina"; Switzerland and others "ciclosporin" (INN name, WHO). The trade name worldwide for cyclosporin A is Sandimmun®, except for North America where it is spelt Sandimmune® (20).

1.3. Production of Cyclosporine

Cyclosporine is produced by fungi of the genus *Tolypocladium* in submerged cultures (1). In normal fermentation broths it is the main component of a group of cyclic undecapeptides. The isolation of these naturally occurring cyclosporines has been reported partly by RUEGGER et al. (22) and by TRABER et al. (23–26). They usually differ in their chemical structure only at one amino acid and their biosynthesis can be directed by externally supplying the corresponding precursor. For example, KOBEL et al. (27) reported a more selective production of (norvaline-2)cyclosporine by adding L-norvaline to the fermentation broths. Addition of D,L-α-aminobutyric acid almost completely suppresses the formation of undesired cyclosporine analogues and leads to a higher percentage of cyclosporine (1) in the crude product (for details on the structure of other metabolites see section 1.6).

1.4. Elucidation of the Structure of Cyclosporine – Conformation of Cyclosporine in the Crystal (for the conformation of cyclosporine in aprotic solvents see section 1.8)

The structure of cyclosporine (1) (Fig. 1) was determined by chemical degradation (22) together with an X-ray crystallographic analysis of the iodo derivative (2). As described by PETCHER et al. (28), this analysis not only established the amino acid sequence of the native peptide – the sequence D-alanine-L-alanine could not be established with certainty by chemical methods – but also gave an insight into its shape. Cyclosporine (1) is a neutral hydrophobic cyclic peptide composed of eleven amino acid residues, all having the L-configuration of the natural amino acids except for the D-alanine (D-Ala) in position

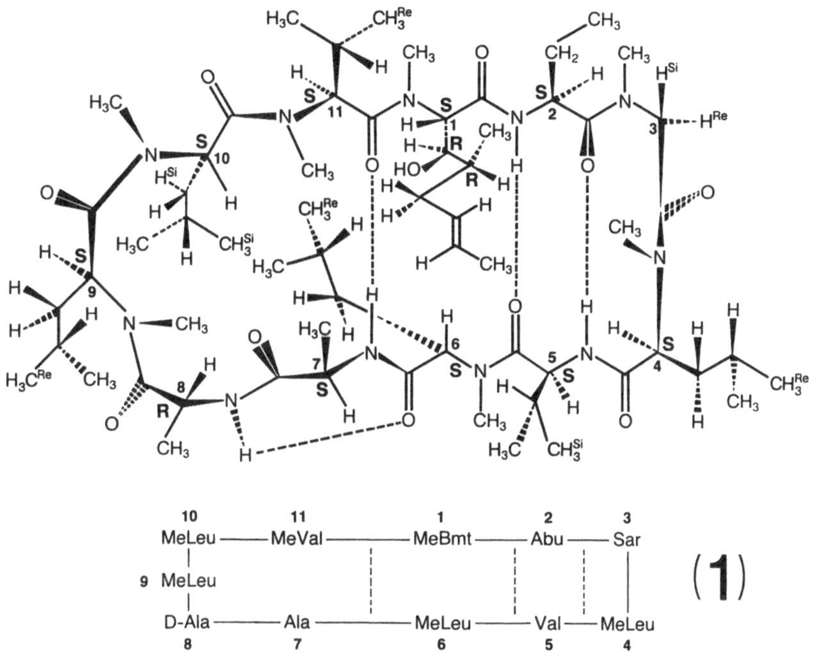

Fig. 1. Structure of cyclosporine (**1**) (schematic) corresponding to the conformation observed in the crystal (see Fig. 2). MeBmt = (4R)-4-[(E)-2-butenyl]-4,N-dimethyl-L-threonine (see text)

8 and the non-chiral sarcosine (Sar, N-methylglycine) in position 3. Ten of the eleven ring members are derivatives of known aliphatic amino acids: they are α-aminobutyric acid (Abu) in position 2, sarcosine (Sar) in position 3, N-methylleucine (MeLeu) in positions 4, 6, 9 and 10, valine (Val) in position 5, alanine (Ala) in position 7, D-alanine (D-Ala) in position 8 and N-methylvaline (MeVal) in position 11. These amino acids were readily characterized following acidic hydrolysis of cyclosporine or were converted to the corresponding N-trifluoroacetyl-O-methylesters.

One amino acid, namely that in position 1, was previously unknown. Being composed of 9 carbon atoms this amino acid was reported under the name "C-9 amino acid" (*22*). It was found to have the structure (2S,3R,4R,6E)-3-hydroxy-4-methyl-2-(methylamino)-6-octenoic acid, and, in accord with amino acid nomenclature, is now designated (4R)-4((E)-2-butenyl)-4,N-dimethyl-L-threonine and abbreviated MeBmt. Thus the novel amino acid has the polar features of an N-methyl-L-threonine which is substituted at the end of the carbon chain by butenyl and methyl groups. This amino acid was previously not known in the free form, since only artifacts and derivatives were obtained from degradation experiments on cyclosporine (*22*). Hydrolysis of cyclosporine (**1**) followed by ion-exchange chromatography affords a mixture of two cyclic tetrahydrofuran derivatives (**3**) and (**4**) of the MeBmt amino acids as the only isolable MeBmt-derivatives, while acid treatment of cyclosporine (**1**) in absence of water (CH$_3$SO$_3$H, CH$_3$OH, 50°) effects an N,O-acyl migration of the methylvaline moiety and furnishes isocyclosporine (**5**). A modified Edman degradation of isocyclosporine (**5**) using methylisothiocyanate in pyridine and then trifluoroacetic acid in 1-chlorbutane at 10 to 15° produces an anhydrothiohydantoinderivative (**6**) as the only isolable MeBmt-derivative and thus established MeBmt as the first amino acid in the peptide sequence. The complete sequence, except for the alanines, was established by repetitive Edman degradation.

In order to determine the absolute configuration of the amino acid in position 1 of cyclosporine (**1**) and in order to have definitive proof of its structure, the preparation of a suitable derivative for X-ray crystallographic analysis was undertaken. Treatment of cyclosporine (**1**) with iodine and thallium acetate, a procedure established by CAMBIE *et al.* (*29*), gave an iodo derivative (**2**) (Fig. 3) in which the MeBmt amino acid was present as an iodinated cyclic derivative. This compound (**2**), on treatment with zinc and acetic acid, regenerated cyclosporine (**1**). X-ray analysis (*28*) established the structure of the iodo derivative (**2**) from which the structure of cyclosporine itself could be deduced, with the exception of the geometry of the double bond of the MeBmt residue. This final problem was solved by measuring the coupling constant (16 Hz) between the vinyl protons in the 360 MHz NMR spectrum of cyclosporine (**1**) in deuterobenzene, which characterized the geometry as trans (E).

The structural assignments have since been confirmed by an X-ray crystallographic analysis (*30*) of cyclosporine (**1**) itself. The X-ray analysis also shows that the backbone of cyclosporine (**1**) has the same conformation as the heavy atom derivative (**2**). This conformational similarity in two different crystal structures can be considered as evidence for a relative high internal stability of the molecule against intermolecular forces (*30*).

The conformation of cyclosporine observed in the crystal obtained from acetone is shown in Fig. 2 and is schematically represented in Fig. 1. The cyclic peptide assumes a rather rigid conformation. A large portion of the backbone (residues 11–7) forms a β-fragment and adopts an antiparallel β-pleated sheet conformation which contains three transannular H-bonds and is markedly twisted (NH of Abu-2 to CO

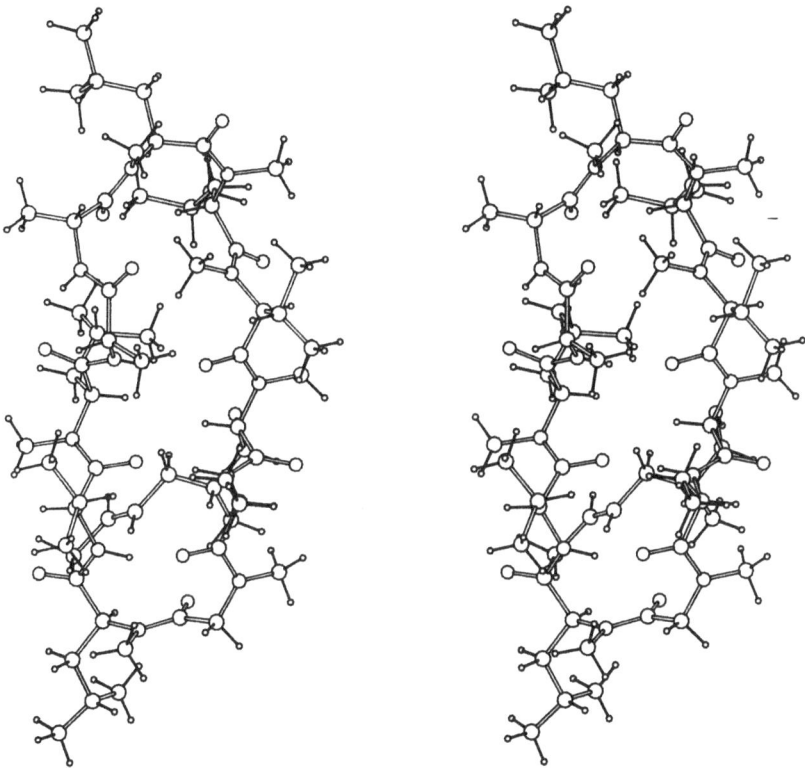

Fig. 2. Conformation of cyclosporine (1) in the crystal (stereograph)

of Val-5, NH of Val-5 to CO of Abu-2 and NH of Ala-7 to CO of MeVal-11). The sarcosine in position 3 and N-methylleucine in position 4 participate in a type II′ β-turn (*31–34*). This means that in the orientations chosen in Figures 1 and 2, the CO group of Abu-2 points up but is bridged with the NH-group of Val-5; the CO group of Sar-3 is down and the N-CH$_3$ groups of MeLeu-4 up. The Si proton of the methylene group of Sar-3 is axial, and the carbon chain of MeLeu-4 equatorial. The twist in the β-sheet is right-handed, as indicated by the virtual torsion angle ($+12°$) formed by the nitrogen of Val-5, the carbon of the CO group of Abu-2, the nitrogen of Abu-2 and the carbon of the CO group of Val-5. The remaining residues 7-11 form an open loop which contains the only cis-amide linkage in the molecule between two adjacent N-methylleucine residues 9 and 10. The fourth intramolecular H-bond (NH of D-Ala-8 to CO of MeLeu-6) is part of a seven-membered ring. This type of conformation for peptides was

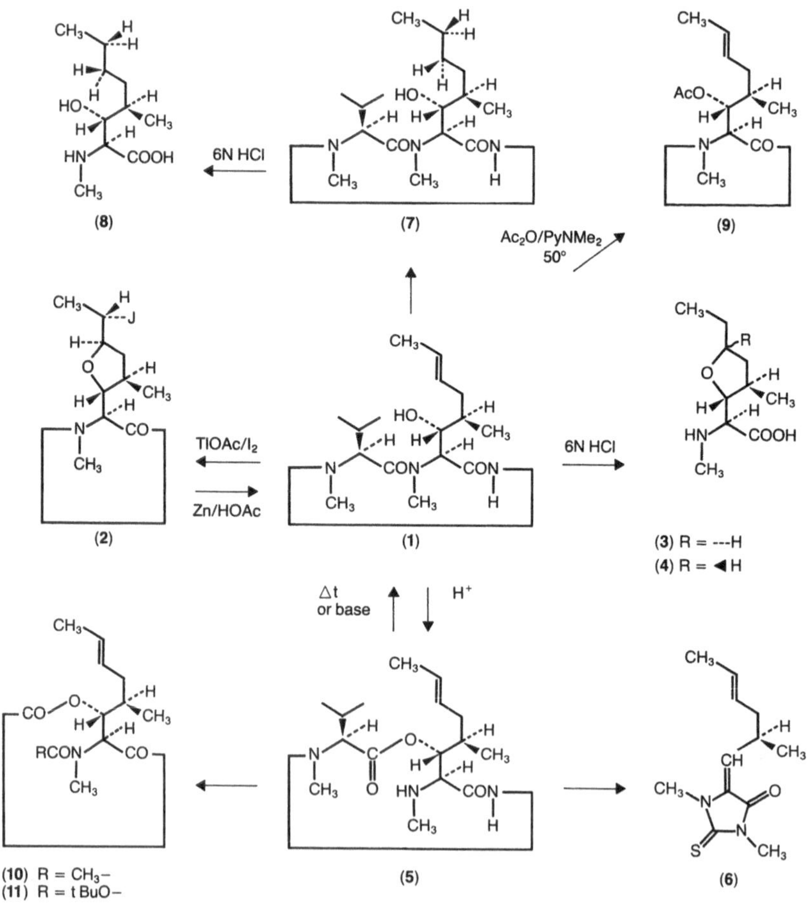

Fig. 3. Some chemical modifications of cyclosporine (1) (partial structures)

predicted by PULLMAN (35) and has also been observed in tight γ-turns of proteins (30 and Ref. 9–12 cited there). Another remarkable feature is the position of the N-methyl group of MeVal-11 which projects into the center of the loop fragment and makes a number of close contacts with backbone atoms. This methyl group together with the four intramolecular hydrogen bonds whose positions are practically forced by the presence of the six other N-methyl groups and the cis amide bond contribute significantly to the rigidity of the cyclosporine skeleton. The carbon chain of MeBmt-1 is neatly folded into the ply of the β-pleated sheet and thus allows the molecule to adopt a globular compact shape. As a consequence of the rather rigid conformation of the cyclosporine skeleton, 6 amino acids have their carbon chains directed quasi perpendicular to the plane of the peptide ring. They are Abu-2, Val-5, Ala-7 and MeVal-11, projecting up and MeBmt-1 and MeLeu-6 projecting down in Fig. 1. The remaining 5 amino acids have their carbon chain more or less in the plane of the peptide ring.

1.5. Some Chemical Modifications of Cyclosporine (Figs. 3 and 4)

Cyclosporine (1) can easily be reduced (H_2/Pd on C/ethanol) to dihydrocyclosporine (7). Hydrolysis of dihydrocyclosporine (7) with 6 N HCl followed by ion exchange chromatography affords a dihydro-MeBmt derivative (8) which can be related to a synthetic reduced MeBmt amino acid (36) (see section 4.2). Cyclosporine (1) can only be acetylated under drastic conditions (4-(dimethylamino)pyridine/acetic anhydride/50°) (25) to acetyl-cyclosporine (9). Unexpectedly cyclosporine (1) does not react with acetyl chloride in pyridine at room temperature. Under these conditions or at room temperature isocyclosporine (5) reacts to form N-acetyl-isocyclosporine (10).

Treatment of isocyclosporine (5) for 15 hours (37) with di-*tert*-butylpyrocarbonate in dioxane-1 N NaOH (9:1) at room temperature affords N-Boc-isocyclosporine (11). Treatment of O-acetyl-cyclosporine (9) in dimethylformamide with methyl iodide in the presence of silveroxide at room temperature (25) gives tetramethyl-O-acetyl-cyclosporine (12) (Fig. 4). This N-permethylation reaction has been used to correlate most of the natural N-desmethylated cyclosporine metabolites (see section 1.6) with cyclosporine (1) or with its analogue (position 2).

Iodination experiments (38) have shown that the formation of the iodo derivative (2) (Fig. 3) is temperature controlled and that depending on the conditions the other possible trans isomer (13) (Fig. 4) is also formed. Treatment of both iodides (2) and (13) with zinc in acetic acid reconverts them back to cyclosporine (1). Formation of (2) and (13) is also observed if cyclosporine (1) is treated with iodine in the presence of potassium carbonated in methylene chloride (38). Selective iodination to the iodide (2) is achieved at −70°.

(Thr^2)cyclosporine (14) (23) has been correlated chemically with cyclosporine (1) *via* selective tosylation ($CH_3C_6H_4SO_2Cl$ in pyridine at room temperature) of the threonine residue, conversion into the corresponding iodide (16) with sodium iodide in acetone followed by reduction to cyclosporine (1) with tri-n-butylstannic hydride. The selective formation of (16) depends on the time of the reaction, longer reaction times affording more of the epimeric iodide (17). Both iodides (16) and (17) can in this way be converted to cyclosporine (1) or used

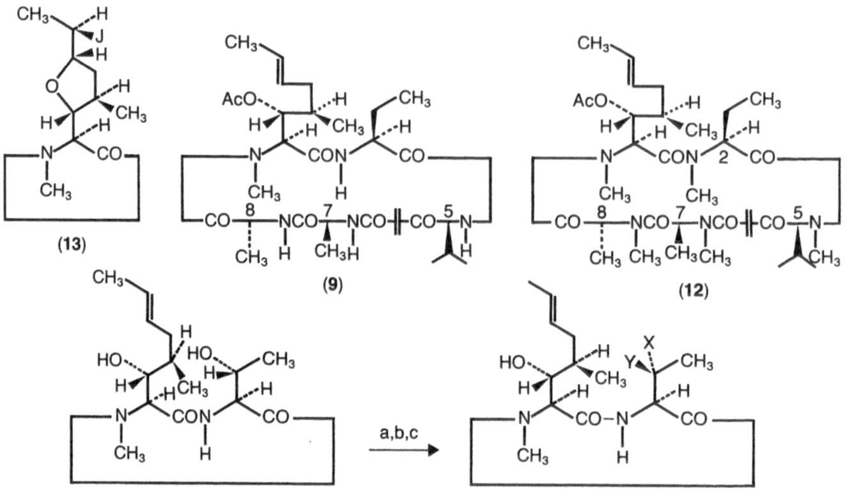

(14) [Thr^2]cyclosporine

a) tosylchloride/pyridine
b) NaJ/acetone
c) (n-Bu)$_3$ SnH

(15) Y = H; X = $-OSO_2C_6H_4CH_3$
(16) Y = J (iodide); X = H
(17) Y = H; X = J
(1) Y = H; X = H
(18) Y = H (T); X = T (H)

Fig. 4. Some chemical reactions with cyclosporine analogues. Conversion of (threonine-2)cyclosporine (**14**) to cyclosporine (**1**) (partial structures)

as starting materials for the preparation of tritiated cyclosporine (**18**) with the aid of tritium gas in the presence of a 10% Pd-$CaCO_3$ catalyst which had been de-activated with a 10-fold molar excess of triethylamine (with respect to Pd). This partial poisoning of the catalyst prevents tritium addition to the double bond of the iodides (**16**) or (**17**) and permits selective deiodination to labelled cyclosporine (**18**) (*39*).

1.6. Natural Cyclosporine Analogues (Table 1)

The fungus giving rise to cyclosporine which was initially identified as *Trichoderma polysporum* (*1*) is now classified under the genus *Tolypocladium* established by GAMS (*40*) and is called *Tolypocladium inflatum Gams* (*27*). This fungus in submerged cultures (see section 1.3) produces in addition to cyclosporine (**1**) a plethora of minor metabolites of the same structural type. A total of twentyfive natural components (**1**) and (**19** to **42**) (Table 1) has been reported as "cyclosporins A to Z" (*22–26*). The most common variations occur at position 2, the α-aminobutyric acid residue Abu of cyclosporine (CS) (**1**) being replaced by alanine (a), threonine (b), valine (c) or norvaline (d). Fourteen

Table 1. *25 Fungal Metabolites or Natural Cyclosporine (CS) Analogues (CS A to CS Z)*

Name	CS Contains	Pos.	Metabolite/Analogue Contains	Lit.	Name
CS A	Abu	2	Abu	22, 28	(1) Cyclosporine = CS
CS B	Abu	2	Ala	24	(19) [Ala^2]CS
CS C	Abu	2	Thr	22, 23	(20) [Thr^2]CS
CS D	Abu	2	Val	24	(21) [Val^2]CS
CS E	MeVal	11	Val	24, 25	(22) [Val^{11}]CS
CS F	MeBmt	1	(3'-Desoxy)MeBmt	25	(23) [3'-DesoxyMe Bmt^1]CS
CS G	Abu	2	Nva	25	(24) [Nva^2]CS
CS H	MeVal	11	D-MeVal	25	(25) [D-MeVal^{11}]CS
CS I	Abu^2, MeLeu^{10}	2, 10	Val^{2}Leu^{10}	25	(26) [Val^2, Leu^{10}]CS
CS K	MeBmt^1, Abu^2	1, 2	(3'-Desoxy)MeBmt^1 Val^2	26	(27) [3'-DesoxyMe Bmt^1-Val^2]CS
CS L	MeBmt	1	Bmt	26	(28) [Bmt^1]CS
CS M	Abu^2, Val^5	2, 5	Nva^2, Nva^5	26	(29) [Nva$^{2, 5}$]CS
CS N	Abu^2, MeLeu^{10}	2, 10	Nva^{2}Leu^{10}	26	(30) [Nva^{2}Leu^{10}]CS
CS O	MeBmt^1, Abu^2	1, 2	MeLeu^1, Nva^2	26	(31) [MeLeu^1, Nva^2]CS
CS P	MeBmt^1, Abu^2	1, 2	Bmt^1, Thr^2	26	(32) [Bmt^1 Thr^2]CS
CS Q	MeLeu	4	Val	26	(33) [Val^4]CS
CS R	MeLeu^{6}MeLeu^{10}	6, 10	Leu^6, Leu^{10}	26	(34) [Leu$^{6, 10}$]CS^a
CS S	Abu^2, MeLeu^4	2, 4	Thr^2, Val^4	26	(35) [Thr^{2}Val^4]CS
CS T	MeLeu	10	Leu	26	(36) [Leu^{10}]CS
CS U	MeLeu	6	Leu	26	(37) [Leu^6]CS
CS V	Ala	7	Abu	26	(38) [Abu^7]CS
CS W	Abu^2, MeVal^{11}	2, 11	Thr^2, Val^{11}	26	(39) [Thr^{2}Val^{11}]CS
CS X	Abu^2, MeLeu^9	2, 9	Nva^{2}Leu^9	26	(40) [Nva^{2}Leu^9]CS
CS Y	Abu^2, MeLeu^6	2, 6	Nva^{2}Leu^6	26	(41) [Nva^2, Leu^6]CS
CS Z	MeBmt	1	(3'-Desoxy-4'-des-methyl)-6',7'(di-hydro)MeBmt	26	(42) [MeAoc^1]CS^b

^{a} Structure not yet firmly established;
^{b} Aoc = L-2-aminooctanoic acid (*19*).

Val^2 means valine in position 2.

Metabolites have such a change; they are respectively: (a) (Ala^2)CS
(**19**); (b) (Thr^2)CS (**20**); (Bmt^1, Thr^2)CS (**32**); (Thr^2, Val^4)CS (**35**);
(Thr^2, Val^{11})CS (**39**); (c) (Val^2)CS (**21**); (Val^2, Leu^{10})CS (**26**); ((3'-des-
oxy)MeBmt^1 Val^2)CS (**27**); (d) (Nva^2)CS (**24**); (Nva^2, Nva^5)CS (**29**);
(Nva^2, Leu^{10})CS (**30**); (MeLeu^1, Nva^2)CS (**31**); (Nva^2, Leu^9)CS (**40**)
and (Nva^2, Leu^6)CS (**41**). Among these, seven analogues (**32**), (**35**),
(**39**), (**26**), (**30**), (**40**) and (**41**) contain a N-desmethylated amino acid
residue as an additional variation in the molecule; two which have
an additional variation in position 1 are ((3'-desoxy)MeBmt^1, Val^2)CS

(27) and (MeLeu^1, Nva^2)CS (31). In (Nva^2, Nva^5)CS (29) Abu-2 and
Val-5 have been replaced by a norvaline residue. Of the remaining
ten cyclosporine analogues containing Abu in position 2, five metabolites contain a N-desmethylated amino acid residue as an additional
variation. They are (Val^{11})CS (22); (Bmt^1)CS (28), (Val^4)CS (33);
(Leu^{10})CS (36) and (Leu^6)CS (37). One metabolite (Leu^6, Leu^{10})CS
(34) contains two N-desmethylated amino acid residues (2 Leu's) probably in positions 6 and 10. This structure remains to be firmly established. Thus six metabolites containing an Abu residue in position 2
are N-desmethylated derivatives of cyclosporine (1). Two of the remaining metabolites have an additional variation in position 1; they are
((3'-desoxy)MeBmt^1)CS (23) and (MeAoc^1)CS [(N-methyl-L-2-amino-
octanoyl^1)cyclosporine] (42), a derivative containing a reduced
"MeBmt" having lost its hydroxy and methyl groups in position 3'
and 4' respectively. In the last two metabolites, (Abu^7)CS (38) and
(D-MeVal^{11})CS (25), alanine in position 7 and N-methylvaline in position 11 have been replaced by Abu and D-MeVal respectively.

From this description it follows that the metabolites produced by
the fungus *Tolypocladium inflatum* mainly differ by a change in position
2 and or by introduction of a N-desmethylated amino acid residue.
Further minor variations are observed in positions 1, 5 and 7. In the
fermentation broths variations at positions 3 and 8 have not yet been
found. MeBmt in position 1 has been replaced by (3'-desoxy)MeBmt
in (desoxy-MeBmt^1)CS (23) and in (desoxy-MeBmt^1, Val^2)CS (27), by
MeLeu in (MeLeu^1, Nva^2)CS (31) and by MeAoc (N-methyl-L-2-ami-
no-octanoyl) in (MeAoc^1)CS (42). In position 5 valine has been replaced by norvaline in (Nva^2, Nva^5)CS (29). In position 7 alanine has
been replaced by Abu as in (Abu^7)CS (38). Both the latter variations
have only been found once among the natural analogues of cyclosporine isolated so far.

Noteworthy also are the variations observed in position 4 where
N-methylleucine has unexpectedly been replaced by a valine residue
as noted in (Val^4)CS (33) and in (Thr^2, Val^4)CS (35). Another striking
variation producing considerable conformational changes has been en-
countered in (D-MeVal^{11})CS (25) which contains N-methyl-D-valine
instead of the natural L-epimer. The biosynthesis of this derivative
requires further study.

1.7. Pharmacokinetics and Metabolism of Cyclosporine (1) (Fig. 5)

Studies of WOOD *et al.* (*41*) showed that after oral administration
of an olive oil-based solution, the bioavailability of cyclosporine is

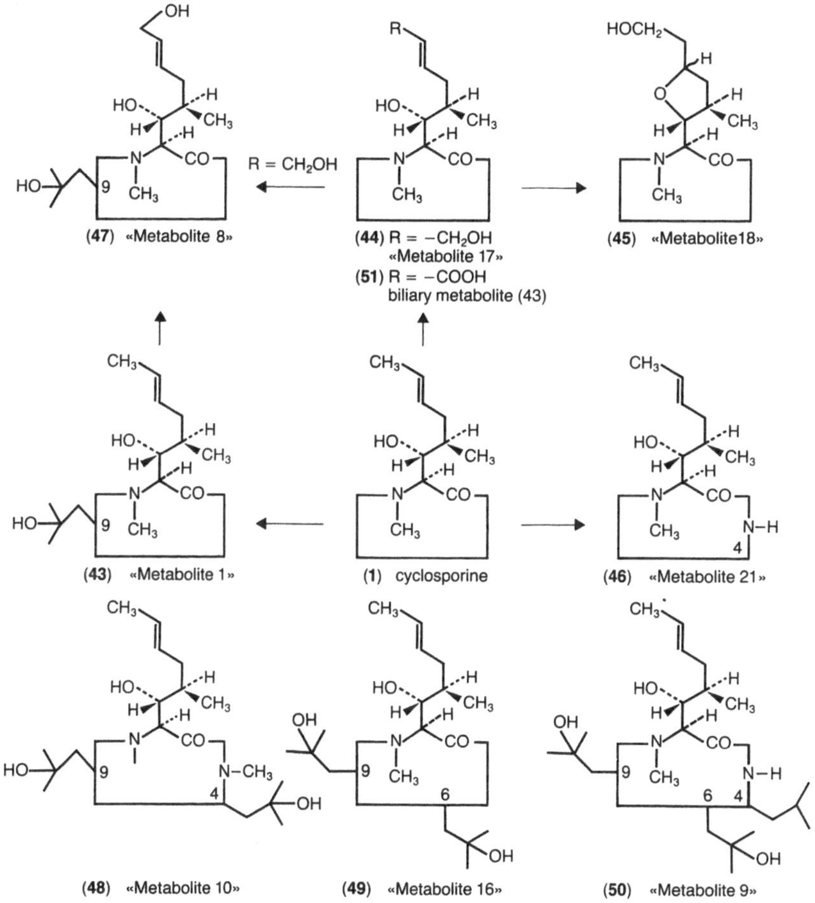

Fig. 5. Proposed pathways (42) for the biotransformation of cyclosporine (1)

20 to 50%. Peak blood concentrations are usually reached after 1–4 hours. Cyclosporine is extensively distributed to extravascular tissues because of its high lipid solubility. In the blood, approximately 50% of the drug is associated with erythrocytes, (0–20% with leucocytes, and of the 30–40% found in plasma, about 90% is bound to lipoproteins. The drug is eliminated primary by biliary mechanisms, the terminal elimination half life being in the range of 14 to 27 hours. Approximately seventeen metabolites are formed from the parent drug, all of which are present in considerably lower concentrations than cyclosporine itself. The metabolism of cyclosporine has been studied in the rat, rabbit, dog and man (42) and the biotransformation pathways

are similar in all species. Nine of the metabolites have been isolated and identified and all retained the intact cyclic peptide structure of the parent drug. Structural modifications originated from oxidation of only four of the eleven amino acids. Hydroxylation reactions appeared to be restricted to the terminal carbon atom (8') of amino acid 1 (MeBmt) and the gamma-position (4') of the N-methyl-leucines 4, 6 and 9. Further oxidation of MeBmt-1 to the carboxylic acid derivative has recently been reported (43). Oxidative N-demethylation, which presumably also proceeds via hydroxylation, appears to mainly occur on N-methylleucine 4.

The primary metabolites are the monohydroxylated cyclosporines (gamma-hydroxy-MeLeu⁹)cyclosporine (43) or *Metabolite 1* (nomenclature according to (42), (8'-hydroxy-MeBmt¹)cyclosporine (44) or *Metabolite 17,* (8'-hydroxy-3'6'-oxy-dihydro-MeBmt¹)cyclosporine (45) or *Metabolite 18* and the N-desmethylated (Leu⁴)cyclosporine (46) or *Metabolite 21.* The monohydroxylated derivatives can serve as substrates for further metabolism, hydroxylation then occurring on amino acids 1, 4 or 6. From (gamma-hydroxy-MeLeu⁹)-cyclosporine (43) the following secondary metabolites are then formed (8'-hydroxy-MeBmt¹, gamma-hydroxy-MeLeu⁹)cyclosporine (47) or *Metabolite 8,* (gamma-hydroxy-MeLeu⁴, gamma-hydroxy-MeLeu⁹)cyclosporine (48) or *Metabolite 10* and (gamma-hydroxy-MeLeu⁶, gamma-hydroxy-MeLeu⁹)-cyclosporine (49) or *Metabolite 16.* The identification of the metabolite (gamma-hydroxy-MeLeu⁶,⁹, Leu⁴)cyclosporine (50) or *Metabolite 9* indicates either that N-demethylation is possible on the metabolite hydroxylated on the gamma-carbon of N-methylleucines 6 and 9 (49) or that the primary N-desmethylated metabolite (Leu⁴)cyclosporine (46) can be oxidized by the enzymes hydroxylating N-methylleucines 6 and 9. The structures of several further urinary metabolites remain incompletely determined and are currently under further investigation in the Sandoz laboratories.

The main metabolite found in human blood is the (8'-hydroxy-MeBmt¹)cyclosporine (44) whereas the main metabolite found in rats is the (gamma-hydroxy-MeLeu⁹)cyclosporine (43). The primary biliary metabolite of cyclosporine from rabbit and human has recently (43) been identified as (7'-carboxy-7'-desmethyl-MeBmt¹)cyclosporine (51), in which the allylic 8'methyl group of MeBmt-1 has been oxidized to an α,β-unsaturated carboxylic acid. Noteworthy is that the cyclosporine metabolite (44) in animals has been found less active *in vitro* and *in vivo* than cyclosporine (1) (44). The unsaturated carboxylic acid metabolite (51) has also been reported to be inactive *in vitro* (43). All other metabolites described in Fig. 5 are also considerably less active *in vitro* than cyclosporine (1).

1.8. Elucidation of the Conformation of Cyclosporine (1) in Aprotic Solvents (Fig. 6 and Fig. 7)

Modern two dimensional NMR techniques permitted complete assignment of all C- and H- and some of the N-signals in the NMR spectra of cyclosporine (1) in CDCl$_3$ and C$_6$D$_6$ (45).

1.8.1. Backbone Conformation

In apolar solvents (CDCl$_3$, C$_6$D$_6$, CD$_2$Cl$_2$) only one conformation of cyclosporine (1) is populated to an extent of about 95% as determined by integration of the CH$_3$N resonances (30). From the NMR parameters it seems that the backbone conformation of cyclosporine (1) in solution in aprotic solvents is very similar to the X-ray structure (Fig. 6 and Fig. 7). A β-pleated sheet involving a βII'-bend about Abu-

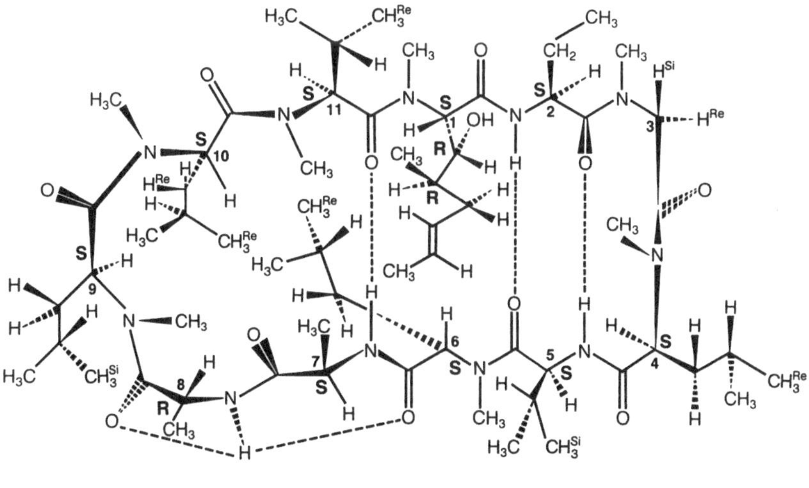

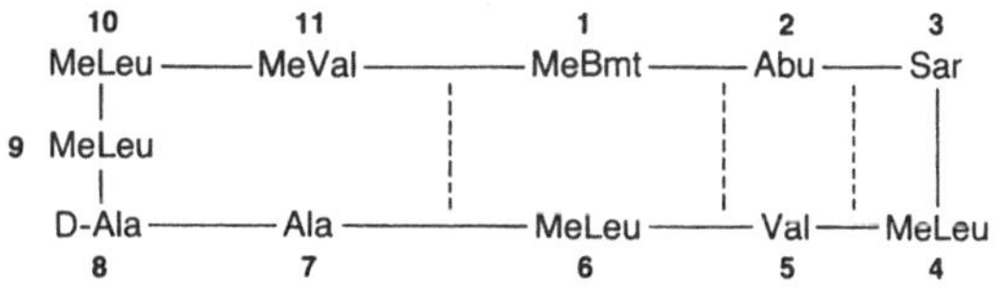

Fig. 6. Structure of cyclosporine (1) (schematic) corresponding to the conformation observed in apolar solvents (30). In solution (3'-OH) of McBmt points up to the carbonyl of MeBmt, NH of D-Ala-8 forms simultaneous bridges to CO-6 and CO-8, and carbon chain of MeLeu-10 has lightly rotated around Cα-Cβ-sigma bond compared to conformation of (1) in crystal (Fig. 1)

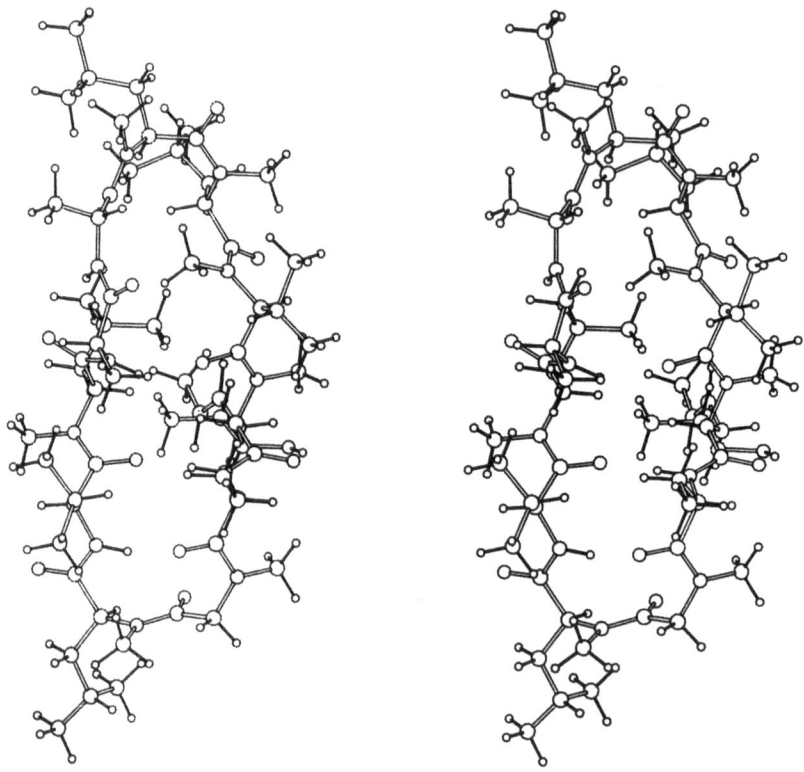

Fig. 7. Cyclosporine (**1**) computer-generated conformation in apolar solvents (in solution the distal atoms of Abu² and MeBmt-1 have a high flexibility)

2 to Val-5 is built up *via* three H-bridges. As in the crystal the peptide bond between MeLeu-9 and MeLeu-10 is in the cis conformation in the most stable conformation in solution. A minor variation has been observed in the backbone conformation in solution compared to that in the crystal. In the region of D-Ala-8 a simultaneous formation of 5-ring hydrogen bond in addition to the γ^i-bend of D-Ala-8(NH) has been found in solution.

1.8.2. Carbon Chains Conformations of Amino Acid Residues of Cyclosporine (**1**)

The main difference between crystal and solution conformations lies in the orientation of the carbon chain of MeBmt-1. Its carbon chain has rotated around the $C_\alpha - C_\beta$ sigma bond by about 120° compared to its conformation in the crystalline state. This rotation allows the formation of an intramolecular H-bond between the β-hydroxy and the carbonyl O-atom of MeBmt itself replacing the intermolecular H-bond found in the crystal.

The carbon chain conformations of MeVal-11 and Val-5 and of 3 of the MeLeu residues (MeLeu-4, MeLeu-6, and MeLeu-9) are virtually identical in crystal and solution. However, MeLeu-10 is rotated by 120° (χ_1). This suggests a rather rigid conformation for these amino acid residues and for the cyclosporine molecule as a whole.

The greatest freedom of movement has been found for the distal atoms of the carbon chain of Abu-2 and MeBmt-1, indicating greater flexibility of there carbon chains (30).

1.8.3. Cyclosporine Conformation in Biological Fluids

To what extent the conformations of cyclosporine (1) observed in the crystalline state (Fig. 1 and Fig. 2) or the conformation obtained by NMR measurements in solution in aprotic solvents (Fig. 6 and Fig. 7) are relevant for biological activity is unclear. Results presented in the following sections however suggest that in aqueous milieu and possibly on the receptor the conformation of cyclosporine resembles more the one found in the crystalline state.

2. Monoclonal Antibodies to Cyclosporine (1)

The rigid ring structure of the cyclosporine (1) molecule provides a conformationally highly restricted antigen. Since many chemically and configurationally defined variants of the molecule have been obtained by synthesis, derivation or isolation (see sections 1 and 4), cyclosporine (1) and its derivatives are interesting models for basic immunochemical studies. Their rigid ring structures make it possible to study the structural parameters that govern epitope-antibody interactions.

2.1. The Antibody Response Induced by Cyclosporine

Cyclosporine coupled to a protein carrier is used to raise antibodies against cyclosporine. Since no adequate reactive group is available on the cyclosporine molecule – the OH of MeBmt in position 1 is not reactive enough (see section 1.5) – a natural cyclosporine analogue possessing a threonine residue at position 2 (Thr^2)cyclosporine (**20**) and a synthetic lysyl cyclosporine analogue with D-Lysine at position 8, (D-Lys^8)cyclosporine (**52**), were used for coupling of cyclosporine to a carrier protein. High anticyclosporine titers were obtained after short immunization protocols with these conjugates independent of the nature of the carrier and the route of immunization (*46, 47, 48*).

2.2. Fine Specificity of Monoclonal Antibodies

A large number of monoclonal antibodies were tested using cyclosporine analogues for which the skeletal configuration was preserved (see section 4.4.3). It thus was possible to determine the influence of amino acid substitutions in each position on antibody recognition. When the capacity of a derivative to inhibit the reactions between cyclosporine and antibody was the same as that of the unmodified cyclosporine, it was concluded that the modified residue was not involved in antibody binding. Conversely when the inhibitory capacity of the derivative was decreased, it was assumed that the modified amino acid was a contact residue of the epitope recognized by the monoclonal antibody. This approach made it possible to propose which residues were part of the epitopes recognized by different antibodies (*47*). Antibodies were specific for small well defined clusters of amino acid residues contiguous on the molecular surface of cyclosporine and they were able to recognize different faces of the cyclosporine molecule.

2.3. The Conformation of Cyclosporine Seen by Antibodies in Biological Fluids Seems to Be Similar to that Observed in Crystal

The fine specificity of monoclonal antibodies for different parts of the amino acid residue MeBmt in position 1 of cyclosporine has been studied using several cyclosporine analogues modified at different positions in the carbon chain of MeBmt (*49*). Some antibodies seemed to recognize specifically changes at the β- and γ-carbons of MeBmt, but not at the end of the carbon chain or by contrast, only at the end of it. In metabolite (**44**) (section 1.7) the additional hydroxy group in position 8' of MeBmt is strongly discriminated by antibodies which recognize "the back" face (*47*) of the cyclosporine molecule, namely amino acid residues 4, 5, 6 and 7, but are unable to see changes in positions 3'(β) or 4'(γ) of MeBmt at the "bottom" face of cyclosporine.

Such results, together with the specific antigenic patterns of the monoclonal antibodies (47) seem to indicate, that most probably the cyclosporine conformation in biological fluids resembles that observed in
the solid state (Fig. 1 and Fig. 2) rather than the conformation observed
in aprotic solvents (Fig. 6 and Fig. 7). In aqueous milieu the carbon
chain of MeBmt might be expected to fold back against the lipophilic
ply of the β-pleated sheet portion of cyclosporine rather than protrud-
ing from the structure as it does in aprotic solvents. Intermolecular
hydrogen bonds between cyclosporine and water could compensate
partly for the loss of the hydrogen bond between the carbonyl and
OH groups of MeBmt seen in aprotic solvents allowing cyclosporine
to adopt a conformation closely resembling that determined by X-ray
crystallography.

3. The Biosynthesis of Cyclosporine (Fig. 8)

In the initial biosynthetic studies of KOBEL et al. (50), ^{3}H and ^{13}C
labelled precursors were fed to the culture and the position of incorporation of the label in isotopic cyclosporine (53) was determined by
NMR spectroscopy. It was demonstrated that the N-methyl groups
of cyclosporine and the methyl group in the gamma-position of the
unsaturated amino acid MeBmt are introduced as intact methyl groups
from methionine and that the remaining carbons of the MeBmt moiety
are derived from the head to tail condensation of four acetate units.
The ^{13}C-NMR spectrum of the enriched cyclosporine derived from
[1-^{13}C]-acetate showed 4 enhanced signals corresponding to carbon
atoms 1, 3, 5 and 7 of the Bmt unit and no ^{13}C incorporation into
any other amino acid.

In an analogous example of non-ribosomal cyclic peptide synthesis
by fungi, ZOCHER and KLEINKAUF (51) showed that in the case of
enniatin a multifunctional enzyme carries out the N-methylation of
the constituent amino acids subsequent to the ATP-dependent activation step when they are bound to the enzyme as thioesters. In studies
on cyclosporine biosynthesis (52) support for a similar mechanism was
implied from the observation that ^{14}C-sarcosine was not incorporated
and that the radioactivity from [Me-^{14}C]methionine incorporated into
each N-methylated amino acid was directly proportional to the number
of residues in cyclosporine indicating that N-methylation of the amino
acids occurred simultaneously.

From the results of KOBEL et al. (50) the site of biosynthesis of
the Bmt unit remains unclear. However, with short-term feeding experi-

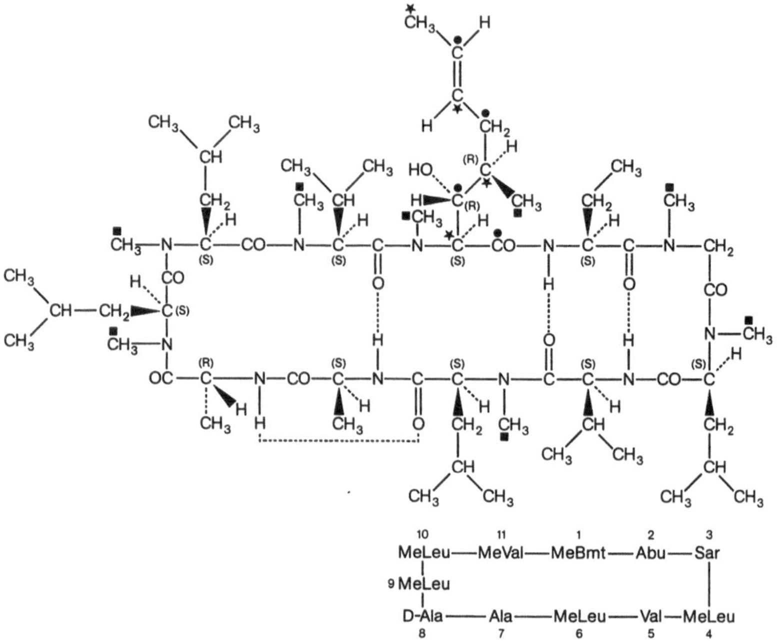

Fig. 8. Structure of isotopic cyclosporine (53) and distribution of ^{13}C and ^{3}H after addition of [1-^{13}C]acetate ●, [2-^{13}C]acetate *, [methyl-^{13}C]methionine ■ and [methyl-^{3}H]methionine ■

ments using labelled acetate and methionine, the KLEINKAUF group (52) only detected incorporation of label into the N-methyl group suggesting that the amino acid Bmt is not biosynthesized on the enzyme matrix but at some other site before incorporation into cyclosporine. Further experiments (53) related to the metabolic regulation of enniatin have shown that the multienzyme enniatin synthetase, which consists of a single polypeptide chain, can be covalently immobilized to activated agarose. This immobilized enzyme catalyzed synthesis of the cyclohexadepsipeptide enniatins (54) involving at least five different reaction steps (54) and thereby indicated that the enniatins (54) were synthesized by a single macromolecule and that interaction with a number of different enzymes was not required.

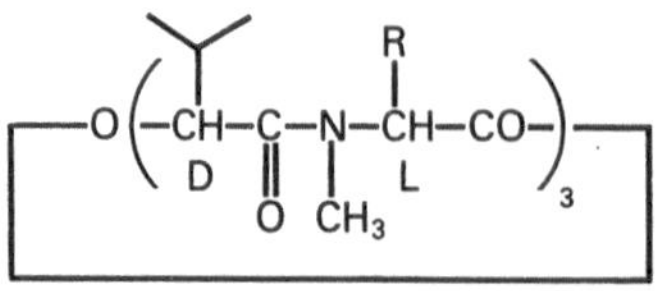

Enniatin A:R = −CH(CH$_3$)CH$_2$CH$_3$
(54)   B:R = −CH(CH$_3$)$_2$
          C:R = −CH$_2$CH(CH$_3$)$_2$

References, pp. 164–168

In similar studies related to the biosynthesis of cyclosporine (**1**) (*55*) a multifunctional enzyme with an apparent molecular weight of about 700000 could be isolated from a *Tolypocladium inflatum* cyclosporine producer. This multienzyme is most probably involved in cyclosporine biosynthesis. Besides its ability to catalyze the ATP-pyrophosphate exchange reactions dependent on the N-unmethylated constituent amino acids of cyclosporine the enzyme is capable of forming covalent enzyme substrate complexes. Evidence has been obtained which suggests that covalent binding of substrate N-unmethylated amino acids occurs via thioester linkages. The same enzyme catalyzes the N-methylation of S-thioesterified valine, leucine and glycine residues, which are present in cyclosporine. N-methylation appears to take place at the stage of the thioesterified amino acid since the N-methylamino acid can be cleaved from the enzyme by performic acid treatment following incubation of the enzyme in the presence of ATP, S-adenosyl-L-methionine and the unmethylated amino acid. The enzyme is not able to synthesize cyclosporine *de novo* but it catalyzes the formation of the dikelopiperazine cyclo(D-Ala-MeLeu) (**55**) from D-alanine and L-leucine consuming ATP and S-adenosyl-L-methionine in the process. The cyclopeptide (**55**) represents a partial sequence of cyclosporine and suggests that the synthesis of cyclosporine (**1**) starts at amino acid D-alanine-8. The reason for formation of (**55**) and not the total cyclosporine peptide ring is still unclear.

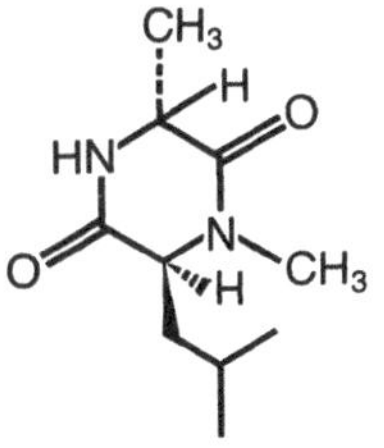

(**55**)

It is noteworthy that the formation of cyclo(D-Ala-MeLeu) (**55**) strongly resembles that of cyclo(D-Phe-Pro) (**56**) in Gramicidin S, cyclo(D-Phe-Pro-Val-L-Orn-L-Leu-D-Phe-Pro-L-Leu-L-Orn-L-Leu) (**57**), when only D-Phe and L-Pro are present as the substrate amino acids of Gramicidin S synthetase. Apparently in the case of "cyclosporine synthetase" a similar situation exists. It is speculated (*55*) that due to the inability of the enzyme to synthesize the whole undecapeptide, synthesis of cyclo(D-Ala-MeLeu) (**55**) takes place as an intramolecular cyclisation of a thioester bound dipeptide. A possible explanation could be the absence of an essential factor (protein substrate) in the cell-free system used.

4. Synthesis of Cyclosporine

4.1. Introduction

Because the natural cyclosporine derivatives contain mainly aliphatic amino acids which are not readily modified chemically, an approach using total synthesis was included among the methods used
for exploring structure-activity relationships for this substance class.
Three major problems had to be overcome: the synthesis of the unusual
amino-acid Bmt in stereochemically pure form, the construction of
the peptide chain containing N-methylated sterically hindered amino
acids without racemisation, and cyclization of the undecapeptide chain.

4.2 Synthesis of the Enantiomerically Pure Amino Acid MeBmt (83) (16)

The basic problems associated with synthesis of $(2S,3R,4R,6E)$-3-
hydroxy-4-methyl-2-(methylamino)-6-octenoic acid (83), also designated (chapter 1.4) as $(4R)$-4-$((E)$-2-butenyl)-4,N-dimethyl-L-threonine
(MeBmt), are those of forming three contiguous asymmetric centres
with the correct configuration and ensuring that only the trans isomer
of the double bond is formed. In MeBmt (83) the MeNH- and OH-
groups are in a *threo*-configuration and the OH- and CH_3-groups in

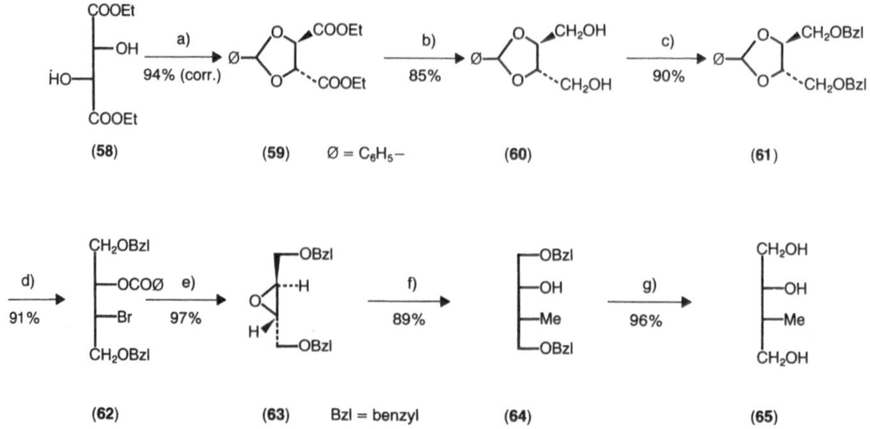

Fig. 9. Synthesis of (R,R)-3-methyl-1,2,4-butanetriol (65). a) PhCHO/HC-(OEt)$_3$/TosOH ·
H$_2$O. b) LiAlH$_4$/tetrahydrofuran (THF). c) BzlBr/toluene/KOH. d) N-bromosuccinimide
(NBS)/CCl$_4$. e) KOH/EtOH. f) 2MeLi(CuI)/Et$_2$O, −15°. g) Pd/H$_2$

an *erythro-configuration*. (R,R)-$(+)$-Tartaric acid was used as the basic chiral building block and modified in three major operations to introduce the features of the target amino acid (**83**). In the first operation, one hydroxy group of the (R,R)-$(+)$-tartaric acid molecule was incorporated with the correct configuration and the other OH-group was replaced by a CH_3-group accompanied by inversion of configuration. This provided asymmetric centers C3 and C4 of the amino acid MeBmt as shown in Fig. 9 by formation of the triol (**65**).

The second operation consisted of introducing the (E)-butenyl moiety (Fig. 10) after the carbon chain was elongated by formation of the aldehyde (**69**). This was the starting material for a Wittig reaction under SCHLOSSER and CHRISTMANN conditions (*56*) to produce diol (**71**).

The third operation consisted of oxidizing diol (**71**) to hydroxyaldehyde (**76**) (Fig. 11) and introducing the MeNH- and COOH-groups *via* a cyclic intermediate (**79**). This permitted stereochemical control during the formation of the third asymmetric centre at C2 (Fig. 12).

4.2.1. Synthesis of $(2R,3R)$-3-methyl-1,2,4-butanetriol (**65**); Formation of the Asymmetric Centers C3 and C4

As shown in Fig. 9, the OH-groups of (R,R)-diethyltartrate (**58**) were first protected by reaction with benzaldehyde to give the acetal (**59**). Reduction with $LiAlH_4$ to give the diol (**60**) and subsequent benzylation then furnished the dioxolane (**61**). For replacement of one of the OH-groups of tartaric acid by a CH_3-group with inversion of configuration, epoxide (**63**) was first prepared from dioxolane (**61**) according to a procedure described by SEELEY and MCELWEE (*57*); because the symmetry of (**61**) it is immaterial on which of the two secondary C-atoms the displacement occurs. Thus, treatment of dioxolane (**61**) with N-bromo-succinimide yielded bromo ester (**62**) which upon alkaline hydrolysis was directly converted into the optically active and therefore trans-disubstituted oxirane (**63**). Methylation of the epoxide (C2-symmetry) using the method of JOHNSON *et al.* (*58*) afforded a single product (**64**). Hydrogenolysis of the benzyl (Bzl)-protecting groups led to formation of the triol (**65**) in 56% overall yield from diethyl tartrate (**58**) (seven steps).

4.2.2. Synthesis of $(2R,3R,5E)$-3-methyl-5-heptene-1,2-diol (**71**); Carbon Chain Elongation with Introduction of the Trans Double Bond

The two vicinal OH-groups of triol (**65**) were, as shown in Fig. 10, selectively protected by preparation of ketal (**66**). The formation of

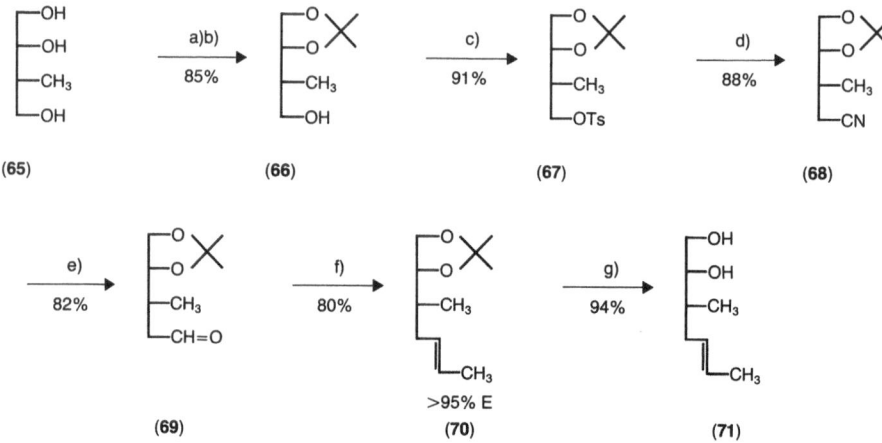

(65)   (66)   (67)   (68)

(69)   (70)   (71)

Fig. 10. Synthesis of (R,R,E)-3-methyl-5-heptene-1,2,-diol (71). a) Me₂C(OMe)₂/TsOH/ C₆H₆, reflux, 2 h. b) Acetone, TsOH, reflux, 15 h. c) TsCl/Py, 35°, 4 h. d) KCN/DMSO, 20°, 3 days. e) DlBAH/hexane, −75°, 2 h. f) Ph₃EtPBr/BuLi, Schlosser conditions (56). g) 1.1 equiv. of 1N HCl, THF/H₂O 4:1, 20°, 2 days

10–15% of the isomeric 1,3-dioxane in addition to the desired dioxolane (66) could not be avoided. Dioxolane (66) was converted into tosylate (67). Subsequent carbon chain elongation with potassium cyanide in dimethyl sulfoxide furnished the nitrile (68) which was converted into the aldehyde (69). The latter was subjected to a Wittig reaction to give the octene (70) (68). After removal of the isopropylidene-protecting group the diol (71) was obtained in 42% overall yield from the triol (65) (six steps).

4.2.3. Synthesis of (2R,3R,5E)-2-hydroxy-3-methyl-5-heptenal (76); Oxidation of Diol (71) to Hydroxyaldehyde (76)

Oxidation of diol (71) (Fig. 11) could be effected by the PFITZNER-MOFFATT method (59) in one step, but only in low yield (25%). To obtain the hydroxyaldehyde (76) in high yield, it was necessary to protect the secondary OH-group of the diol (71). This was done by mono-benzoylation to the monobenzoate (72) followed by protection of the secondary OH-group as the ethoxyethyl derivative (73) and alkaline hydrolysis of the benzoate to give the primary alcohol (74). Using the PFITZNER-MOFFATT method, (74) could then be oxidized to aldehyde (75) in 95% yield. Removal of the ethoxyethyl protecting group furnished hydroxyaldehyde (76) in a total yield of 70% (five steps).

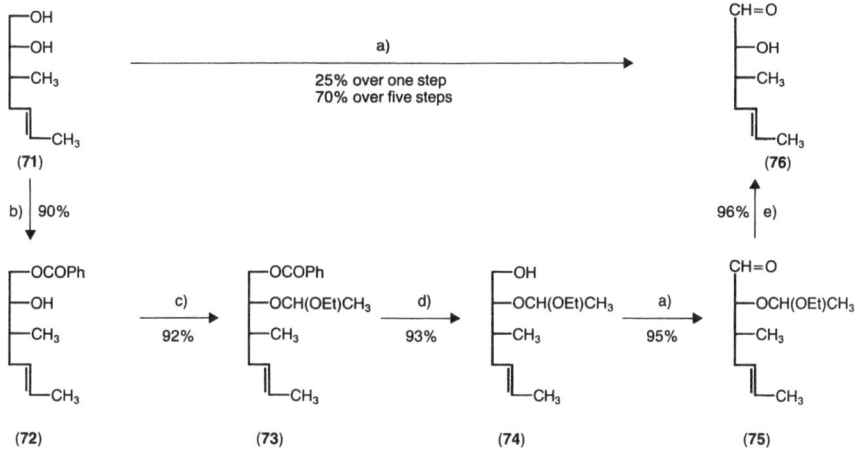

Fig. 11. Synthesis of (R,R,E)-2-Hydroxy-3-methyl-5-heptenal (75). a) DCCI/DMSO/ C₆H₆/Py/TFA, 20°, 2 h. b) PhCOCl/Py, 20°, 1 h. c) CH₂=CHOEt/TFA, 20°, 1–3 days. d) 10 N KOH/EtOH, 20°, 1.5 h. e) 1 N HCl/THF, 20°, 2 h

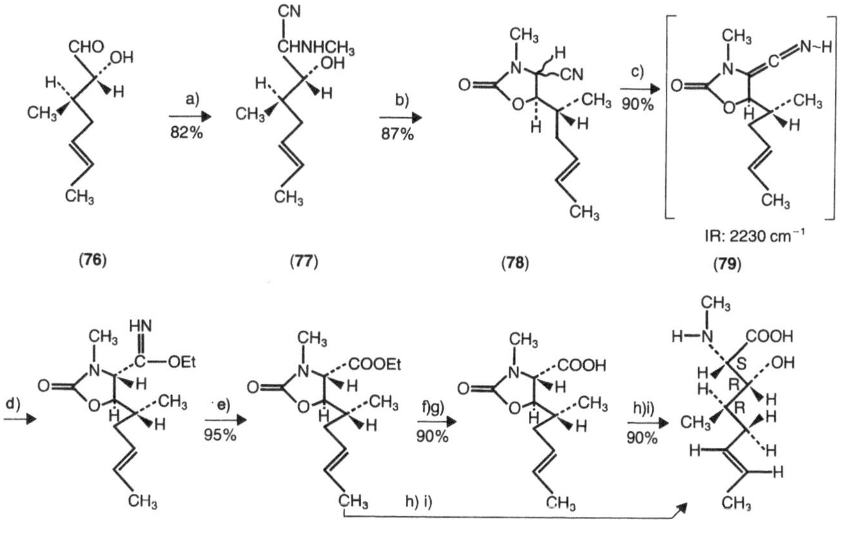

Fig. 12. Synthesis of (2S,3R,4R,6E)-3-hydroxy-4-methyl-2-methylamino-6-octenoic acid (83). a) KCN/MeNH₂·HCl/MeOH/H₂O, 20°, 2 h. b) 1,1′-Carbonyldiimidazol/CH₂Cl₂, 20°, 16 h. c) K₂CO₃/EtOH, 20°, 6 h. d) EtOH. e) 1 equiv. of 1 N HCl/EtOH, 20°, 1.5 h. f) 0.1 N KOH/dioxane, 20°, 1 h. g) HCl (pH 2). h) 2 N KOH/H₂O, 80°, 3 h. i) HCl (pH 5)

4.2.4. Synthesis of (2S,3R,4R,6E)-3-hydroxy-4-methyl-
2-(methylamino)-6-octenoic acid (MeBmt) (83); Introduction
of the Methylamino Group and the Carboxy Group

Fig. 12 outlines the last steps of the MeBmt (**83**) synthesis. Reaction of freshly prepared hydroxyaldehyde (**76**) with potassium cyanide and methylammonium chloride afforded cyanohydrin (**77**) as a mixture of diastereomers. The mixture (**77**) was converted into the 2-oxazolidin-ones (**78**) (6:1/cis:trans relative to the ring). This oxazolidinone mixture (**78**) could be converted in high yield into imidate (**80**) by treatment with potassium carbonate in ethanol. The intermediate iminomethylene derivative (**79**) (IR band at 2230 cm^{-1} in ethanol) reacted stereospecifi-cally to give the thermodynamically more stable trans-imidate (**80**). Hydrolysis of the imidate (**80**) furnished the enantiomerically pure N-methylamino acid derivative (**81**) with the O- and N-functional groups in the desired *threo* configuration. Both protecting groups of the N-methylamino acid derivative (**81**) could be removed in one step, or stepwise via the acid derivative (**82**) using warm aqueous potassium hydroxide solution or cold potassium hydroxide in dioxane respective-ly. The desired N-methylamino acid (**83**) crystallized from the reaction mixture following acidification to pH 5 with 1N HCl. The yield of free MeBmt (**83**) was 48% after six steps from the hydroxyaldehyde (**76**) or 7.8% after 24 steps from (*R,R*)-diethyltartrate (**58**). This corre-sponds to an average yield of 90% per step.

The stereospecific synthesis described here and reported in detail in (*16*) allowed characterization of the new N-methylamino acid MeBmt (**83**) for the first time and thus opened the way for a total synthesis of cyclosporine.

4.3. Strategy Used for the Synthesis of Cyclosporine (1) (*17, 60*)

For the synthesis of cyclosporine (**1**) (Fig. 1), the peptide bond be-tween the L-alanine in position 7 and the D-alanine in position 8 was chosen for the cyclization step. There were two main reasons for choos-ing this strategy: 1. The intramolecular hydrogen bonds between the amide groups of this linear peptide could operate so as to stabilize the open chain in a folded conformation which approximates the cyclic structure of cyclosporine and thus assist cyclization. 2. Bond formation between N-methylated amino acids is more difficult than between non-methylated amino acids (for further details see (*60, 61, 62*)); therefore, bond formation between the only consecutive pair of non-methylated amino acids in cyclosporine appeared to be the logical choice for the cyclization step.

For the synthesis of the undecapeptide, a fragment-condensation technique introducing the amino acid MeBmt at the end of the synthesis was used. In this way, the number of steps after the introduction of this amino acid was minimized. Fig. 13 shows the sequence of steps used for joining the (protected) peptide fragments. The amino groups of the amino acids and peptides being reacted were protected with a *tert*-butoxycarbonyl group (Boc) and the carboxy groups with a benzyloxy group (benzyl ester; OBzl). The carboxy groups were generally activated using a variation of the mixed pivalic anhydride method reported by ZAORAL (63) and adapted by us for N-methylamino acid derivatives (60). This strategy allowed slow anhydride formation in chloroform at −20 °C with pivaloyl chloride in the presence of 2 equivalents of a tertiary base such as N-methylmorpholine before adding the amino acid or peptide esters to be coupled in the form of their free bases. The free bases were obtained by removal of the Boc-protecting groups with trifluoroacetic acid at −20 °C and subsequent neutralization with $NaHCO_3$; the benzyloxy-protecting groups of the peptide intermediates were removed hydrogenolytically with palladium on charcoal as catalyst in ethanol. The peptides were built up in the direction shown by the horizontal arrows in Fig. 13 using the step sequence which is indicated numerically. Bonds 1, 2 and 3 were made first and the tetrapeptide Boc-D-Ala-MeLeu-MeLeu-MeVal-OH (89) thus synthesized. The tetrapeptide (88) could not be made directly starting from the right by making bond 3 first because of instantaneous formation of the diketopiperazine (105) from the dipeptide H-MeLeu-MeVal-OBzl. Carefully controlled reaction conditions permitted formation of the dipeptides Boc-D-Ala-MeLeu-OBzl (84) and Boc-D-Ala-MeLeu-OH (85), then of the tripeptides Boc-D-Ala-MeLeu-MeLeu-OBzl (86) and Boc-D-Ala-MeLeu-MeLeu-OH (87), and finally synthesis of the tetrapeptides Boc-D-Ala-MeLeu-MeLeu-MeVal-OBzl (88) and Boc-D-Ala-MeLeu-MeLeu-MeVal-OH (89) in good yields. Bond 4 was made and the dipeptides Boc-Abu-Sar-OBzl (90) and Boc-Abu-Sar-OH (91) synthesized. The tetrapeptide Boc-MeLeu-Val-MeLeu-Ala-OBzl (96) was synthesized from the right to the left by forming bonds 5, 6 and 7 in that order, by elongating the peptide chain from the carboxy toward the amino end thereby minimising racemization since in that way only anhydrides of amino or N-methylamino acids protected on the nitrogen by an urethane group were prepared. In the synthesis of the peptides Boc-MeLeu-Ala-OBzl (92), Boc-Val-MeLeu-Ala-OBzl (94), Boc-MeLeu-Val-MeLeu-Ala-OBzl (96) and for the dipeptide Boc-Abu-Sar-OBzl (90), there was no coupling of consecutive N-methylated amino acids. This was reflected in the good yield obtained for each coupling reaction. The Boc-protecting groups of the peptide intermediates (92),

step sequence for the peptide bond formation

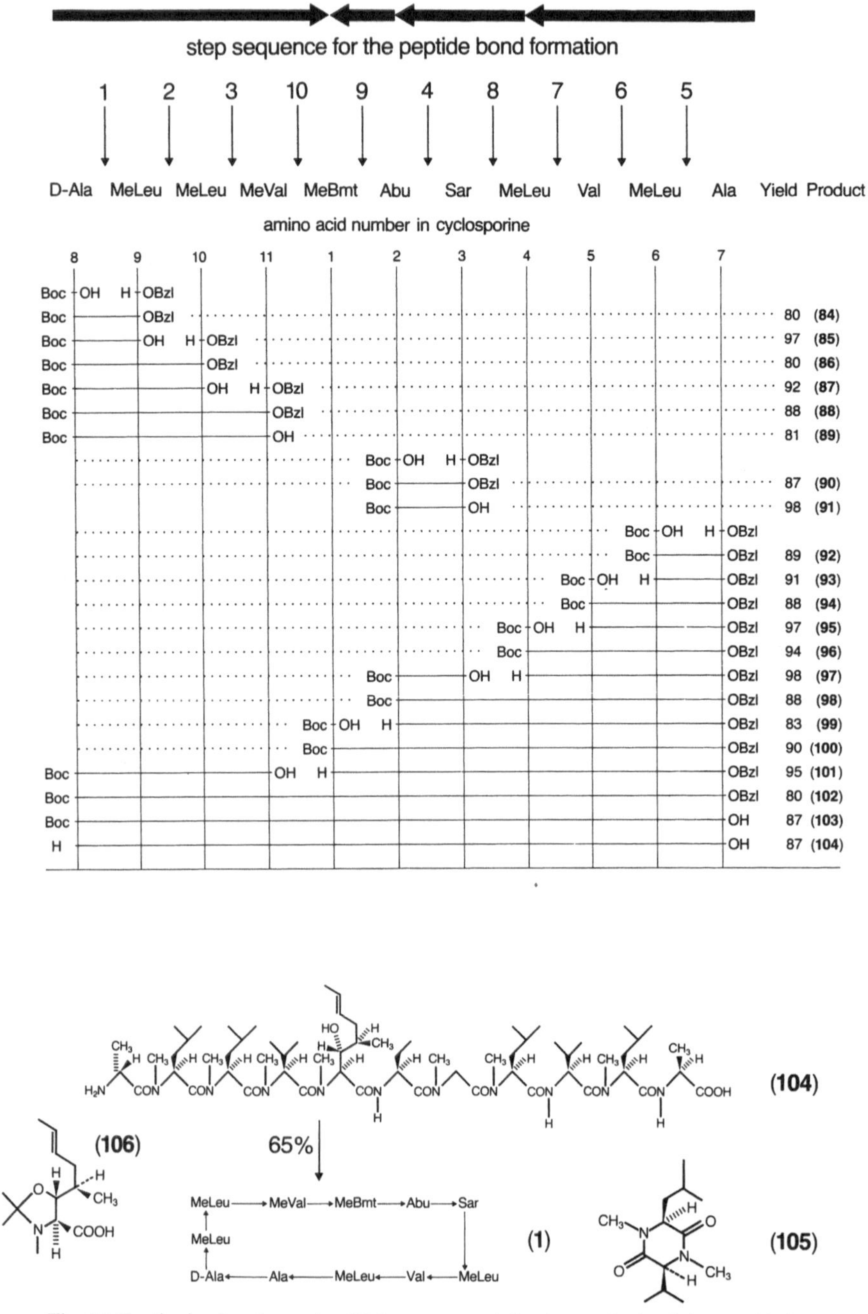

1	2	3	10	9	4	8	7	6	5

| D-Ala | MeLeu | MeLeu | MeVal | MeBmt | Abu | Sar | MeLeu | Val | MeLeu | Ala | Yield | Product |

amino acid number in cyclosporine

8	9	10	11	1	2	3	4	5	6	7	Yield	Product
Boc–OH	H–OBzl											
Boc	OBzl										80	(84)
Boc	OH	H–OBzl									97	(85)
Boc		OBzl									80	(86)
Boc		OH	H–OBzl								92	(87)
Boc			OBzl								88	(88)
Boc			OH								81	(89)
				Boc–OH	H–OBzl							
				Boc	OBzl						87	(90)
				Boc		OH					98	(91)
									Boc–OH	H–OBzl		
									Boc	OBzl	89	(92)
								Boc–OH	H	OBzl	91	(93)
								Boc		OBzl	88	(94)
							Boc–OH	H		OBzl	97	(95)
							Boc			OBzl	94	(96)
						Boc–OH	H			OBzl	98	(97)
						Boc				OBzl	88	(98)
				Boc–OH	H					OBzl	83	(99)
				Boc						OBzl	90	(100)
Boc			OH	H						OBzl	95	(101)
Boc										OBzl	80	(102)
Boc										OH	87	(103)
H										OH	87	(104)

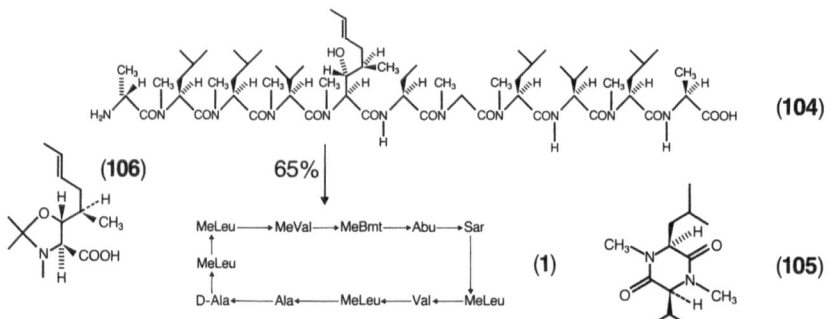

(104)

(106)   65%   **(1)**   **(105)**

MeLeu⟶MeVal⟶MeBmt⟶Abu⟶Sar

MeLeu

D-Ala←Ala←MeLeu←Val←MeLeu

Fig. 13. Synthesis of cyclosporine (1) (strategy used for the synthesis of the peptides)

References, pp. 164–168

(94) and (96) were removed with CF_3COOH at -20 °C, the acid neutralized with $NaHCO_3$, and the peptide bases H-MeLeu-Ala-OBzl (93), H-Val-MeLeu-Ala-OBzl (95) and H-MeLeu-Val-MeLeu-Ala-OBzl (97) isolated. The dipeptide Boc-Abu-Sar-OH (91) and tetrapeptide H-MeLeu-Val-MeLeu-Ala-OBzl (97) were coupled using the mixed pivalic anhydride method described above and so forming the bond 8 and the hexapeptide (98) in good (88%) yield. To prepare the protected heptapeptide (100) from the protected MeBmt amino acid and the hexapeptide H-Abu-Sar-MeLeu-Val-MeLeu-Ala-OBzl (99) (bond 9), the dicyclohexylcarbodiimide (DCCI) coupling method was used in the presence of N-hydroxybenztriazol (BtOH) (64). The Boc-protecting group of the heptapeptide (100) was removed with CF_3COOH at -20 °C, the acid neutralized with $NaHCO_3$, and the heptapeptide base H-MeBmt-Abu-Sar-MeLeu-Val-MeLeu-Ala-OBzl (101) isolated. For incorporating of the N-methylamino acid MeBmt (83), the hydroxy- and N-methylamino functions could also be protected in the form of a dimethyloxazolidine derivative (106) (17, 18). The final amide linkage 10 was made to produce the undecapeptide Boc-D-Ala-MeLeu-MeLeu-MeVal-MeBmt-Abu-Sar-MeLeu-Val-MeLeu-Ala-OBzl (102) by coupling Boc-D-Ala-MeLeu-MeLeu-MeVal-OH (89) with the heptapeptide (101) with the aid of the reagent (1H-benzo[d][1,2,3]triazol-1-yloxy)-tris-(dimethylamino)phosphonium hexafluorophosphate ($BtOP(NMe_2)_3^+PF_6^-$) (65) in the presence of N-methylmorpholine in CH_2Cl_2 at room temperature. The ester group of the undecapeptide (102) was removed by hydrolysis with NaOH at 0 °C to yield the Boc-protected undecapeptide acid (103). The Boc-group of the undecapeptide (103) was removed with CF_3COOH at -20 °C, then the unprotected undecapeptide (104) was cyclized in CH_2Cl_2 (0.0002 M) using 4 equivalents of propanephosphonic acid anhydride ($CH_3CH_2CH_2PO_2)_n$ (66) in the presence of 5 equivalents of 4-(dimethylamino)pyridine (1 day at room temperature) to yield crystalline cyclosporine (1), isolated in 65% yield. Similar cyclisation yields were obtained by using ($BtOP(NMe_2)_3^+PF_6^-$) (65) or the pentafluorophenol-DCCI complex (67) as coupling reagents for the last step of the cyclosporine synthesis (17). Using the fragment-condensation technique described here, it is now possible to synthesize very efficiently cyclosporine (1) in 30% yield with respect to the N-methylamino acid MeBmt (83). Thus, due to the molecular weight increase during this synthesis, 1.7 g of cyclosporine (1) could be produced starting from 1 g of the N-methylamino acid MeBmt (83).

The synthesis of cyclosporine (1) described in detail in (17) and (60) has been applied to the preparation of analogs needed to attack the many unanswered problems concerning the structure-activity relationships of this drug.

4.4. Cyclosporine Analogues

Some naturally occurring modifications of the cyclosporine peptide structure have been described in section 1.6 and these natural cyclosporine analogues have provided the basis of our initial understanding of the structure activity relationships that will be described below. Using the synthetic approach potentially any amino acid of the peptide chain of cyclosporine (1) can be modified and specific aspects of the structure-activity relationships examined one after the other.

4.4.1. Synthetic Analogues

The synthetic derivatives described here cover three aspects: 1) the importance of MeBmt (83) in position 1 of cyclosporine (1) for activity, 2) the effect of varying the amino acids adjacent to the MeBmt unit and 3) some derivatives in which the loop formed by amino acids 7 to 11 (Fig. 1) is modified. Fourteen such variations will be described here together with the synthesis of (N-methyl-D-valine-11)cyclosporine (25), a naturally occurring isomer of cyclosporine (1). These synthetic analogues and the strategy for their synthesis are described in Fig. 14.

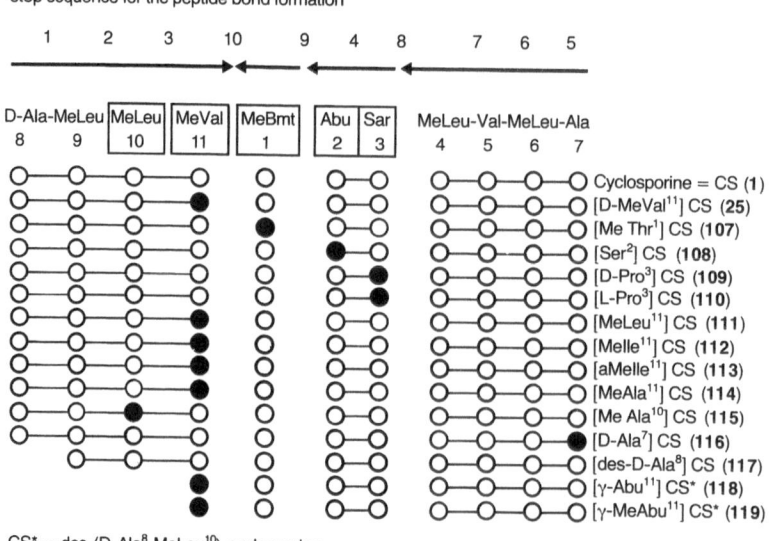

Fig. 14. Synthetic analogues of cyclosporine (1) and strategy for their synthesis

In order to study the importance of the unusual carbon chain of the MeBmt in position 1 of cyclosporine (1), this amino acid was replaced by N-methyl-threonine (MeThr) which retains the polar features of MeBmt (83) but not the extended lipophilic carbon chain and (MeThr¹)cyclosporine (107) was synthesized (68).

(Threonine-2)cyclosporine (20) [(Thr²)CS] is a naturally occurring analogue of cyclosporine (1) and a potent immunosuppressant but unlike the other natural variations of the amino acid in position 2 of cyclosporine (1), carries a hydroxy group. To examine whether or not this hydroxy group is involved in some special receptor interaction, a modified cyclosporine (108) with a serine in position 2, replacing Abu in cyclosporine (1) and threonine in (Thr²)CS (20) was prepared.

X-ray analysis of cyclosporine (1) (Fig. 2) indicates that the sarcosine in position 3 participates in a type II′ β-turn (see chapter 1.4). Replacement of sarcosine by D-proline as in (D-Pro³)CS (109) would be expected to stabilize this structural feature and make it more rigid. The proline ring would be expected to take an equatorial position relative to the peptide ring and should not alter the conformation of the peptide backbone relative to that of cyclosporine (1). The (L-Proline-3)cyclosporine (110) was also made for comparison as it would be predicted that in this case the strain introduced by the proline ring should prevent the formation of the type II′ β-turn at position 3 observed in cyclosporine (1).

To relieve steric crowding and to introduce more flexibility into the peptide ring at position 11, MeVal was replaced by MeLeu, a N-methylamino acid residue having one additional methylene group on its carbon chain, and (N-methyl-leucine-11)cyclosporine (111) was prepared. Other variations in position 11 nearer to that of the natural cyclosporine (1) were made by replacing MeVal by MeIle as in (N-methyl-L-isoleucine-11)cyclosporine (112) and by aMeIle as in (N-methyl-L-alloisoleucine-11)cyclosporine (113). These two N-methylamino acids have one additional methyl group on their carbon chain. In (MeIle¹¹)CS (112) a methyl group replaces a hydrogen atom of the Si-methyl of MeVal of cyclosporine (1) and in (aMeIle)CS (113) a methyl group replaces a hydrogen atom of the Re-methyl of MeVal of cyclosporine (1).

A further variation was the introduction in position 11 of a less hindered smaller N-methylamino acid residue, namely MeAla instead of MeVal, thus replacing the two methyl groups of MeVal by two hydrogens as is the case in (N-methylalanine-11)cyclosporine (114). A similar variation was the introduction in position 10 of MeAla instead of MeLeu, thus replacing the isopropyl group of MeLeu by a hydrogen as in (N-methyl-alanine-10)-cyclosporine (115). To examine whether or not a D-alanine residue in position 7 could be introduced without provoking an important change in the conformation of the peptide backbone of cyclosporine (1), a modified cyclosporine (116) with a D-alanine residue replacing the L-alanyl residue in cyclosporine (1) was prepared.

If the H-bond between alanine-7 and N-methylvaline-11 is considered as a ring-forming bond, then the amino acid residues 7 to 11 of cyclosporine form a 17-membered ring as an open loop. To examine the importance of the size of this loop a modified cyclosporine with 14-membered ring was prepared by removing the amino acid residue in position 8 and preparing a (des-8-D-alanine)cyclosporine (117) as a cyclodecapeptide analogue.

Following the same approach, two cyclooctapeptides containing 10-membered ring analogues of the loop were synthesized. The amino acid residues 8 to 11 D-Ala-MeLeu-MeLeu-MeVal were replaced by 4-amino-butanoic acid and 4-methylamino-butanoic acid residues producing   [11-(4-amino-butanoyl)]-des-8-D-alanyl-des-9-(N-methyl)leucyl-des-10-(N-methyl)leucyl-cyclosporine (118) and [11-(4-(methylamino)-butanoyl)]-des-8-D-alanyl-des-9-(N-methyl)leucyl-des-10-(N-methyl)leucyl-cyclosporine (119). These two cyclosporine derivatives (118) and (119) could have the amino acid residues 1 to 6 in a β-pleated sheet conformation as in cyclosporine (1) held together by three transannular H-bonds. The only difference between the two cases is the smaller ring size of the loop composed of Ala and γ-(Me)Abu only.

The above modifications of cyclosporine were expected to supply information on the sensitivity of the biological activity to minor changes in the cyclosporine molecule. At the same time it should be possible to assess the likelihood of interactions between the receptor and each modified amino acid residue. The contracted cyclosporine derivatives would be expected to provide information about the importance of the loop in the natural substance.

4.4.2. Semi-synthetic Analogues

The first examples of this group of cyclosporine (1) derivatives which were made by chemical derivatization of natural cyclosporines have already been described in section 1.5 and Figs. 3 and 4 and will be cited below for discussion or structure activity relationships. New semi-synthetic cyclosporine analogues have been obtained by SEEBACH et al. (69) using lithium-diisopropylamine in excess to generate a polylithiated cyclosporine species in tetrahydrofuran solution at −78° and then adding electrophilic agents in excess (5–6 eq.) at this temperature leading to selective alkylation of the cyclosporine molecule from the *Re-side* and formation of D-(MeAax^3) cyclosporine analogues. By thus using methyliodide, benzylbromide or dimethyldisulfide as alkylating agents (D-MeAla^3)cyclosporine (**120**), (D-MePhe^3)cyclosporine (**121**) and [D-(α-methylthio)Sar^3]cyclosporine (**122**) could be prepared in 30 to 50% yield. In order to confirm the structure of (D-MeAla^3)cyclosporine (**120**) the analogue (L-MeAla^3)cyclosporine (**123**) was synthesized following the same strategy used earlier for the synthesis of cyclosporine. Synthesis of (D-MeAla^3)cyclosporine gave a cyclosporine derivative which was identical with the semi-synthetic substance (**120**).

4.4.3. Conformation of Cyclosporine Analogues

In order to interpret the structure activity relationships obtained with the new cyclosporine derivatives it is important to know whether or not the structural changes have induced significant changes in the conformation of the peptide chain. The structural assigments for these cyclosporine analogues (Fig. 14 and analogues (**120**), (**121**), (**122**) and (**123**) have been supported by nuclear magnetic resonance spectroscopy (NMR) in deuterochloroform. Except for the (D-MeVal^{11})cyclosporine (**25**) and the (des-8-D-alanine)cyclosporine (**117**), which showed complicated NMR spectra indicating at least seven peptide ring conformations, most derivatives gave NMR spectra similar to that of cyclosporine (**1**) with essentially only one conformation in CDCl$_3$ and only a few changes in chemical shifts of some resonances consistent with what may be expected for each structural change. This is an indication that the peptide conformation for these new cyclosporine analogues is similar to that of cyclosporine (**1**) itself. Two exceptions are (L-proline-3)cyclosporine (**110**) and (L-(N-methyl)alanine-3)cyclosporine (**123**) which might be expected to contain a β-turn of another type involving N-methylamino acids 3 and 4, and for which resonances attributed to amino acid residues 1, 2, 3, 4, 5 and 7 showed appreciable differences. Another exception is (D-alanine-7)cyclosporine (**116**), the NMR spectrum of which indicated the presence of two unequally populated peptide ring conformations. The main conformation was similar to that of natural cyclosporine, but the minor one showed differences for most amino acid resonances indicating an appreciable destabilization of the natural cyclosporine peptide conformation for this derivative (**115**). The structural assignment for (MeLeu^{11})cyclosporine (**111**) has been further supported by an X-ray crystallographic analysis which confirmed the structure predicted by synthesis and physical methods. It also showed that the backbone of (MeLeu^{11})cyclosporine (**111**) has practically the same conformation as cyclosporine (**1**) itself in the solid state. The β-carbons of the N-methylamino acids methylvaline and methylleucine in position 11 are both in the same position in the respective molecules.

5. Biological Activity and Structure-activity Relationships of Cyclosporine Analogues

The relative potencies of some synthetic or semi-synthetic and naturally occurring analogues of cyclosporine (**1**) are shown in Table 2 ([a] no significant activity, [b] average activity and [c] indicates potent immunosuppressive activity). The compounds were characterized using pharmacological models previously used to test cyclosporine (**1**) (2–6). Suppression of the humoral immune response was determined by measuring the antibody-forming cells in the spleens of mice immunized against sheep erythrocytes (Jerne's plaque-forming test). Suppression of the cell-mediated immune response was quantified in an immune delayed-type hypersensitivity reaction (DTH) by measuring suppression of the oxazolone-induced skin reaction in mice. The suppressive index *in vitro* could be detected by quantifying the suppression of the proliferative response of lymphoid cells which are activated by mitogens or alloantigens. Anti-inflammatory activity was quantified by measuring the suppression of the Freund's adjuvant induced arthritis in the rat, both when administered preventively at the time of sensitization or when given therapeutically during the established disease.

From the results presented in Table 2 it is evident that the carbon chain of amino acid 1 (MeBmt) is important for biological activity. Removal of the non-polar part of the carbon chain as in (N-methyl-threonine-1)cyclosporine (**107**) reduces the immunosuppressive activity dramatically as do modifications of the hydroxy group as seen in ((OAc)MeBmt[1])cyclosporine (**9**) and [(3′-desoxy)MeBmt[1]]cyclosporine (**23**). Even the double bond makes a contribution to the biological activity since dihydro compounds like (Dihydro-MeBmt[1])cyclosporine (**7**) are generally less potent as immunosuppressants. Removal of the methyl group of the nitrogen of MeBmt also reduces the immunosuppressive activity as seen in (Bmt[1])cyclosporine (**26**). The free N-methyl-amino acid H-MeBmt-OH (**83**) alone is not sufficient for immunosuppressive activity.

Amino acid 2 permits some variation with alkyl chains of 2–3 carbons resulting in good activity. (Alanine-2)cyclosporine (**19**) with only a methyl group in position 2 is less active. (Threonine-2)cyclosporine

Table 2. *Immunosuppressive Activity of Some Cyclosporine Analogues and Derivatives.* Irrespective of its Source, the Cyclosporine **(1)** (CS) has Strong Immunosuppressive Activity ([c]); H-MeBmt-OH **(83)** has no activity ([a])

CS Contains	Pos.	Analogue/Derivative Contains	Source	Nb.	Activity
MeBmt	1	(3'-desoxy)MeBmt	n	**(23)**	[a]
MeBmt	1	(Dihydro)MeBmt	ss	**(7)**	[b]
MeBmt	1	(OAc)MeBmt	ss	**(9)**	[a]
MeBmt	1	MeThr	s	**(107)**	[a]
MeBmt	1	Bmt	n	**(26)**	[b]
Abu	2	Ser	s	**(108)**	[b]
Abu	2	Ala	n	**(19)**	[b]
Abu	2	Thr	n	**(20)**	[c]
Abu	2	Val	n	**(21)**	[b]
Abu	2	Nva	n	**(22)**	[c]
Sar	3	D-Pro	s	**(109)**	[a]
Sar	3	L-Pro	s	**(110)**	[a]
Sar	3	D-Ala	ss, s	**(120)**	[b]
Sar	3	L-Ala	ss, s	**(123)**	[a]
Sar	3	D-Phe	ss	**(121)**	[a]
MeLeu	4	Val	n	**(33)**	[a]
Abu^2, Val^5	2, 5	Nva^{2}Nva^5	n	**(29)**	[b]
MeLeu	6	Leu	n	**(37)**	[a]
Ala	7	Abu	n	**(38)**	[b]
Ala	7	D-Ala	s	**(116)**	[a]
D-Ala	8	des-8-D-Ala	s	**(117)**	[a]
Abu^2, MeLeu^9	2, 9	Nva^2, Leu^9	n	**(40)**	[a]
MeLeu	10	Leu	n	**(36)**	[a]
MeVal	11	MeLeu	s	**(111)**	[a]
MeVal	11	D-MeVal	n, s	**(25)**	[a]
MeVal	11	MeIle	s	**(112)**	[b]
MeVal	11	aMeIle	s	**(113)**	[a]
MeVal	11	MeAla	s	**(114)**	[a]
MeVal	11	Val	n	**(22)**	[a]
8-9-10-MeVal^{11}	11	des(8-9-10) − (γ-Abu^{11})	s	**(118)**	[a]
8-9-10-MeVal^{11}	11	des(8-9-10)-(γ-MeAbu^{11})	s	**(119)**	[a]

n = natural; ss = semisynthetic; s = synthetic.

[a] No significant activity.

[b] Average activity.

[c] Strong immunosuppressive activity in pharmacological models previously used to test cyclosporine (*2–6*).

(20) and (norvaline-2)cyclosporine **(22)** are potent immunosuppressants. (Valine-2)cyclosporine **(21)** is slightly less potent. Although (Thr^2)cyclosporine **(20)** is a potent immunosuppressant, (Ser^2)cyclosporine **(108)** is significantly less active which suggests that the hydroxy group is not involved in a strong interaction with the receptor and that the hydrophobic interactions between the alkyl chain of amino acid 2 and the receptor are more important. Although (D-proline-3)cyclosporine (109) should have almost the same conformation as cyclosporine (1) itself, the NMR spectrum being similar to that of cyclosporine (1), it is a negative variant suggesting that in this case the increased steric bulk of the proline ring in position 3 probably prevents efficient binding of this derivative (109) to the cyclosporine receptor. The loss of activity observed with (L-proline-3)cyclosporine (110), a derivative in which the β-turn in positions 3 and 4 is different (NMR) from that of cyclosporine (1), is not surprising if this region of the cyclosporine molecule is involved in interactions with the receptor. This variation introduces not only an increase in steric bulk but also a change in overall shape in an apparently sensitive region. N-Methylamino acid 3 permits some variations with small side chains showing relatively good activity. (D-(N-Methyl)alanine-3)cyclosporine (120) is slightly less active than cyclosporine (1), but (D-(N-methyl)-phenylalanine-3)cyclosporine (121) and (L-(N-methyl)alanine-3)cyclosporine (123) have reduced activity. The former substance (121) has probably lost activity because of the increased steric bulk of the additional benzyl group; the latter (123) has the β-turn in position 3 and 4 modified (NMR) in a similar manner to (L-Pro³)cyclosporine (110), suggesting that in this case, compared with (D-MeAla³)cyclosporine (121), modification of the β-turn is sufficient for loss of immunosuppressive activity. In positions 4, 6, 9, 10 and 11 removal of the methyl group of the nitrogen of the N-methylamino acids reduces the immunosuppressive activity dramatically as seen in (valine-4)cyclosporine (33) (substitution of MeLeu by Val), (leucine-4)cyclosporine (46) (substitution of MeLeu by Leu), (leucine-6)cyclosporine (37) (substitution of MeLeu by Leu), (norvaline-2-leucine-9)cyclosporine (40) (substitution of MeLeu⁹ by Leu), (leucine-10)cyclosporine (36) (substitution of Me-Leu by Leu) and (valine-11)cyclosporine (22) (substitution of MeVal by Val). In these cases loss of activity is probably due to additional H-bonds which induce conformation disturbances as observed in the solid state by crystallographic X-ray analysis of an iodo derivative of (Val¹¹)cyclosporine (22) (25).

(Norvaline-2-norvaline-5)cyclosporine (29) represents a variation at amino-acid 5 which retains immunosuppressive activity. This is a single and very conservative variation. It would therefore be premature to draw general conclusions as to whether or not residue 5 plays an important role in determining immunosuppressive potency.

The interpretation of the loss of activity observed with (D-Me-Val¹¹)cyclosporine (25), a derivative in which the conformation of the peptide ring at this position is certainly different (X-ray, NMR) from that of cyclosporine (1), must be reduced to stating that the configuration of the carbon chain of amino acid 11 is important for maintaining the active conformation. The great change in the overall shape of the molecule induced by this variation does not allow more detailed conclusions as to structure-activity relationships to be drawn. From the results obtained from the other variations in position 11 it is clear that minor variations in structure of N-methylamino acid 11 can reduce the immunosuppressive activity of cyclosporine analogues. Although (Me-Leu[11])cyclosporine (111) should have almost the same conformation as cyclosporine (1) itself, based on similar NMR spectra and X-ray analysis, it is inactive, suggesting that in this case the increased steric bulk of the carbon chain in position 11 probably prevents efficient binding of this derivative to the cyclosporine receptor. From the low immunosuppressive activity of (N-methylalloisoleucine-11)cyclosporine (112), it can be stated that the Re-methyl group of N-methylvaline, which based on the X-ray crystal structure of cyclosporine (1) would appear to be closer to the N-methyl group of MeBmt, is somewhat more sensitive to variation than the Si-methyl group.

The lower activity of the MeAla analogue (114) could be due to the failure to fill a critical hydrophobic cavity on the receptor or to reduced conformational rigidity of the molecule. Immunosuppressive activity is also reduced by major conformational and structural changes in the loop of cyclosporine (1) as shown by the results obtained with (Val[11])cyclosporine (22), (D-Ala[7])cyclosporine (116), (Abu[7])cyclosporine (38), (des-D-Ala[8])cyclosporine (117), (des-D-Ala[8]-des-MeLeu[9]-des-MeLeu[10]-γ-Abu-[11])cyclosporine (118) and (des-D-Ala[8]-des-Me-Leu[9]-des-MeLeu[10]-γ-MeAbu[11])cyclosporine (119). The introduction of an additional H-bond as seen above between amino acids 8 and 11 (X-ray) as in the case of (valine-11)-cyclosporine (22) or the replacement of L-alanine in position 7 by D-alanine or α-aminobutyric acid residues induces disturbances in the loop which suppress the immunosuppressive activity of these derivatives. The same is true if the size of the loop is reduced. Removal of the D-alanine residue in position 8 or replacement of the peptide segment D-Ala-MeLeu-MeLeu-MeVal of cyclosporine (1) by 4-aminobutanoyl or 4-(methylamino)butanoyl residues produced more constrained cyclosporine analogues which were all inactive as immunosuppressants although they might be expected to have maintained the β-pleated sheet portion of the cyclosporine structure. However the latter two derivatives (118) and (119) also lack the isopropyl group of MeVal[11] which may also account for their lack of immunosuppressive activity. The loop obviously is of some importance for biological activity of cyclosporine but to determine the features which are essential requires further investigation.

5.1. Conclusions

Some structure-activity relationships for cyclosporine (1) are now emerging from the study of natural cyclosporine analogues and specifically modified derivatives. However, much remains to be done in order to obtain a more precise image and then to prepare analogues with an improved profile of activity. Nevertheless, it is already clear that the unusual MeBmt chain is intimately involved in the biological actions of the cyclosporine molecule but that it alone is not sufficient for immunosuppressive activity. The biological activity is associated with a larger portion of the cyclosporine structure and the results presented above indicate that this probably includes amino acid residues 1, 2, 3, 10 and 11. The synthesis of further derivatives should help answer many of the remaining questions concerning the structure-activity relationships of this drug.

6. Mechanism of Action of Cyclosporin A

6.1. Introduction

The pioneering studies of BOREL et al. (2) in 1976 established the lymphocyte specificity of cyclosporine and the conclusions drawn at that time remain fundamental to an understanding of the mechanism of action of the drug even today. Simple experimental systems indicated that cyclosporine was the first agent to exhibit a selective effect on lymphocytes; it suppressed humoral immunity to T-cell-dependent antigens in a dose-dependent fashion; it was most active if given at the time of immunisation or antigenic challenge but the suppressive effect was reversible. It was an effective prophylactic agent in models of autoimmunity if given during the sensitisation phase and was therapeutically useful if administered at the onset of symptoms; it had no effect against acute inflammatory responses but reduced chronic inflammation in several models; it acted specifically and reversibly on lymphocytes, preferentially inhibiting T-cells though at higher concentrations also depressing B-cells. The T-helper cell was the main target and T-effector cells were inhibited; its inhibitory effect was time-dependent because cyclosporine affected rather the induction phase than the proliferative phase of lymphoid cell populations. Cyclosporine had no functional effects on hemopoietic or phagocytic cells and it did not lower resistance to bacterial or fungal infections. These unusual properties of cyclosporine triggered not only the extensive chemical investigations and studies on structure-activity relationships outlined in the previous sections, but also led to an extensive investigation on the mode of action of this immunodepressive drug. The advances in understanding and dissecting the individual steps of the immune response allowed far more detailed studies to define the site of action of cyclosporine. This topic has been comprehensively reviewed recently (21, 70–73); here we shall attempt to concentrate on the main steps of the mechanism of action of cyclosporine and tentatively formulate in a few pages what was suggested by the results of numerous studies of the individual processes in the cascade of events of lymphoid cell activation.

6.2. Lymphokines and the Immune System

The interaction of antigen with surface antibodies on B-lymphocytes is not a sufficient signal to stimulate a resting B lymphocyte to replicate and mature into an antibody-secreting plasma cell. Rather what is required is a complex interaction between antigen-specific T- and B-lymphocytes and non-specific cells such as macrophages and other antigen-presenting cells. A subpopulation of T-lymphocytes, called helper T-cells, aids not only B-lymphocytes but also other T-cell subpopulations, namely the cytolytic (killer T-cells) and suppressor T-cells, to divide and mature into effector cells of the immune response (74).

For helper T-cells to be activated in their turn, it is necessary for the antigen to be correctly presented by a specialized accessory cell. The antigen undergoes a preliminary processing and is then presented to the T-helper cell in association with Class II histocompatibility signals given by glycoproteins (encoded in man by human leukocyte antigen (HLA) genes (75)) on the surface of the antigen-presenting accessory cell (i.e. macrophage). Following correct presentation, the interacting T-helper cell sees the antigen in the context of (compatible) Class II HLA molecules (=MHC, major histocompatibility complex) and is thereby triggered to proliferate (activated from the resting G_0 phase of the cell cycle). The proliferation is accompanied in the G_1 phase of the cell cycle by synthesis of an array of soluble factors called lymphokines which induce other cells of the immune system to differentiate and thus to become activated by antigen. One of these lymphokines found in the supernatants of mitogen-stimulated lymphocyte cultures allows T-cells from normal human bone marrow to be cultivated for months in vitro (76). The underlying activity of these T-cells supernatants was subsumed under the name T-cell growth factor. Later this growth factor was renamed interleukin-2 (IL-2).

6.3. Importance of Interleukin-2 for Lymphocytes (74)

If a resting T-lymphocyte encounters a foreign antigen which is recognized by its specific receptor structure in association with histocompatibility molecules on antigen-presenting cells, this T-lymphocyte expresses within 6 to 12 hours a new receptor, the IL-2 receptor. This receptor expression induced by antigen or experimentally also by T-cell mitogens, is called stage 1 of T-cell activation.

In the second stage of the T-cell response, the same antigen or mitogen stimulus promotes IL-2 production in at least one T-cell subpopulation, the helper T-cells. For the production of IL-2 another lymphokine interleukin-1 (IL-1) secreted by macrophages or the presence of adherent cells seems to be necessary. The third stage of the T-cell response, the T-cell replication, is the exclusive result of the interaction of IL-2 and the IL-2 receptor, irrespective of the subpopulation to which the IL-2 receptor expressing T-cell belongs. This mechanism of action differs from classical hormone-receptor concept only in that the same cell can produce and consume the growth factor. IL-2 is the only one of at least 10 lymphokines which influence the growth and maturation of other cells such as B- and T-cells and myeloid cells.

Whenever there is a change in the T-cell response, whether of genetic origin or due to immune modulators, it is now possible to say whether the steps IL-2 receptor expression (1), IL-2 production (2) or IL-2-induced T-cell replication (3) are involved. The increasing understanding of the role of interleukin-2 in the immune response has also afforded new insight into the mechanism of action of cyclosporine. Its selective immunodepressive action is based on inhibition of lymphokine production (step 2). In supernatants of mitogen-stimulated T-cell cultures there is no IL-2 if cyclosporine is added in an immunosuppressive concentration (0.2–1.0 µg/ml) at the time of induction (77, 78). Studies on the expression of IL-2 receptor with monoclonal antibodies, which bind selectively to the IL-2 receptor on human lymphocytes (anti-TAC-antibody) have shown that IL-2 receptors were expressed in the presence of cyclosporine and that addition of IL-2-containing supernatants to a large extent restored the proliferative responses of T-lymphocytes (79).

The inhibition of interleukin-2 synthesis by cyclosporine has obvious consequences in the sequence of events which follows T-cell activation. Proliferation of cytolytic T-lymphocytes is abrogated in the presence of effective concentrations of cyclosporine, while their cytolytic effect is unaffected. This holds both for specific, Class I (MHC-restricted cytolytic T-cells (killer cells) which can produce their own IL-2 (helper-independent killer cells) and for cytolytic T-cell precursors which clonally expand in response to IL-2 provided by helper T-cells. The effect of cyclosporine on suppressor T-lymphocytes is not fully understood at present. *In vivo* experimental models strongly suggest that cyclosporine selectively spares or even favours the activation of T-suppressor cells. This topic has been comprehensively reviewed recently (*80*).

6.4. Cyclosporine and Transcriptional Control

Studies on IL-2 gene expression with the cloned leukemic T-cell line Jurkat and a cDNA probe for human IL-2 have shown (*81*) that cyclosporine inhibits the expression of mRNA for IL-2 and several other lymphokines, while the expression of other inducible genes is not affected.

6.5. Membrane and Cellular Receptors

A cytosolic receptor protein, designated cyclophilin (*82*, *83*) has recently been described. Cyclophilin specifically binds cyclosporine with a dissociation constant of about 2×10^{-7} M/l. The affinity of a series of cyclosporine derivatives to cyclophilin correlates with their relative immunosuppressive activity. Whether cyclophilin binds cyclosporine on the membrane and is then internalized or cyclosporine simply diffuses through the membrane and binds to cyclophilin on the cytosolic side of the membrane is still unclear.

Recent studies (*84*) have shown that cyclosporine binds to calmodulin in the presence of calcium with binding kinetics similar to the binding of cyclosporine to intact cells. Direct evidence for binding of cyclosporine to calmodulin in T-lymphocytes was provided (*85*) using a radiolabelled, ultraviolet photoactive derivative of cyclosporine. However the binding affinity of cyclosporine to the calmodulin protein is about one hundred times less than that of cyclosporine to cyclophilin (*86*). Calmodulin binding also provides no explanation for the cellular specificity of cyclosporine. Another hypothesis on the mode of action of cyclosporine is based on findings (*72*, *73*) which showed that cyclosporine can displace prolactin from its receptor present on lymphocytes. Cyclosporine also blocked the prolactin-induced ornithine decarboxylate activity in lymphocytes and kidney cells. The role of prolactin and other proteins in lymphocyte activation is currently under investigation.

6.6. Conclusions

In our present understanding of the mode of action of cyclosporine suppression of the production of interleukin-2 and other lymphokines appears to account for the major effect of immunosuppressive activity. Which cyclosporine protein complex is transported into the cell nucleus and how this complex blocks a step at or before initiation of transcription of the genes activated by antigen-lymphocyte interaction is unclear but such questions are likely to be answered in the near future.

It is a pleasure to acknowledge with sincere thanks the encouragement of Dr. T. Payne and his help in improving the manuscript.

References

1. DREYFUSS, M., E. HAERRI, H. HOFFMANN, H. KOBEL, W. PACHE, and H. TSCHERTER: Cyclosporin A and C. New Metabolites from *Trichoderma polysporum* (Link ex Pers.) Rifai. Eur. J. Appl. Microbiol. **3**, 125 (1976).
2. BOREL, J.F., C. FEURER, H.U. GUBLER, and H. STAEHELIN: Biological Effects of Cyclosporin A: A New Antilymphocytic Agent. Agents Actions **6**, 468 (1976).
3. BOREL, J.F.: Comparative Study of *in vitro* and *in vivo* Drug Effects on Cell-Mediated Cytotoxicity. Immunology **31**, 631 (1976).
4. BOREL, J.F., C. FEURER, C. MAGNEE, and H. STAEHELIN: Effects of the New Anti-Lymphocytic Peptide Cyclosporin A in Animals. Immunology **32**, 1017 (1977).
5. BOREL, J.F., D. WIESINGER, and H.U. GUBLER: Effects of the Antilymphocytic Agent Cyclosporin A in Chronic Inflammation. Europ. J. Rheumatol. Inflammation **1**, 237 (1978).
6. WIESINGER, D., and J.F. BOREL: Studies on the Mechanism of Action of Cyclosporin A. Immunobiol. **156**, 454 (1979).
7. BUEDING, E., J. HAWKINS, and Y.N. CHA: Antischistosomal Effects of Cyclosporin A. Agents Actions **11**, 380 (1981).
8. THOMMEN-SCOTT, K.: Antimalarial Activity of Cyclosporin A. Agents Actions **11**, 770 (1981).
9. MORRIS, P.J.: Cyclosporin A Overview. Transplantation **32**, 349 (1981).
10. MORRIS, P.J.: The Impact of Cyclosporin A on Transplantation. Advances in Surgery **17**, 99 (1984).
11. CALNE, R.Y., D.J.G. WHITE, S. THIRU, D.B. EVANS, P. McMASTER, D.C. DUNN, G.N. CRADDOCK, D.B. PENTLOW, and K. ROLLES: Cyclosporin A in Patients Receiving Renal Allografts from Cadaver Donors. Lancet ii, 1323 (1978).
12. POWLES, R.L., A.J. BARRETT, H. CLINK, H.E.M. KAY, J. SLOANE, and T.J. McELWAIN: Cyclosporin A for the Treatment of GVH-Disease in Man. Lancet ii, 1327 (1978).
13. KAHAN, B.D., ed.: Cyclosporine, Biological and Clinical Applications. Orlando: Grune and Stratton 1984; or In: Transplantation Proceedings, Vol. XV, No. 4, suppl. 1 (December) p. 2207 (1983).
14. SCHINDLER, R., ed.: Cyclosporin in Autoimmune Diseases. Berlin-Heidelberg: Springer. 1985.

15. WENGER, R.M.: Chemistry of Cyclosporin. In: Cyclosporin A. p. 19. Amsterdam: Elsevier Biomed. 1982.
16. WENGER, R.M.: Synthesis of Cyclosporine. I. Synthesis of Enantiomerically Pure (2*S*,3*R*,4*R*,6*E*)-3-hydroxy-4-methyl-2-(methylamino)-6-octenoic Acid Starting from Tartaric Acid. Helv. Chim. Acta 66, 2308 (1983).
17. WENGER, R.M.: Synthesis of Cyclosporine. Helv. Chim. Acta 67, 502 (1984).
18. WENGER, R.M.: Synthesis of Cyclosporine and Analogues: Structural Requirements for Immunosuppressive Activity. Angew. Chem. Intern. Ed. Engl. 24, 77 (1985).
19. IUPAC-IUB Joint Commission on Biochemical Nomenclature: Nomenclature and Symbolism for Amino Acids and Peptides, Recommendations 1983. Eur. J. Biochem. 138, 9 (1984).
20. BOREL, J.F., ed.: Ciclosporin. Progress in Allergy, Vol. 38, p. 9. Basel: Karger. 1986.
21. WENGER, R.M., T.G. PAYNE, and M.H. SCHREIER: Cyclosporine: Chemistry, Structure-Activity Relationships and Mode of Action. In: Progress in Clinical Biochemistry and Medicine, p. 157. Berlin-Heidelberg: Springer. 1986.
22. RÜEGGER, A., M. KUHN, H. LICHTI, H.R. LOOSLI, R. HUGUENIN, CH. QUIQUEREZ, and A. VON WARTBURG: Cyclosporin A, ein immunosuppressiv wirksamer Peptidmetabolit aus *Trichoderma polysporum* (Link ex Pers.) Rifai. Helv. Chim. Acta 59, 1075 (1976).
23. TRABER, R., M. KUHN, A. RÜEGGER, H. LICHTI, H.R. LOOSLI, and A. VON WARTBURG: Die Struktur von Cyclosporin C. Helv. Chim. Acta 60, 1247 (1977).
24. TRABER, R., M. KUHN, H.R. LOOSLI, W. PACHE, and A. VON WARTBURG: Neue Cyclopeptide aus *Trichoderma polysporum* (Link ex Pers.) Rifai: die Cyclosporine B, D und E. Helv. Chim. Acta 60, 1568 (1977).
25. TRABER, R., H.R. LOOSLI, H. HOFMANN, M. KUHN, and A. VON WARTBURG: Isolierung und Strukturvermittlung der neuen Cyclosporine E, F, G, H und I. Helv. Chim. Acta 65, 1655 (1982).
26.*a)* TRABER, R., H. Hofmann, H.R. Loosli, and A. von Wartburg: Cyclosporins, a New Group of Potent Immunoregulators – Structure Activity Relationships. 24th Interscience Conference on Antimicrobial Agents and Chemotherapy, Washington D.C., 8–10 Oct. 1984, Poster Presentation, Abstract No. 900.
26.*b)* VON WARTBURG A., and R. TRABER: Chemistry of the Natural Cyclosporine Metabolites in Progress in Allergy, Vol. 38, p. 28. Basel: Karger. 1986.
27. KOBEL, H., and R. TRABER: Directed Biosynthesis of Cyclosporins. Eur. J. Appl. Microbiol. Biotechnol. 14, 237 (1982).
28. PETCHER, T.J., H.P. WEBER, and A. RÜEGGER: Crystal and Molecular Structure of an Iodo-Derivative of the Cyclic Undecapeptide Cyclosporin A. Helv. Chim. Acta 59, 1480 (1976).
29. CAMBIE, R.C., R.C. HAYWARD, J.L. ROBERTS, P.S. RUTLEDGE: Diterpenes. Iodocarboxylation of Phyllocladene and Isophyllocladene. J. Chem. Soc. Perkin I, 1, 1120 (1974).
30. LOOSLI, H.R., H. KESSLER, H. OSCHKINAT, H.P. WEBER, T.J. PETCHER, and A. WIDMER: The Conformation of Cyclosporin A in the Crystal and in Solution. Helv. Chim. Acta 68, 682 (1985).
31. CHOU, P.Y., and G.D. FASMAN: β-Turns in Proteins. J. Mol. Biol. 115, 135 (1977).
32. SMITH, J.A., and L.G. PEASE: Reverse Turns in Peptides and Proteins. CRC Critical Reviews in Biochemistry 8, 315 (1980).
33. RICHARDSON, J.S.: The Anatomy and Taxonomy of Protein Structure. Adv. Protein Chem. 34, 167 (1981).
34. RICHARDSON, J.S.: A New Twist for Hairpin Turns. Nature 316, 102 (1985).
35. PULLMAN, B., and B. MAIGRET: In: Conformation of Biological Molecules and Polymers, p. 13. New York: Academic Press. 1973.

36. KUHN, M., and R.M. WENGER: Dihydro-MeBmt Natural and Synthetic. Unpublished results.
37. WENGER, R.M.: N-BOC- and N-Z-isocyclosporine. Unpublished results.
38. BOLLINGER, P.: Iodination of Cyclosporine. Unpublished results.
39. VOGES, R., B. VON WARTBURG, H.R. LOOSLI: Tritiated Compounds for in vivo Investigations. Second Intern. Symposium on the Synthesis and Application of Isotopically Labelled Compounds, Kansas City, Sept. 1985.
40. GAMS, W.: Tolypocladium, eine Hyphomycetengattung mit geschwollenen Phialiden. Personia 6, 185 (1971).
41. WOOD, A.J., G. MAURER, W. NIEDERBERGER, and T. BEVERIDGE: Cyclosporine: Pharmacokinetics, Metabolism and Drug Interaction. Transplantation Proceedings 15 (Suppl. 1), 2409 (1983); (or Ref. 13, p. 193).
42. MAURER, G., H.R. LOOSLI, E. SCHREIER, and B. KELLER: Disposition of Cyclosporine in Several Animal Species and Man. I. Structural Elucidation of its Metabolites. Drug Metabolism and Disposition 12, 120 (1984).
43. HARTMAN, N.R., L.A. TRIMBLE, J.C. VEDERAS, and I. JARDINE: An Acid Metabolite of Cyclosporine. Biochem. Biophys. Res. Comm. 133, 964 (1985).
44. HIESTAND, P., and B. RYFFEL: Pharmacological Activity of Cyclosporine Metabolites. Unpublished results.
45. KESSLER, H., H.R. LOOSLI, and H. OSCHKINAT: Peptide Conformations: Assignment of ^{1}H, ^{13}C, and ^{15}N-NMR Spectra of Cyclosporin A in CDCl$_3$ and C$_6$D$_6$ by a Combination of Homo- and Heteronuclear Two-Dimensional Techniques. Helv. Chim. Acta 68, 661 (1985).
46. DONATSCH, P., E. ABISCH, M. HOMBERGER, R. TRABER, M. TRAPP, and R. VOGES: A Radioimmunoassay to Measure Cyclosporin A in Plasma and Serum Samples. J. Immunoassay 2, 19 (1981).
47. QUESNIAUX, V., R. TEES, M.H. SCHREIER, R.M. WENGER, P. DONATSCH, M.H.V. VAN REGENMORTEL: Monoclonal Antibodies to Ciclosporin. In: Progress in Allergy, Vol. 38, p. 108 (1986).
48. ROSENTHALER, J., P. BALL, and H. MÜNZER: Monoclonal Antibodies to Ciclosporin. Manuscript in preparation.
49. QUESNIAUX, V., R.M. WENGER, M.H. SCHREIER, and M.H.V. VAN REGENMORTEL: Manuscript in preparation.
50. KOBEL, H., H.R. LOOSLI, and R. VOGES: Contribution to Knowledge of the Biosynthesis of Cyclosporin A. Experientia 39, 873 (1983).
51. ZOCHER, R., and H. KLEINKAUF: Biosynthesis of Enniatin B. Partial Purification and Characterization of the Synthesizing Enzyme and Studies of the Biosynthesis. Biochem. Biophys. Res. Comm. 81, 1162 (1978).
52. ZOCHER, R., N. MADRY, H. PEETERS, and H. KLEINKAUF: Biosynthesis of Cyclosporin A. Phytochemistry 23, 549 (1984).
53. SIEGBAHN, N., K. MOSBACH, K. GRODZKI, R. ZOCHER, N. MADRY, and H. KLEINKAUF: Covalent Immobilization of the Multienzyme Enniatin Synthetase. Biotechnology Letters 7, 297 (1985).
54. ZOCHER, R., U. KELLER, and H. KLEINKAUF: Enniatin Synthetase, a Novel Type of Multifunctional Enzyme Catalysing Depsipeptide Synthesis in Fusarium oxysporum. Biochemistry 21, 43 (1982).
55. ZOCHER, R., T. NIHIRA, E. PAUL, N. MADRY, H. PEETERS, H. KLEINKAUF, and U. KELLER: Biosynthesis of Cyclosporin A: Partial purification and Properties of a Multifunctional Enzyme from: Tolypocladium inflatum. Biochemistry 25, 550 (1986).
56. SCHLOSSER, M., and K.F. CHRISTMANN: Mechanismus und Stereochemie der Wittig-Reaktion. Justus Liebigs Ann. Chem. 708, 1 (1967).

57. SEELEY, D.A., J. McELWEE: Stereospecific Synthesis of cis and trans Epoxides from the Same Diol. J. Org. Chem. **38**, 1691 (1973).

58. HERR, R.W., D.M. WIELAND, and C.R. JOHNSON: Reactions of Organocopper Reagents with Oxiranes. J. Amer. Chem. Soc. **92**, 3813 (1970).

59. PFITZNER, K.E., and J.G. MOFFATT: Sulfoxide-Carbodiimide Reactions. Scope of the Oxidation Reaction. J. Amer. Chem. Soc. **87**, 5670 (1965).

60. WENGER, R.M.: Synthesis of Cyclosporine. Helv. Chim. Acta **66**, 2672 (1983).

61. McDERMOTT, J.R., and N.L. BENOITON: (a) N-methylamino Acids in Peptide Synthesis. Razemization During Deprotection by Saponification and Acidolysis. Can. J. Chem. **51**, 2551 (1973); (b) Razemization and Yields in Peptide-Bond Formation. Can. J. Chem. **51**, 2562 (1973).

62. TUNG, R.D., and D.H. RICH: Bis(2-oxo-3-oxazolidinyl)phosphinic Chloride as a Coupling Reagent for N-alkyl Amino Acids. J. Amer. Chem. Soc. **107**, 4342 (1985).

63. ZAORAL, M.: Amino Acids and Peptides. Pivaloyl Chloride as a Reagent in the Mixed Anhydride Snthesis of Peptides. Collect. Czech. Chem. Commun. **27**, 1273 (1962).

64. KOENIG, W., and R. GEIGER: Eine neue Methode zur Synthese von Peptiden. Aktivierung der Carboxylgruppe mit DCCI unter Zusatz von 1-Hydroxybenztriazolen. Chem. Ber. **103**, 788 (1970).

65. CASTRO, B., J.R. DORMOY, G. EVIN, and C. SELVE: Réactifs de couplage peptidique. L'hexafluorophosphate de benzotriazolyl-N-oxytrisdimethyl-aminophonium (BOP). Tetrahedron Lett. **14**, 1219 (1975).

66. WISSMANN, H., and H.J. KLEINER: Neue Peptidsynthese. Angew. Chem. **92**, 129 (1980); Angew. Chem. Int. Ed. Engl. **19**, 133 (1980).

67. KOVACS, J.: In: The Peptides, Vol. 2, p. 485. New York: Academic Press. 1980.

68. WENGER, R.M.: Synthesis of Cyclosporine and Analogues: Structure Activity Relationships of New Cyclosporine Derivatives. Transplantation Proceedings, Vol. XV, No. 4, Suppl. 1, 2230 (1983).

69. SEEBACH, D., W. MURTIASHAW, R. NAEF, S.I. SHODA (ETH Zürich), and M. KRIEGER, P. BOLLINGER, A. LEUTWILER, and R. WENGER (Sandoz Ltd.): Aktive Cyclosporin-Derivate durch C-Alkylierung unter Ersatz von H^{Re} der Sarkosin-Einheit. Erzeugung Polythiierter Peptide. Work presented at the autumn session of the Swiss Chemical Society in Berne (October 18, 1985). Manuscript in preparation.

70. BOREL, J.F., and B. RYFFEL: The Mechanism of Action of Ciclosporin, a Continuing Puzzle. In: Ciclosporin in Autoimmune Diseases, p. 24. Berlin-Heidelberg-New York: Springer. 1985.

71. HESS, A.D., and P.M. COLOMBANI: Mechanism of Action, in vitro Studies. In: Progress in Allergy, Vol. 38, p. 198. Basel: Karger. 1986.

72. LARSON, D.F.: Mechanism of Action, Antagonism of the Prolactin Receptor. In: Progress in Allergy, Vol. 38, p. 222. Basel: Karger. 1986.

73. HIESTAND, P.C., and P. MECKLER: Mechanism of Action, Ciclosporin- and Prolactin-Mediated Control of Immunity. In: Progress in Allergy, Vol. 38, p. 239. Basel: Karger. 1986.

74. SCHREIER, M.H.: Interleukin-2 and its Role in the Immune System. Triangle **23**, 141 (1984).

75. BATCHELOR, J.R.: Genetic Role in Autoimmunity. In: Ciclosporin in Autoimmune Diseases, p. 24. Berlin-Heidelberg-New York: Springer. 1985.

76. MORGAN, D.A., F.W. RUSCETTI, and R. GALLO: Selective in vitro Growth of T-Lymphocytes from Normal Human Bone Marrow. Science **193**, 1007 (1976).

77. BUNJES, D., C. HARDT, M. ROELLINGHOFF, and H. WAGNER: Cyclosporin A Mediates Immunosuppression of Primary Cytotoxic T Cell Responses by Impairing the Release of Interleukin 1 and Interleukin 2. Europ. J. Immunol. **11**, 657 (1981).

78. Ryffel, B., U. Goetz, and B. Heuberger: Cyclosporin Receptors on Human Lymphocytes. J. Immunol. **129**, 1978 (1982).

79. MIYAWAKI, T., A. YACHIE, S. OHZEKI, T. NAGAOKI, and N. TANIGUCHI: Cyclosporin A Does not Prevent Expression of Tac Antigen, a Probable TCGF Receptor Molecule, on Mitogen-Stimulated Human T Cells. J. Immunol. **130**, 2737 (1983).

80. KUNKL, A., and G.G.B. KLAUS: Selective Effects of Cyclosporin A on Functional B Cell Subsets in the Mouse. J. Immunol. **125**, 2526 (1980).

81. KROENKE, M., W.J. LEONHARD, J.M. DEPPER, S.K. ARYA, F. WONG-STAAL, R.C. GALLO, T.A. WALDMANN, and W.C. GREENE: Cyclosporin A Inhibits T-Cell Growth Factor Gene Expression at the Level of mRNA Transcription. Proc. Nat. Acad. Sci. USA **81**, 5214 (1984).

82. HANDSCHUMACHER, R.E., M.W. HARDY, J. RICH, R.J. DRUGGE, and D.W. SPEICHER: Cyclosphilin: a Specific Cytosolic Binding Protein for Cyclosporin A. Science **226**, 544 (1984).

83. HARDING, M.W., R.E. HANDSCHUMACHER, and D.W. SPEICHER: Isolation and Amino Acid Sequence of Cyclophilin. J. Biol. Chem. **261**, 8547 (1986).

84. COLOMBANI, P.M., A. ROBB, and A.D. HESS: Cyclosporin A binding to Calmodulin: a Possible Site of Action on T-Lymphocytes. Science **228**, 337 (1985).

85. HESS, A.D., T. TUSZYNSKI, P. ENGEL, P.M. COLOMBANI, J. FARRINGTON, R. WENGER, and B. RYFFEL: Intracellular and Nuclear Localization of Cyclosporine and Peripheral Blood Mononuclear Cells. Transplant. Proc. **18**, 861 (1986).

86. HIESTAND, P. (Sandoz Basel), A.D. HESS (Baltimore), and R.E. HANDSCHUMACHER (New Haven): Unpublished Results, personal communication.

(Received July 22, 1986)

Biosynthesis of Iridoids and Secoiridoids

H. INOUYE and S. UESATO, Faculty of Pharmaceutical Sciences,
Kyoto University, Sakyo-ku, Kyoto 606, Japan

With 26 Figures

Contents

I. Introduction

Research on iridoids began with the discovery of asperuloside (**1**)
(an iridoid glucoside) in 1848 and gentiopicroside (**2**) (a secoiridoid

glucoside) in 1862. However, it was only in the late 1950's that the structures of the compounds of this series began to be elucidated. Thus structure elucidation of the non-glycosidic iridoids iridomyrmecin (3) (1, 2, 2a) isoiridomyrmecin (4) (3, 3a, 4) and nepetalactone (5) (5) and that of the iridoid glucoside plumieride (6) (6) gradually led to the unravelling of the structures of other then unknown iridoids. Since then, large numbers of diverse new iridoids have been found and their structures have been elucidated as time progressed.

As of today, ca. 300 iridoid glucosides, ca. 60 secoiridoid glucosides and ca. 90 non-glycosidic iridoids are known. Thus, the total number of these compounds amounts to ca. 450 whereas according to a 1980 survey (7) these numbers were 158, 38 and 61 respectively. These figures indicate that the number of new compounds is still expanding rapidly. In addition, ca. 40 alkaloidal glucosides and seven hydrangenosides related to the secoiridoid type are known. The reasons for this progress are as follows: 1) From the 1960's to the early 1970's chemists paid attention to the finding that the non-tryptophan moiety of indole alkaloids is of iridoid-secoiridoid origin. 2) Most iridoids are formed by dicotyledons, especially in sympetalous plants, and hence are widely distributed in medicinal plants belonging to this subclass which have been used as crude drugs from ancient times. Many of them have been isolated as the result of pharmacognostic studies. 3) The wide distribution of iridoids in sympetalous plants led chemists to pay attention to them as chemotaxonomic markers.

The controversy on the biosynthetic origin of the non-tryptophan moiety of indole alkaloids, which was a prelude to the biosynthetic study of the iridoids, was initiated with the phenylalanine hypothesis (8, 9) proposed by BARGER and HAHN in the 1930's. Later, WENKERT and BRINGI (in 1959) proposed the prephenic acid hypothesis (10, 11) based on evidence that despite the great number of stereoisomers among the indole alkaloids H-15 always is α-orientated. Subsequently, taking a hint from the newly elucidated structures of gentiopicroside (2), swertiamarin (7), gentianine (8), etc., THOMAS in 1961 (12) and WENKERT in 1962 (13) independently proposed the terpenoid hypothesis, whereas a hypothesis based on an acetate-malonate pathway was advanced by LEETE et al. in 1962 (14, 15). The conflicting hypotheses began to be resolved in 1965 by the groups of SCOTT (16, 17, 18) and ARIGONI (19), who demonstrated that mevalonic acid (MVA) is incorporated into vindoline (9) by Catharanthus roseus (= Vinca rosea). Subsequently, incorporation of MVA and geraniol into various indole alkaloids was demonstrated by many research groups (20–25) and incorporation of variously-labelled loganin (10) into indole alkaloids was demonstrated by BATTERSBY et al. (26, 26a, 27) and by

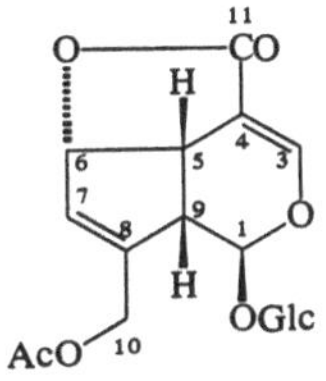

(1) 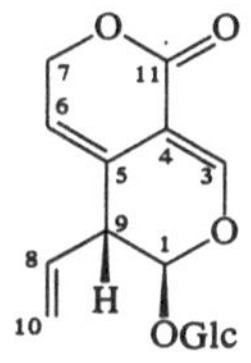　　　　　　(2) 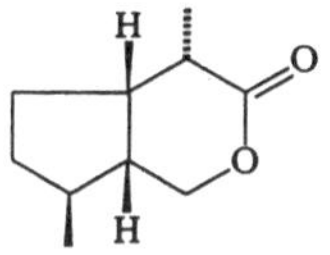　　　　　　(3)

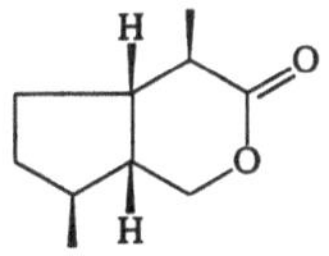

(4) 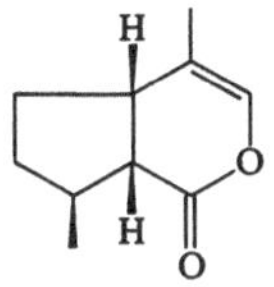　　　　　　(5) 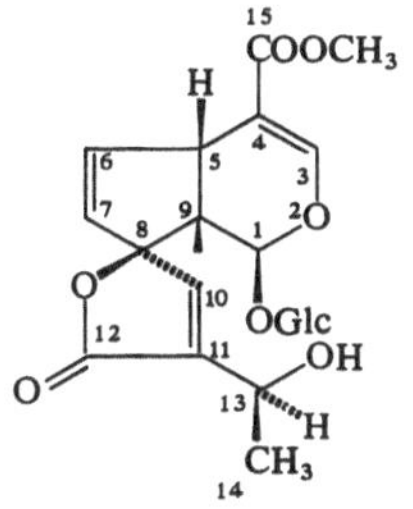　　　　　　(6)

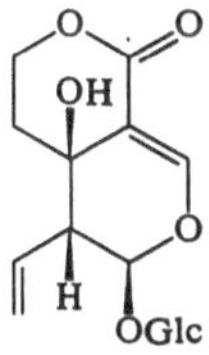

(7) 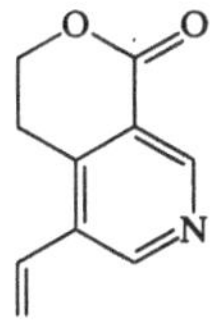　　　　　　(8)

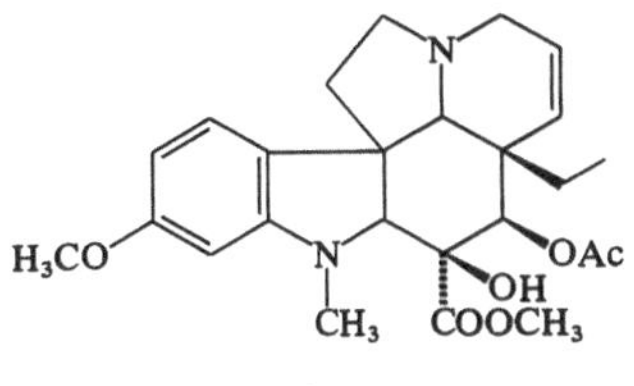

(9) 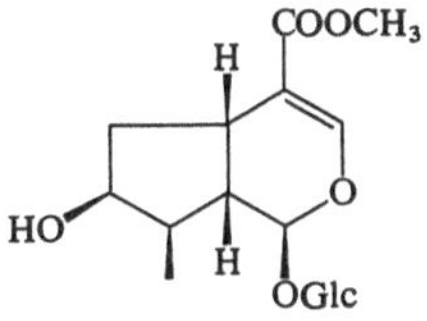　　　　　　(10)

ARGONI and LOEW (28). The origin of the non-tryptophan moiety of the indole alkaloids was thus established.

With the pioneer work cited in the previous paragraphs as a background work on the biosynthesis of the iridoids has been carried on since the early 1960's. Although in the interim a few reviews of this field have been published (29–36) they seem not to cover all aspects. In this article we therefore wish to present a comprehensive review of studies on the biosynthesis of iridoids and secoiridoids.

II. Classification of Iridoids and Secoiridoids

Iridoids and secoiridoids (both non-glycosidic and glycosidic compounds) ought to be classified in this review on the basis of their biosynthesis. At the present time, however, only secoiridoid glycosides can be classified in this manner. For iridoid glycosides which possess a cyclopentane ring, detailed biosynthetic pathways have been established in only a few cases (37–49). Moreover, as will be mentioned in Chapter IV, in one instance the same iridoid glycoside is biosynthesized by different pathways by different plants (49). Thus, an exact understanding of the biosynthesis of these compounds sometimes necessitates the detailed examination even of individual plants.

On the other hand, recent proposals for the biosynthesis of some non-glycosidic iridoids such as dolichodial (11), dolicholactone (12) and nepetalactone (5) (50–55) involve pathways rather different from those usually taken by glycoside iridoids and secoiridoids. However, an iridoid glucoside 1,5,9-epideoxyloganic acid (13) (56) has been isolated recently from *Nepeta cataria* which contains nepetalactone.

Furthermore, both iridomyrmecin (3) and isoiridomyrmecin (4) of the same non-glycosidic lactone type are supposed to be formed by the hydrolysis of an iridoid glycoside, iridodialogentiobioside (14), in *Actinida polygama (57)*. These facts indicate the complexity and diversity of the biosynthetic pathways of these compounds.

For the reasons mentioned above, the classification of iridoids based on their biosynthesis such as the one used by JENSEN *et al.* (58), is not fully satisfactory. Therefore, in this review iridoids are classified conventionally as 1) non-glycosidic iridoids, 2) iridoid glycosides and

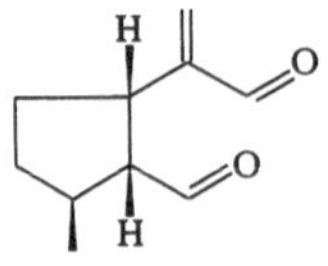

(11) 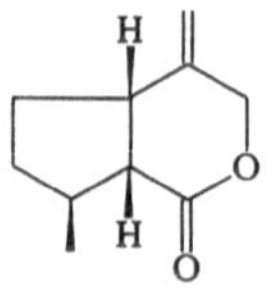   (12)

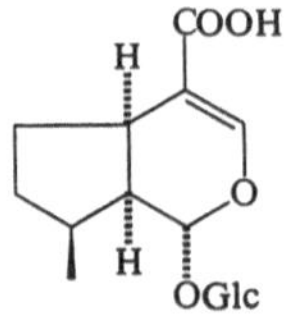

(13) 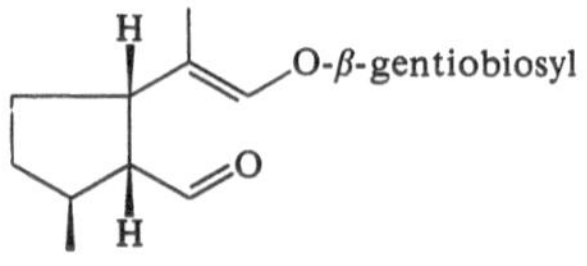   (14)

3) secoiridoid glycosides. The first two groups are not subdivided further, whereas the third group is subdivided on the basis of biosynthetic pathways and structural similarities into the following four subgroups. For each subgroup, representative compounds are given for the reader's convenience.

1. Non-glycosidic iridoids:
Iridodial (15), iridomyrmecin (3), nepetalactone (5), β-skytanthine (16).

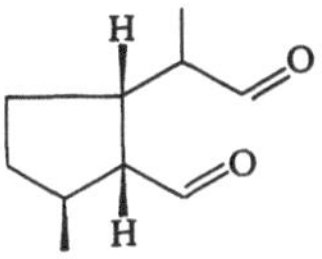

(15) 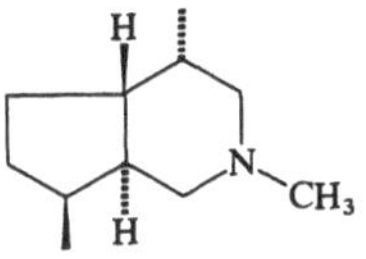   (16)

2. Iridoid glycosides:
Asperuloside (1), loganin (10), deutzioside (17), verbenalin (18), lamiide (19), paederoside (20), monotropein (21), gardenoside (22).

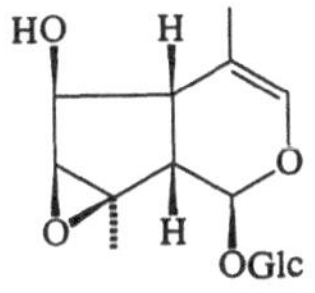

(17) 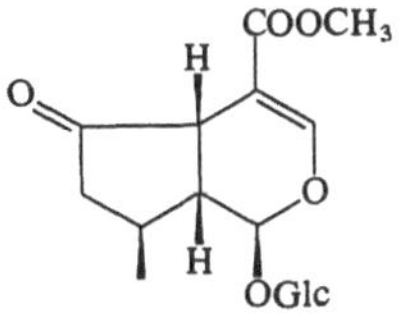   (18) 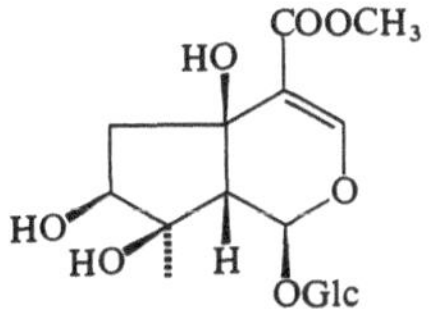   (19)

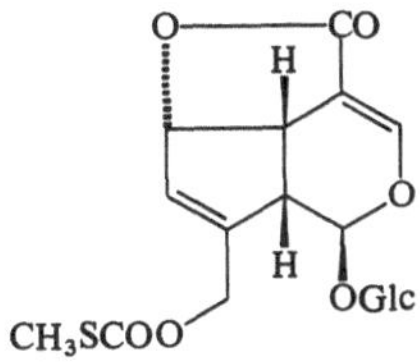

(20)

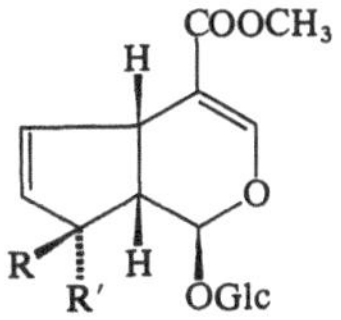

(21) R = CH₂OH   R' = OH
(22) R = OH   R' = CH₂OH

3. Secoiridoid glycosides:
a) Sweroside-morroniside type: secologanin (23), sweroside (24), gentiopicroside (2), foliamenthin (25), morroniside (26), kingiside (27), secogalioside (28).

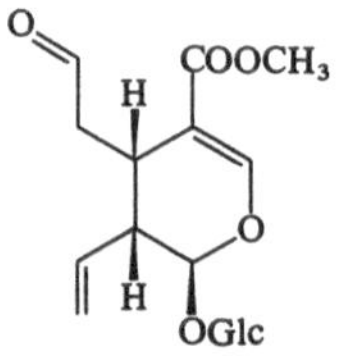

(23) 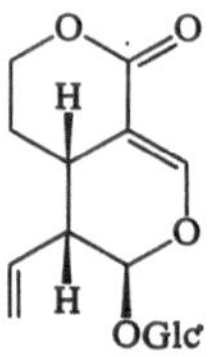 (24) 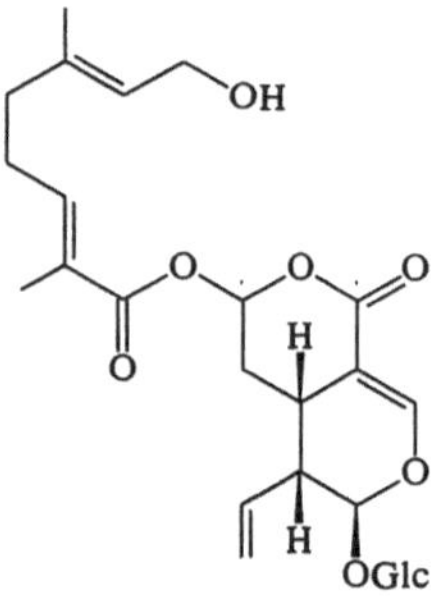 (25)

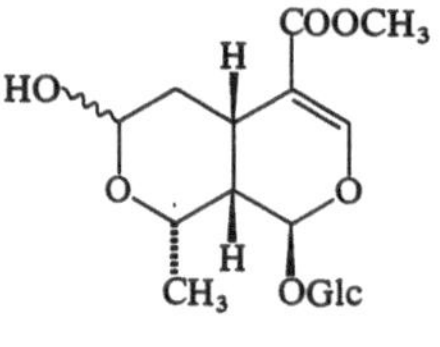

(26) 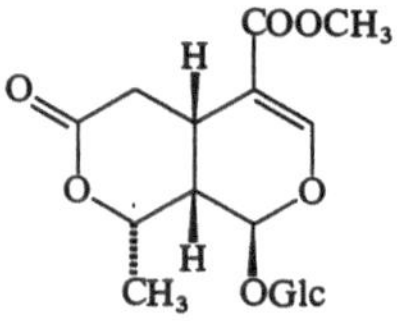 (27) 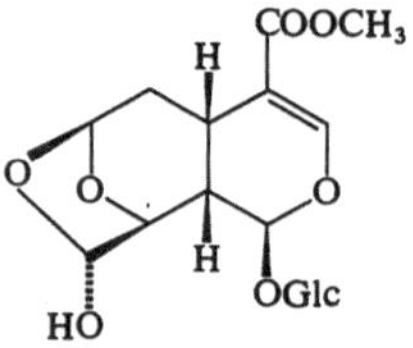 (28)

b) Oleoside-10-hydroxyoleoside type: oleuropein (29), 10-hydroxy-ligustroside (30), ligustarosides A (31) and B (32), nüzhenide (33).

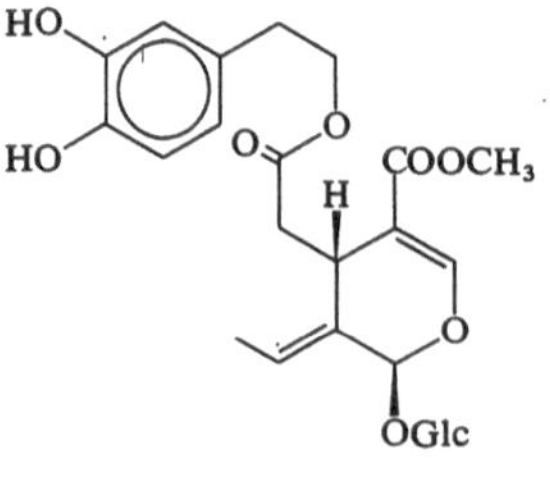

(29) 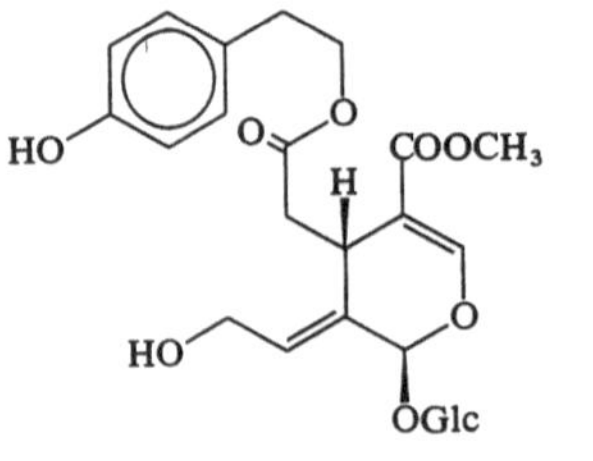 (30)

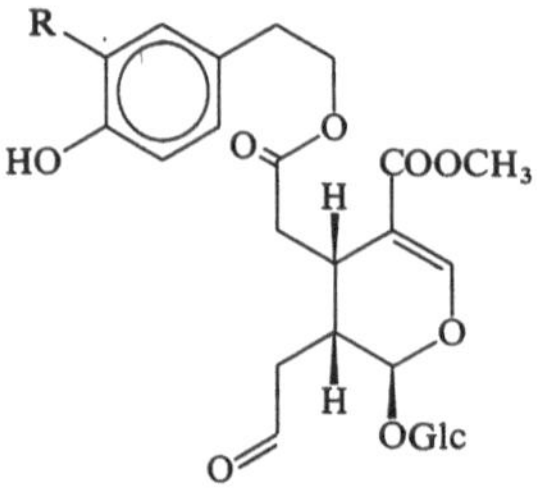

(31) R = OH
(32) R = H

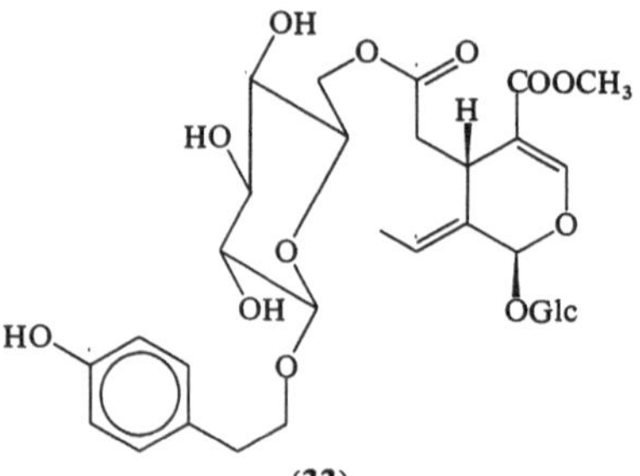

(33)

c) Alkaloidal glycosides containing a secoiridoid skeleton: strictosidine (**34**), vincoside (**35**), ipecoside (**36**).

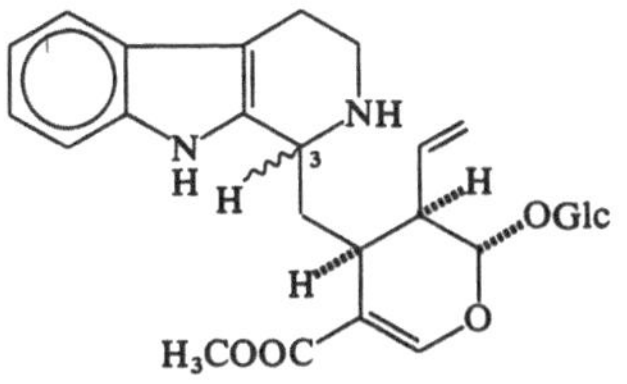

(**34**) 3····H
(**35**) 3—H

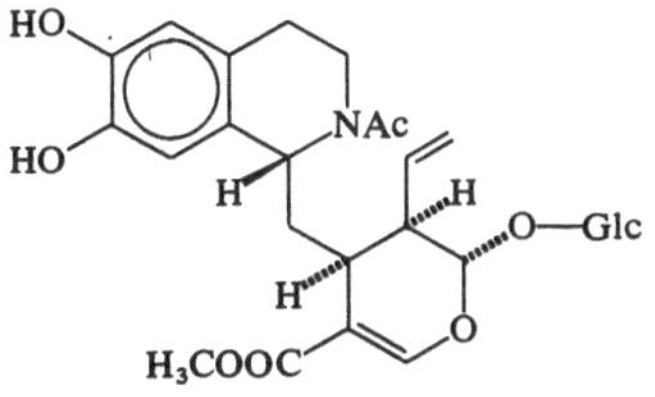

(**36**)

d) Hydrangenosides: hydrangenosides A (**37**), E (**38**) and G (**39**).

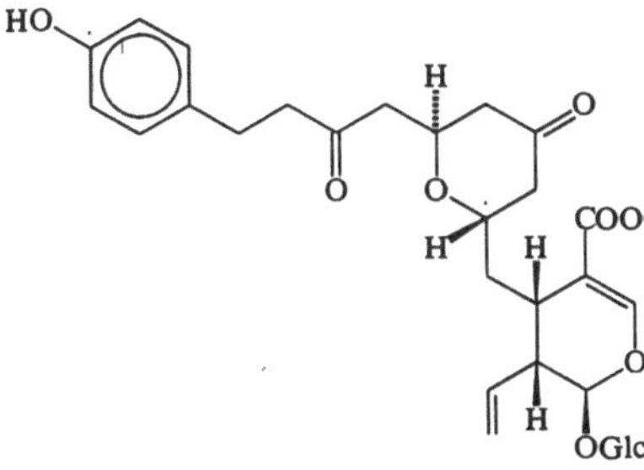

(**37**)

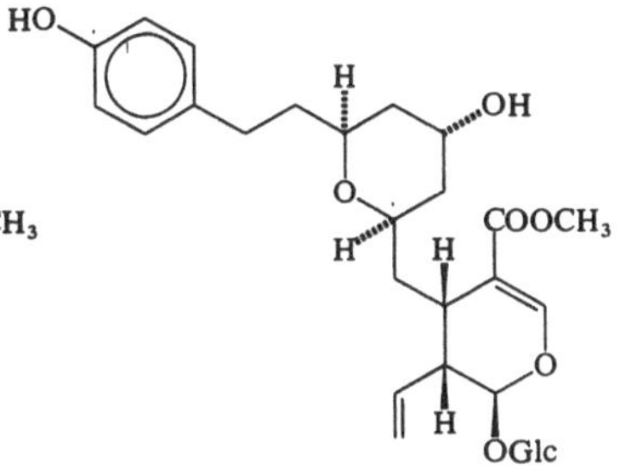

(**38**)

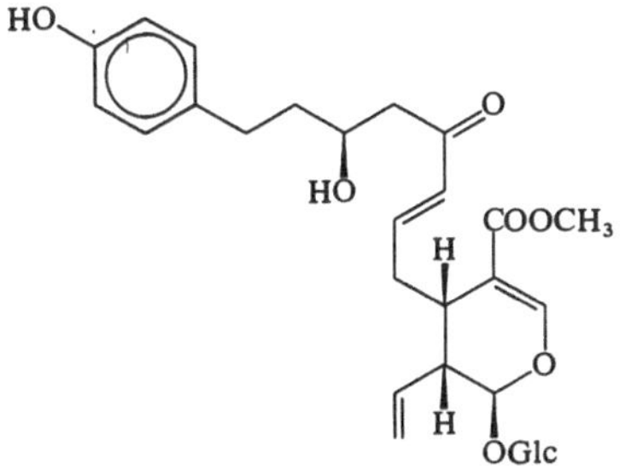

(**39**)

III. Mevalonoid Origin of Iridoids and Secoiridoids and Mechanism of Formation of Iridane Skeleton from Acyclic Monoterpenes

1. Mevalonoid Origin of Iridoids

The first substantial biosynthetic experiments on iridoid glycosides were carried out with plumieride (6) by the group of SCHMID (59) who established its structure.

In 1964, it was found through tracer studies using *Plumiera acutifolia* that the ^{14}C label of [2-^{14}C] MVA was incorporated into C-7, C-3 and C-15 of plumieride (6), whereas the label of [1-^{14}C]acetate was incorporated into the C-12 and C-13 of (6) and that the label from the former precursor was equally distributed between C-3 and C-15. Based on these findings, they inferred, as shown in Fig. 1, that plumieride (6) is biosynthesized through a Michael-type cyclization of 10-oxocitronelall (40) to iridoidial (15), further oxidation to a substance like iridotrial (41), at which stage randomization takes place between C-3 and C-15, and further elaboration involving introduction of two acetate units (59).

The mechanism for the formation of iridodial described above had already been proposed earlier by ROBINSON et al. (60) who synthesized (15) and isoiridomyrmecin (4) from (S)-(−)-citronelall ethylene acetal (42) by this route.

Since then iridodial (15) was long thought to be the key intermediate for the biosynthesis of all iridoids; this is the reason why BRIGGS et al. designated the series of compounds with which this review is concerned as iridoids (29). As mentioned already in Chapter 1, incorporation of MVA into indole alkaloids was demonstrated in 1965 (16, 17, 19) and again considerable randomization of those carbons of the indole alkaloids which correspond to C-3 and C-11 of an iridoid was demonstrated as had already been found in the plumieride (6) biosynthesis (59).

Prompted by research on indole alkaloids, biosynthetic studies on secoridoid glucosides which are biosynthetically related to the non-tryptophan moiety of the indole alkaloids were also being carried out. In 1967, COSCIA et al. (61, 62) found incorporation of [2-^{14}C]MVA into gentiopicroside (2) of *Swertia caroliniensis*. Soon after publication of this work, INOUYE et al. also reported incorporation of [2-^{14}C]MVA into sweroside (24) and swertiamarin (7) of *Swertia japonica* as well as into gentiopicroside (2) of *Gentiana triflora* var. *japonica*. Furthermore, they deduced the sequence sweroside (24) → swertiamarin (7) → gentiopicroside (2) from a comparison of the specific activities of

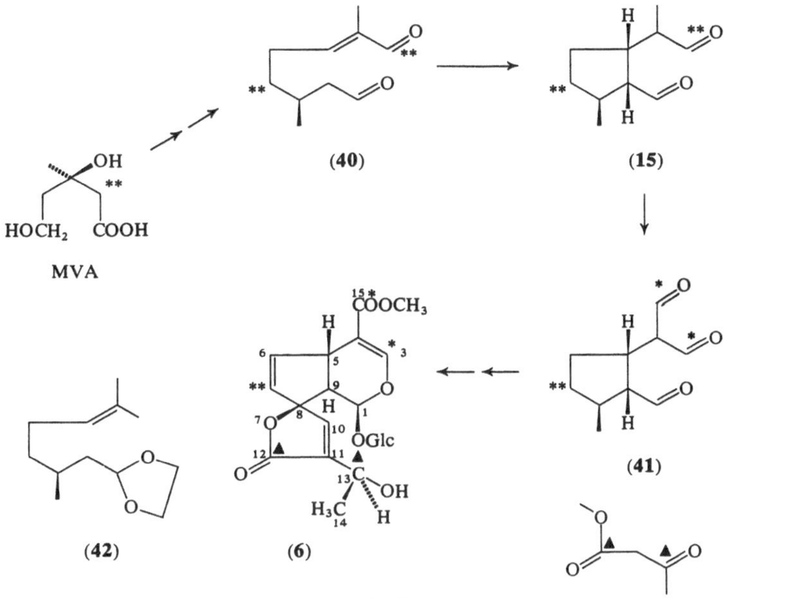

Fig. 1

(24), (7) and (2) and from the structures of these glucosides (63a). Moreover, both groups found that the label from [2-^{14}C]MVA was equally distributed between C-3 and C-11 of these secoiridoid glucosides (Fig. 2).

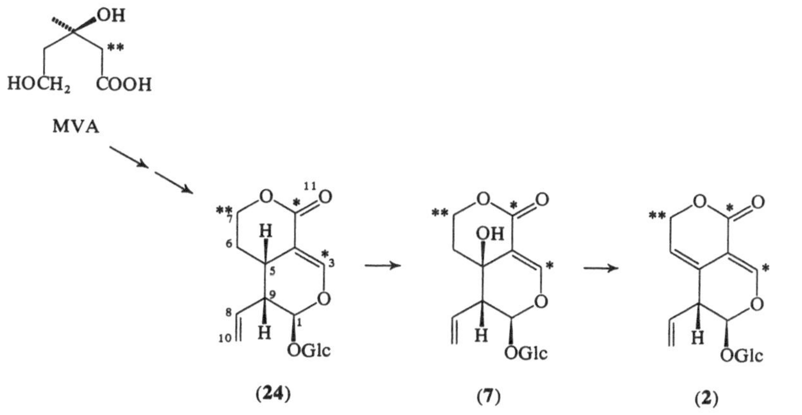

Fig. 2

Later, Coscia et al. (64, 65) administered [4R-4-³H]-, [4S-4-³H]-
or [2S-2-³H]MVA together with [2-¹⁴C]MVA to S. caroliniensis and
obtained the following findings through chemical degradation of isolated loganic acid (43) and gentiopicroside (2): 1) both C-4 pro-R hydrogens of two MVA molecules were retained in loganic acid (43),
whereas both C-4 pro-S hydrogens disappeared in (43). 2) Only one
C-4 pro-R hydrogen of the two MVA molecules was retained in gentiopicroside (2), whereas neither of the two C-4 pro-S hydrogens was
present. 3) The C-2 pro-R hydrogens of MVA were found at C-3 and
C-7 of (43), whereas one pro-S hydrogen was at C-3, the other having
been essentially eliminated from the hydroxylated C-7 position. 4) Gentiopicroside (2) obtained in the [2S-2-³H,2-¹⁴C]MVA incorporation
had the same ³H/¹⁴C ratio as loganic acid (43).

Based on these findings, Coscia et al. inferred the steric course
of the biosynthesis of gentiopicroside (2) from MVA shown in Fig. 3.
Furthermore, they isolated acidic glucosides such as loganic acid (43),

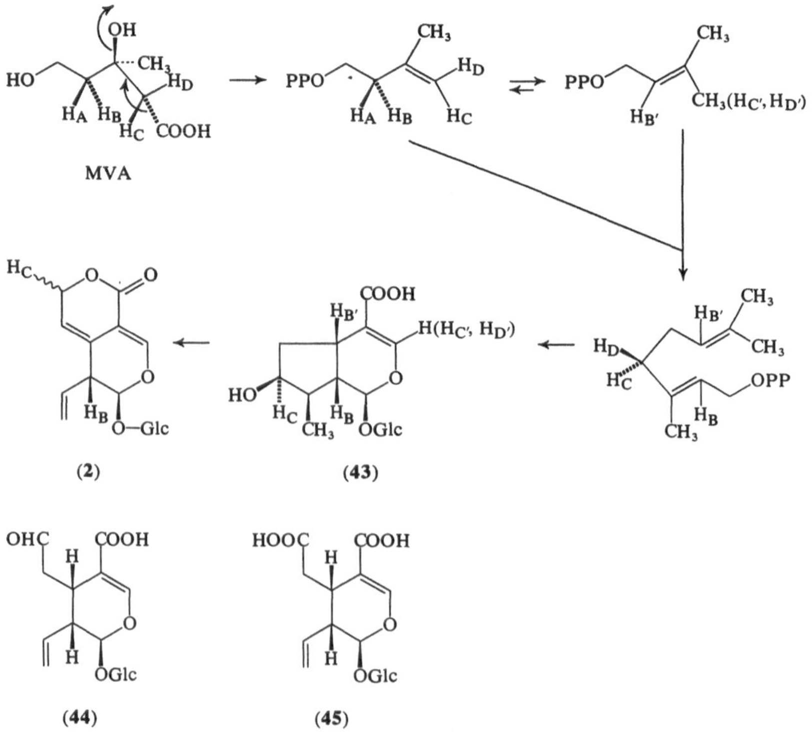

Fig. 3

secologanic acid (44) and secologanoside (45) along with secologanin (23) from *Catharanthus roseus*, and clarified, through administration of [2-^{3}H, 2-^{14}C] MVA to this plant, that the ^{3}H label was retained
on C-3 and C-7 of 43, 44 and 45 (*66, 67*).

In the foregoing paragraphs the evidence for the mevalonoid origin
of iridoid- and secoiridoid glycosides as well as some findings on the
steric course of the biosynthesis of these compounds have been described. In these experiments, randomization of the label from
[2-^{14}C] MVA between C-3 and C-11 was always observed in the iridoid-
and secoiridoid glycosides and in the indole alkaloids.

On the other hand, the unusual distribution of the ^{14}C label from
[2-^{14}C] MVA was recognized in some iridoids. WALLER *et al.* demonstrated incorporation of [2-^{14}C] MVA into β-skytanthine (16) (*68*) of
Skythantus acutus, actinidine (46) (*69*) of *Actinidia polygama* and verbenalin (18) (*70*) of *Verbena officinalis*. In the experiments with *S. acutus* and *V. officinalis*, the ratios of the ^{14}C label from [2-^{14}C] MVA

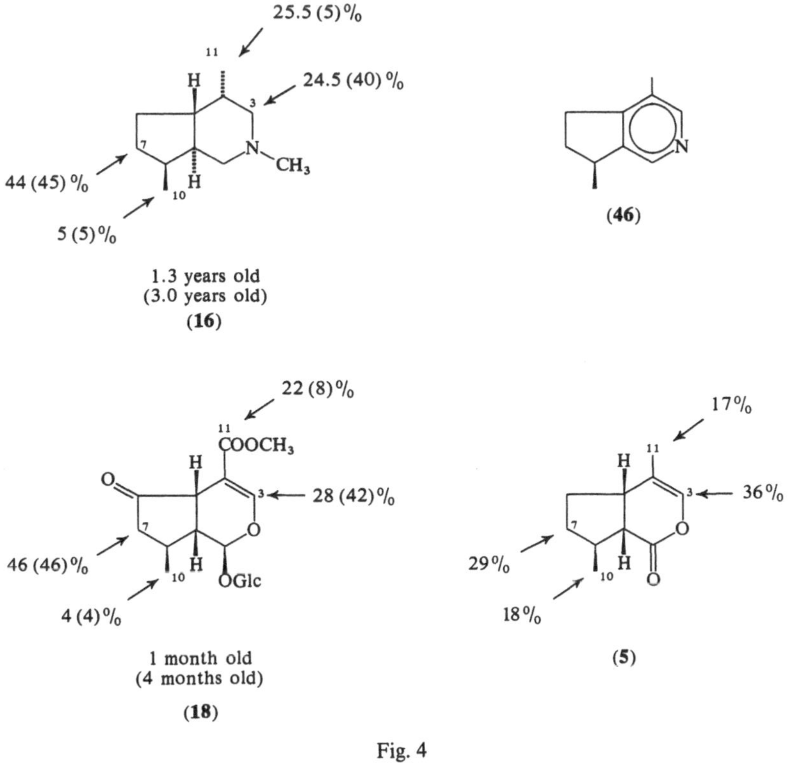

Fig. 4

in C-3 and C-11 of (16) and (18) varied with plant age as illustrated in Fig. 4. These phenomena were explained as follows: Since, in young plants, the pool of iridoids is small, whereas the pool of acyclic monoterpenes such as geraniol and its pyrophosphate is large, there is a much greater chance for randomization in the terminal dimethyl group. By contrast, in old plants the situation is reversed, the pool of iridoids being large and the pool of the acyclic monoterpenes being small; hence there is little chance of randomization in the terminal dimethyl groups. No research has so far been done on the stereochemical course of randomization of the terminal group or on the enzyme system concerned.

In the case of verbenalin (18), SCHMID et al. (71) also observed complete randomization of the label between C-3 and C-11 when [2-^{14}C]MVA was fed to small shoots of V. officinalis. WALLER et al. (72) noted that in Nepeta cataria randomization of the label from [2-^{14}C]MVA took place not only between C-3 and C-11 but also between C-7 and C-10 of nepetalactone (5) and inferred that this occurred at the stage of isopentenyl prophosphate, since the distribution rates of the ^{14}C label at C-11 and C-10 were almost the same.

To explain the unusual phenomena described above they assumed that the randomization between C-3 and C-11 and/or between C-7 and C-10 of iridoids occurred at a stage before cyclopentane ring formation, thus contradicting the opinion of SCHMID that in the biosynthesis of plumieride (6) randomization of C-3 and C-15 takes place after ring closure to iridodial (15) (59).

2. Mechanism of Formation of Iridane Skeleton from Acyclic Monoterpenes

Shortly after the mevalonoid origin of the non-tryptophan moiety of the indole alkaloids was established, several groups also demonstrated that geraniol was a precursor for these alkaloids (21–25). With respect to the iridoid and secoiridoid glycosides, ARIGONI et al. (73) showed incorporation of [4-^{14}C]geraniol into both the acyclic monoterpenoid and secoiridoid portions of foliamenthin (25) in Menyanthes trifoliata in 1968. At the same time, BATTERSBY et al. (74) observed incorporation of [2-^{14}C]geraniol into both the acyclic monoterpenoid and secoiridoid portions of dihydrofoliamenthin (47) in the same plant. BATTERSBY (75) also demonstrated incorporation of [2-^{3}H]geraniol into the isoquinoline alkaloid glucoside ipecoside (36) in Cephaelis ipecacuanha and COSCIA (76) showed incorporation of [1-^{14}C]geranyl pyrophosphate into loganic acid (43) in Swertia caroliniensis.

(25)   (47)

The logical precursors in the sequence following geraniol are 10-hydroxy derivatives of nerol or geraniol.

In 1970, the groups of both ARIGONI (77) and BATTERSBY (78) showed through feeding of various ^{3}H or ^{14}C-labelled acyclic monoterpenes to *Catharanthus roseus* that 10-hydoxynerol (48) and 10-hydroxygeraniol (49) were incorporated into loganin (10) and indole alkaloids such as ajmalicine (50), vindoline (9) and catharanthine (51), whereas citronellol, citronellal, 10-hydroxycitronellol (52), 10-hydroxycitronellal (53) or 10-oxocitronellol (54) and 10-hydroxylinalool (55) were not incorporated into these alkaloids. ARIGONI et al. also observed extensive randomization of the label from both [9-^{14}C]-10-hydroxynerol (48) or [9-^{14}C]-10-hydroxygeraniol (49) between C-3 and C-11 of (10) and between the corresponding positions of the alkaloids and proposed the cyclization mechanism illustrated in Fig. 5 (77). This involves cyclization of 9,10-dioxoneral (56), formed *via* (48) from geraniol, to iridotrial (41), at which stage the randomization occurs; (41) is further metabolized *via* deoxyloganin (57), loganin (10) and secologanin (23) to indole alkaloids. The intermediacy of the 10-hydroxy or 9,10-dioxo derivative of nerol, but not of geraniol, was proposed based on the recognition that the acyclic terpenic acid part of foliamenthin (25) is a derivative of nerol and that [9-^{14}C]-10-hydroxynerol (48) was incorporated into loganin (10) and indole alkaloids more efficiently than [9-^{14}C]-10-hydroxygeraniol (49). However, no reason was given for the suggestion that oxidation of both terminal methyl groups of nerol takes place before cyclopentane ring formation[1]. Nevertheless, this mechanism is noteworthy for its difference from the so far accepted pathway (59, 60) involving Michael-type cyclization of (S)-(−)-10-oxocitronellal (40) to iridodial (15).

---

[1] Shortly before the publication of these articles, BOWMAN and LEETE reported that [4-^{14}C]iridodial (15) was not incorporated into the indole alkaloids of *C. roseus* (80).

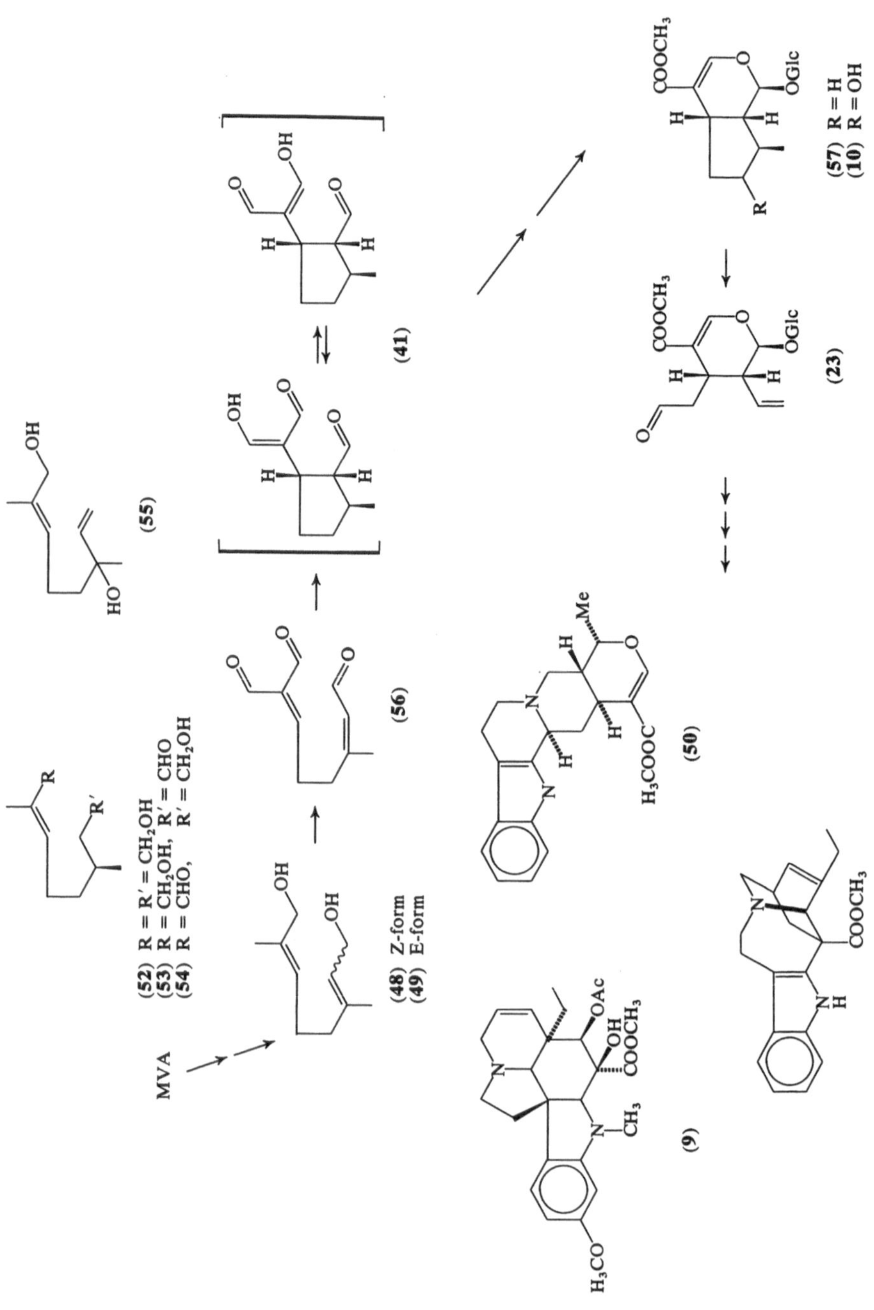

Fig. 5

Subsequently, COSCIA (79) found that an enzyme in the microsomal fraction prepared from the seedlings of *Catharanthus roseus* catalyzes hydroxylation of geraniol or nerol to 10-hydroxygeraniol (49) or 10-hydroxynerol (48) in the presence of $NADP^+$. Since the oxidase which exhibits a typical CO-reduced-P-450 binding spectrum was dependent upon oxygen and NADPH, it was thought to be a cytochrome P-450 type monooxygenase.

Independently, INOUYE and coworkers have studied the mechanism of formation of the iridane skeleton from acyclic monoterpenes in the biosynthesis of iridoid glucosides possessing a highly oxidized cyclopentane ring since the middle 1970's, and have so far found two cyclization modes.

Feeding experiments with [2-^{14}C]MVA indicated incorporation of MVA into lamioside (58), ipolamiide (59), lamiide (19) in *Lamium amplexicaule,* deutzioside (17) in *Deutzia crenata* and asperuloside (1) in *Galium spurium* var. *echinospermon,* and retention of the ^{14}C label only at C-3 and C-7 of these glucosides. It was thus suggested that in these plants the glucosides (58), (59), (19), (17) and (1) are biosynthesized by cyclization of 10-oxogeranial (60)/10-oxoneral (61) to iridodial (15) and by further elaboration without scrambling of C-3 and C-11 to the iridoids (39, 41). This suggestion was confirmed by the finding that [10-^{3}H]-10-hydroxygeraniol (49) was incorporated into deutzioside (17), [10-^{3}H]iridodial (15) into lamioside (58), ipolamiide (59), lamiide (19), deutzioside (17) and asperuloside (1), and [10-^{3}H]iridodial glucoside (62) into lamioside (58) and deutzioside (17) (40, 42) (Fig. 6).

A little later, iridane skeleton formation in the biosynthesis of tarennoside (63) and gardenoside (22) of *Gardenia jasminoides* f. *grandiflora* cell suspension cultures (81) was examined in detail by administering various combinations of the following ^{13}C-labelled putative precursors (82): [9-^{13}C]-10-hydroxygeraniol (49), [2-^{13}C]-9,10-dihydroxygeraniol (64)/[2-^{13}C]-9,10-dihydroxynerol (65) (4:1), (3R)-(+)- and (3S)-(−)-[9-^{13}C]-10-hydroxycitronellol (52) and (3S)-(−)-[8-^{13}C]-9,10-dihydroxycitronellol (66). Compound (49) was incorporated into tarennoside (63) in high yield, but compounds (52) and (66) belonging to the citronellol series were not; also the ^{13}C label of (49) was equally distributed between C-3 and C-11 of (63). Furthermore, [2-^{13}C]-(64)/[2-^{13}C]-(65) was incorporated into (63) much less than [9-^{13}C]-(49), although the incorporation of (64)/(65) was not negligible. It was therefore inferred that in the *G. jasminoides* cell cultures tarennoside (63) and gardenoside (22) are also formed by cyclization of 10-oxogeranial (60) or 10-oxoneral (61) to iridodial (15) and oxidation to iridotrial (41) (45, 47). The aforementioned 9,10-dihydroxygeraniol (64)/9,10-dihydroxynerol (65) and their oxo derivatives, 9,10-dioxoger-

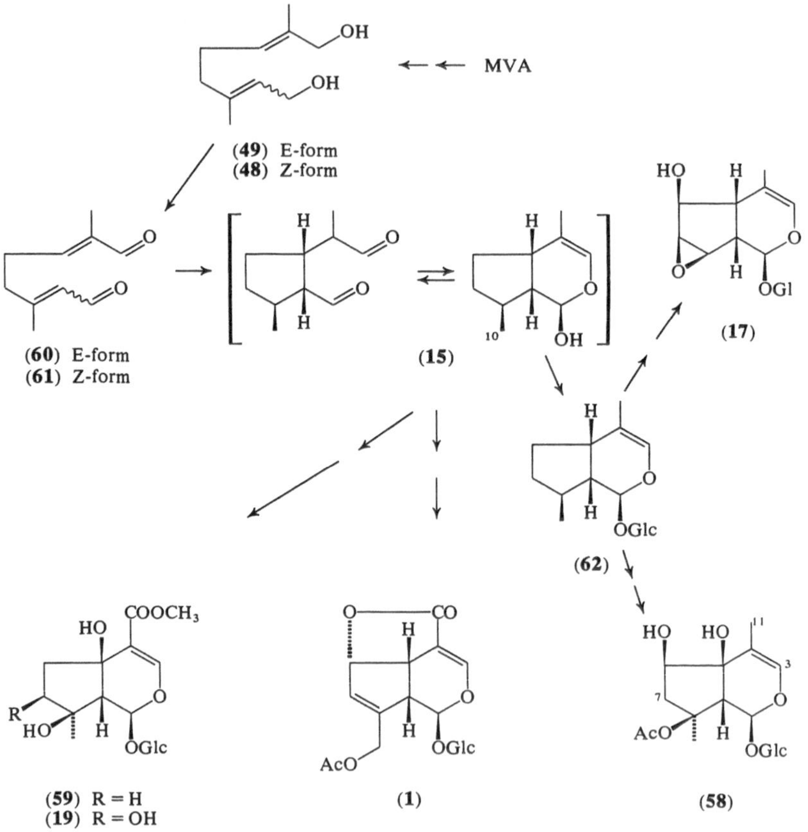

Fig. 6

anial (67)/9,10-dioxoneral (56), presumably are not on the main biosynthetic pathway of tarennoside (63) and gardenoside (22). Subsequently, in order to detect a glucosylated trial derivative following iridotrial (41), dilution analyses of the putative precursors iridotrial glucoside (68) and 7,8-dehydroiridotrial glucoside (69) were attempted after administration of [4-^{13}C]-(49)/[4-^{13}C]-(48) (4:1) to the cell cultures. [10-^{13}C] Boschnaloside (70) was unexpectedly detected by examining the ^{13}C NMR spectrum of the acetate of reisolated mixture of (68) and (69), which did not show any enrichment at the 10β-methyl carbon, but showed a signal at δ 16.10 due to the 10α-methyl carbon of the contaminating [10-^{13}C]-(70). It could therefore be deduced that the cyclization product of the acyclic monoterpene is not iridodial (15), but 8-epiiridodial (71) (46, 48). This deduction was substantiated by

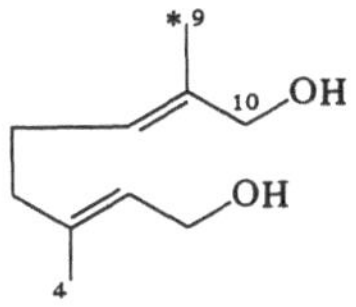

(49)

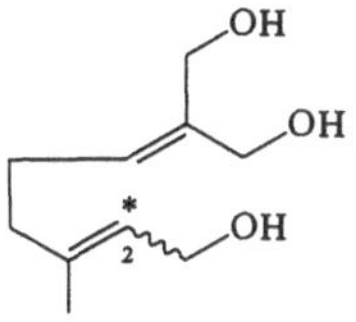

(64) E-form
(65) Z-form

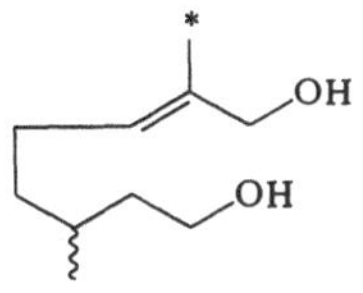

(52)

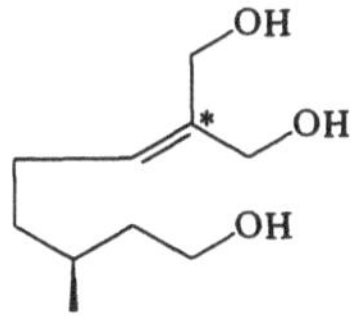

(66)

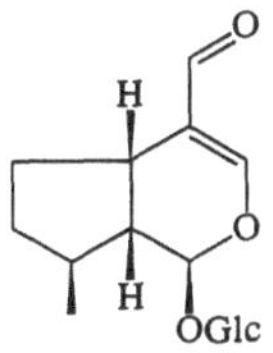

(68)

the high incorporation of [10-^{2}H$_3$]-8-epiiridodial (71) (13.2%) and [11-^{2}H] boschnaloside (70) (41.7%) into tarennoside (63). The intermediate subsequent to (70) was demonstrated to be 7,8-dehydroiridotrial glucoside (69) because of a high incorporation of (69) (70,4%) into tarennoside (63) (Fig. 7).

Shortly before publication of this work, DAMTOFT also reported that 8-epideoxyloganin (72) (43) was a precursor for ipolamiide (59) and lamiide (19) in *Hebenstreitia dentata* and that 8-epideoxyloganic acid (73) (44) was a precursor for aucubin (74) in *Plantago major* and *Scrophularia racemosa* and for antirrhinoside (75) in *Antirrhinum majus*, thus demonstrating indirectly the intermediacy of 8-epiiridodial (71) in the biosynthesis (Fig. 8).

For a long time following the articles of the groups of ARIGONI (77) and BATTERSBY (78) there were no significant further reports on the cyclization mechanism for the biosynthesis of secoiridoids and the related indole alkaloids. Although TIETZE *et al.* in 1981 (83) demonstrated that 10-oxogeraniol (76)/10-oxonerol (77), 10-hydroxygeranial (78)/10-hydroxyneral (79) and 10-oxogeranial (60)/10-oxoneral (61)

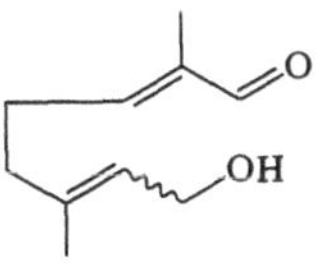

(76) E-form
(77) Z-form

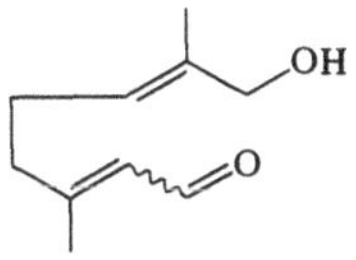

(78) E-form
(79) Z-form

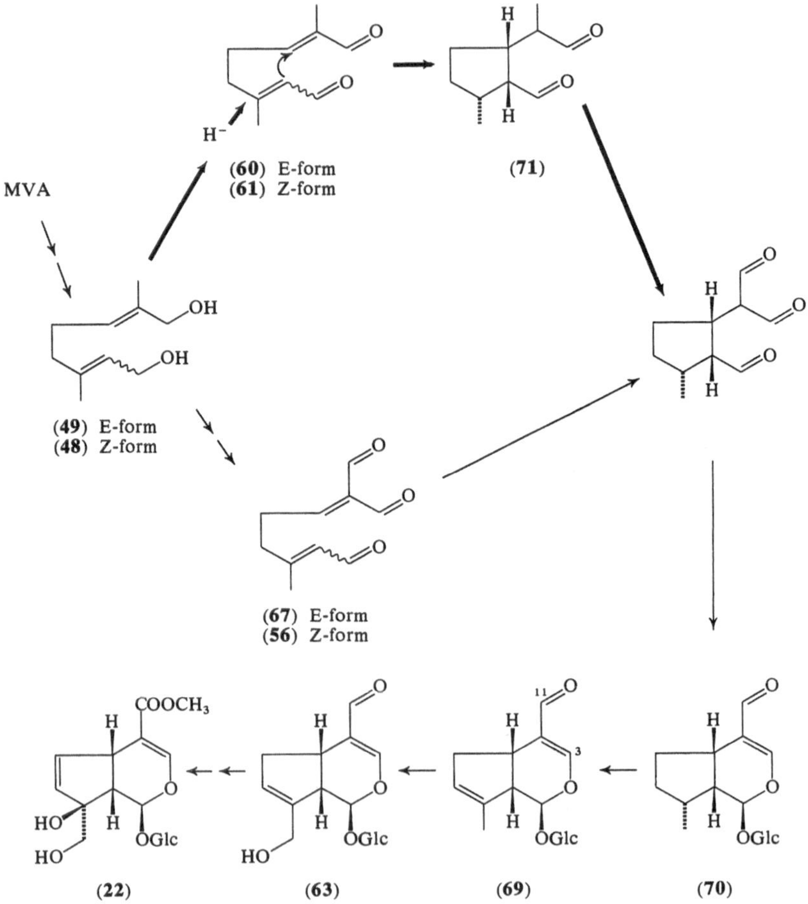
MVA
H⁻
(60) E-form
(61) Z-form
(71)
OH
OH
(49) E-form
(48) Z-form
(67) E-form
(56) Z-form
COOCH₃
HO
HO
OGlc
(22)
(63)
(69)
(70)

Fig. 7

were incorporated into loganin (**10**), vindoline (**9**) and catharanthine (**51**) in *C. roseus*, they did not discuss which of **60**/**61** or the corresponding 9,10-dioxo derivatives are directly responsible for the cyclization. Thus, there still remained unsolved problems regarding the mechanism of formation of the iridane skeleton from acyclic monoterpenes.

In 1983, BALSEVICH and KURZ (*84*) proposed the cyclization pathway depicted in Fig. 9 for secologanin (**23**) and ajmalicine (**50**) in *Catharanthus roseus* cell suspension cultures on the basis of evidence that ^{2}H-labelled acyclic monoterpenes including 9,10-dihydroxygeraniol (**64**), 9-oxo-10-hydroxygeraniol (**80**) and 9,9,10,10-tetraethoxyge-

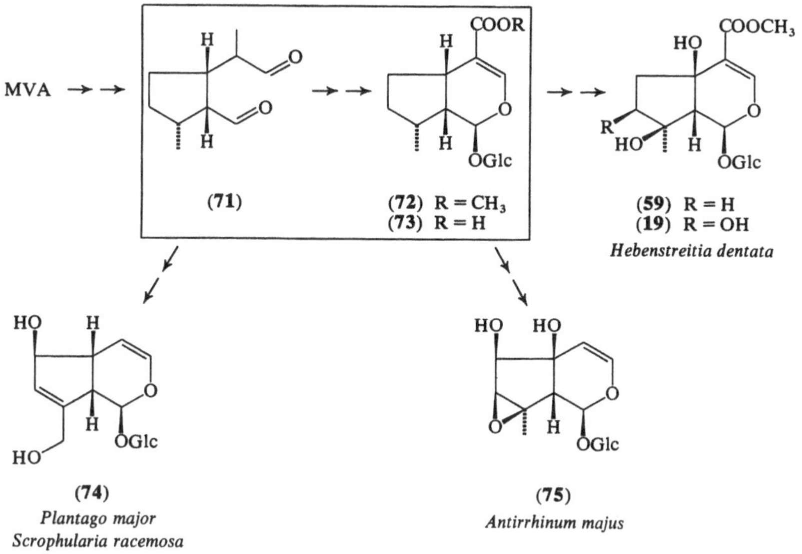

(71)

(72) R = CH₃
(73) R = H

(59) R = H
(19) R = OH
Hebenstreitia dentata

(74)
Plantago major
Scrophularia racemosa

(75)
Antirrhinum majus

Fig. 8

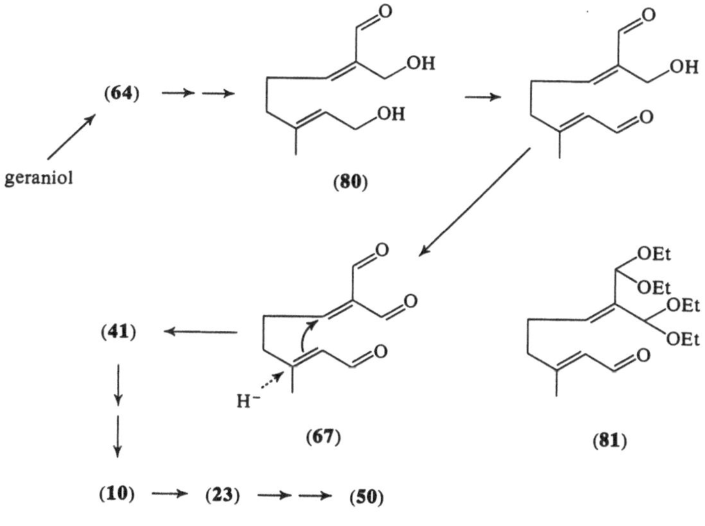

geraniol

(64)

(80)

(41)

(67)

(81)

(10) ⟶ (23) ⟶⟶ (50)

Fig. 9

raniol (81) were incorporated into (23) and (50) in high yields. This pathway is in accordance with that postulated by ARIGONI et al. (77).

Independently, INOUYE and coworkers obtained the following findings. Both [1-³H]-10-hydroxygeraniol (49) and [10-³H]iridodial (15) were incorporated into vindoline (9) of C. roseus and secologanin (23) of Lonicera morrowii, whereas (R,S)-(±)-[10-³H]-hydroxycitronellol (52) was not. Additionally, the respective incorporation rates of [1-³H]-9,10-dihydroxygeraniol (64) and a mixture of [1-³H]-9-hydroxy-10-oxo-geranial (82) and [1-³H]-9-oxo-10-hydroxygeranial (83) into these com-

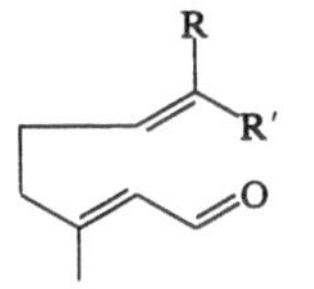

(82) R = CH₂OH, R' = CHO
(83) R = CHO,   R' = CH₂OH

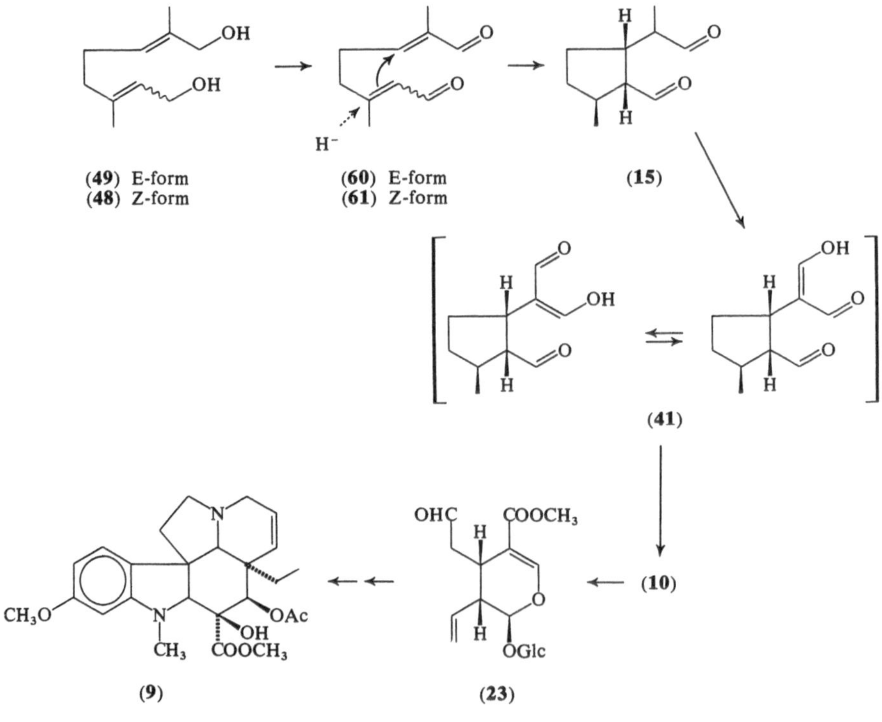

(49) E-form
(48) Z-form

(60) E-form
(61) Z-form

(15)

(41)

(9)

(23)

(10)

Fig. 10

pounds were much less than the incorporation rates of $[1-^3H]-(49)$ and $[10-^3H]-(15)$. It was therefore presumed that secologanin (23) and the indole alkaloids of these plants are biosynthesized *via* cyclization of 10-oxogeranial (60)/10-oxoneral (61) to iridodial (15) in the same way as asperuloside (1) and deutzioside (17) (Fig. 10) (85, 86). This presumption contradicts the proposal by BALSEVICH and KURZ (84) for the biosynthesis of (23) and the indole alkaloids.

Subsequently, in order to examine the applicability of the above mechanism to other plants, $[1-^3H]$-10-hydroxygeraniol (49) and $[1-^3H]$-9,10-dihydroxygeraniol (64) were fed to *Lonicera tatarica* which contains secologanin (23) and $[4-^{13}C]$-49/$[4-^{13}C]$-(48) (4:1) (82), $[9-^{13}C]$-10-hydroxynerol (48), $[2-^{13}C]$-9,10-dihydroxygeraniol (64)/$[2-^{13}C]$-9,10-dihydroxynerol (65) (4:1) and $[10-^2H_2]$iridodial (15) were fed to *Rauwolfia serpentina* cell suspension cultures (87) producing vomilenine (84), ajmaline (85), etc. Higher incorporation yield of (48), (49) and (15) into secologanin (23) and (84) and (85), in comparison with those of (64), clearly indicated that (23), (84) and (85) of these plants and cell cultures are also biosynthesized *via* the same cyclization pathway (88, 89).

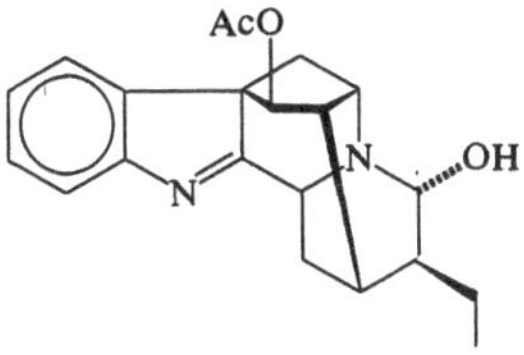

(84) 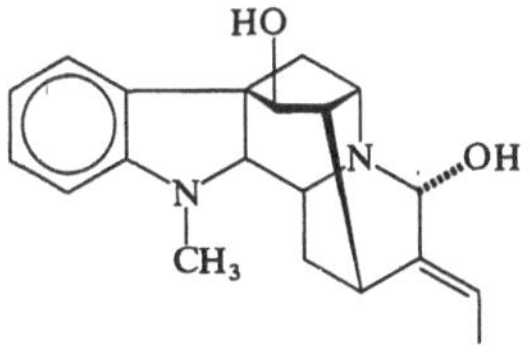   (85)

To provide an unequivocal understanding of iridane skeleton formation in the biosynthesis of secoiridoids and indole alkaloids, enzymatic studies were performed jointly by the groups of INOUYE and ZENK, using cell-free extracts prepared from *R. serpentina* cell suspension cultures (90). The cultured cells obtained 10 days after inoculation were homogenized in 0.1 M tris buffer (pH 7.5, containing 5 mM mercaptoethanol). After centrifugation, the supernatant was saturated with ammonium sulfate and the precipitate obtained by the 70% ammonium sulfate cut was dissolved in the above buffer and desalted by dialysis, leading to the cell free extracts. Then, $[1-^3H]$-10-hydroxygeraniol (49) was incubated with the cell free extract in the presence of NAD/NADH and NADP/NADPH at 25 °C for 20 h. Chromatography of the CH_2Cl_2 soluble portion of the incubation mixture over silica gel plates gave a radioactive conversion product (Fig. 11). In order to confirm that this product was an intermediate in secoiridoid-indole alkaloid

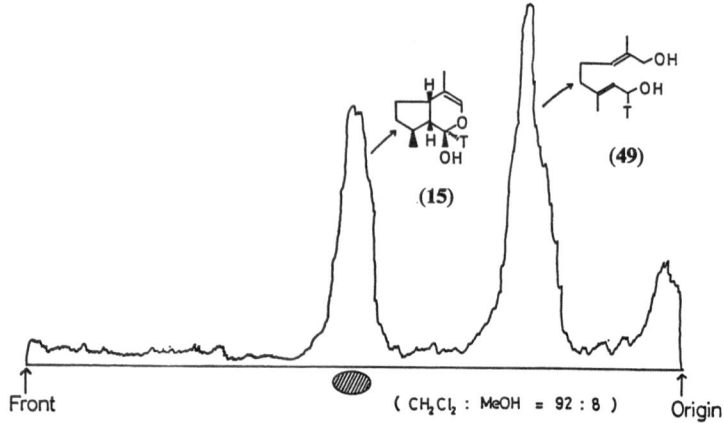

Fig. 11. Radiochromatogram of [1-³H]-10-hydroxygeraniol (49) incubation mixture

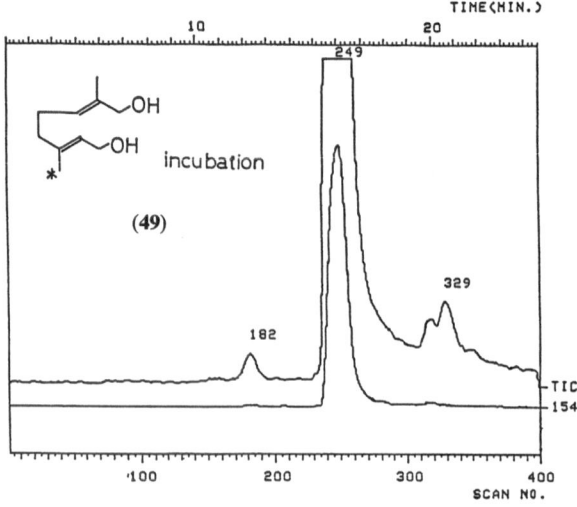

Fig. 12. TIC of the CH₂Cl₂ soluble portion of [4-¹³C]-10-hydroxygeraniol (49) incubation mixture

biosynthesis, it was administered to *R. serpentina* cell suspension cultures which resulted in incorporation into vomilenine (84) (4.6%) and ajmaline (85) (0.7%), respectively.

The structure of the conversion product was established by means of GC-Mass spectrometry in the following way. [4-¹³C]-10-Hydroxygeraniol (49)/[4-¹³C]-10-hydroxynerol (48) (4:1) or [9-¹³C]-10-hydroxy-

nerol (**48**) was incubated with the cell free system in the manner described above. Fig. 12 shows the TIM chromatogram of the CH_2Cl_2
soluble portion of the [4-^{13}C]-(**49**)/[4-^{13}C]-(**48**) incubation mixture. The
mass spectra of the peak at scan no. 182 (a) and the corresponding
peak (b) originating from the [9-^{13}C]-(**48**) incubation mixture had parent peaks at m/z 169.1202 and 169.1183, respectively, whereas that
of a synthetic enol-hemiacetal form (*42*) of iridodial (**15**) exhibited
parent peak (c) at m/z 168.1166 ($C_{10}H_{16}O_2$, Er. 1.6 M). Moreover,
the fragmentation patterns of a and b are in good accord with that
of c, and also correspond with the fragmentation mechanism postulated
for synthetic iridodial (**15**) (Fig. 13). Therefore, the substance of scan
no. 182 was deduced to be the enol-hemiacetal form of ^{13}C-labelled
iridodial (**15**) of empirical formula $C_9^{13}CH_{16}O_2$. This deduction was
confirmed by measuring the ^{13}C NMR spectrum of the CH_2Cl_2 soluble
portion of the [4-^{13}C]-(**49**)/[4-^{13}C]-(**48**) incubation mixture; a ^{13}C-en-
riched signal appeared at 20.60 ppm, corresponding to the C-10 chemi-
cal shift of the enol-hemiacetal form of (**15**). Therefore, it was established that iridodial (**15**) is the intermediate precursor for vomilenine
(**84**) and ajmaline (**85**) in *R. serpentina* cell suspension cultures.

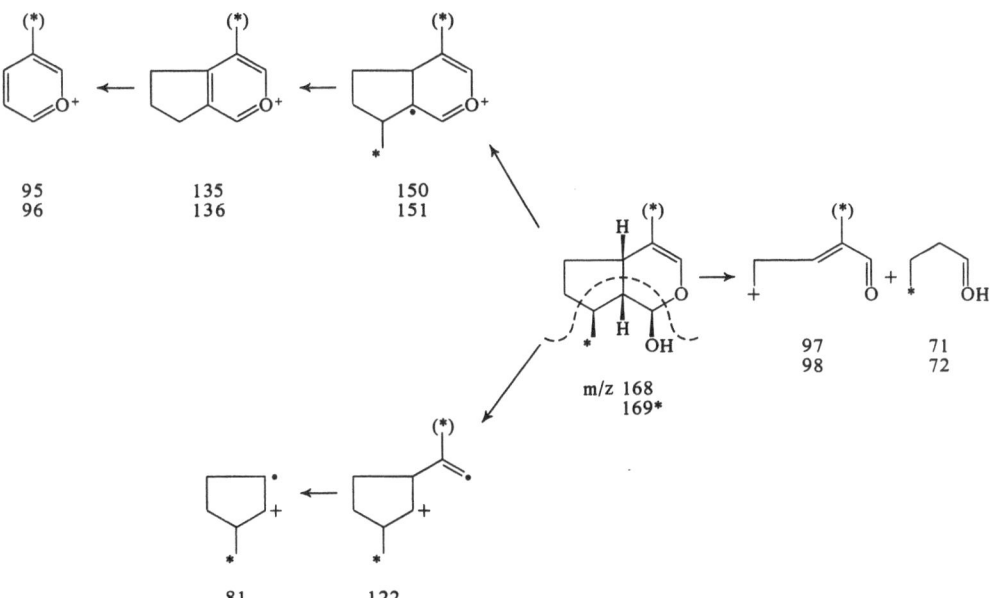

Fig. 13. Postulated fragmentation mechanism for iridodial (**15**)

In subsequent incubation experiments, even in the absence of ATP/ Mg^{2+} and with addition of ATPase to the usual incubation mixture (to decompose traces of endogenous ATP that may be present), enzymatic conversion of (49) or (48) to (15) still proceeded, suggesting the intervention of 10-oxogeranial (60) or 10-oxoneral (61) between (49) or (48) and (15). This was proved by actual conversion of [4-^{13}C]-(60) or [9-^{13}C]-(61) into labelled (15) in the ordinary incubation system. Furthermore, neither (S)-$(-)$-[9-^{13}C]-10-hydroxycitronellol (52) nor [2-^{13}C]-9,10-dihydroxygeraniol (64)/[2-^{13}C]-9,10-dihydroxynerol (65) (4:1) was converted into monocyclic monoterpenes such as iridodial (15) and iridotrial (41), in accordance with the evidence provided by the previous *in vivo* tracer studies (*85, 86, 87, 89*). The cyclase involved was therefore concluded to be a new type of monoterpene cyclase, differing from those so far known in that the former converts the oxo derivatives, but not the pyrophosphates of the acyclic monoterpenes (48) and (49) into the cyclic monoterpene (15).

It appears, therefore, that the cyclization pathway involving conversion of 10-oxogeranial (60)/10-oxoneral (61) is fairly common not only in plants producing iridoid glucosides possessing a highly oxidized cyclopentane ring, but in plants producing secoiridoid glucosides and indole alkaloids.

Recently, however, PAGNONI *et al.* found that (S)-$(-)$-citronellol and (S)-$(-)$-10-hydroxycitronellol (52), but not the corresponding 3,4-unsaturated analogues, can serve as intermediates for dolichodial (11) (*50*), dolicholactone (12) (*51–53*) and teucrein (86) (*52*) in *Teucrium marum* as well as for nepetalactone (5) (*54*) and dihydronepetalactone (87) (*54*) in *Nepeta cataria* (Fig. 14). For dolichodial (11) and teucrein (86) of *T. marum,* they postulated a biosynthesis involving cyclization of 9-hydroxy-10-oxocitronellal (88) to 11-hydroxyiridodial (89), based on the demonstration that the ^{3}H label of [10-^{3}H]-10-hydroxycitronellol (52) was incorporated into C-3 of (11) without scrambling, whereas [10-^{3}H]-iridodial (15) was not.

For dolicholactone (12) of *T. marum* and nepetalactone (5) of *N. cataria,* they deduced a route involving cyclization of (S)-$(-)$-10-oxocitronellal (40) to iridodial (15) and a subsequent 1,5-hydride shift to hydroxy acid (90) from the finding that the ^{3}H label of both [3-^{3}H]iridodial (15) and [10-^{3}H]-10-hydroxycitronellol (52) was retained on C-3 of (12) and (5), whereas that of [1-^{3}H]nerol and [1-^{3}H]-(52) was transferred to C-3. They also assumed that dihydronepetalactone (87) is derived directly from (15) by a 1,5-hydride shift and lactonization of the resulting (90). The possibility that photocitral A (91), easily obtainable by photolysis of citral (*91, 92*), may serve as a biosynthetic intermediate of the non-glycosidic iridoids dolichodial (11) and dolicholactone

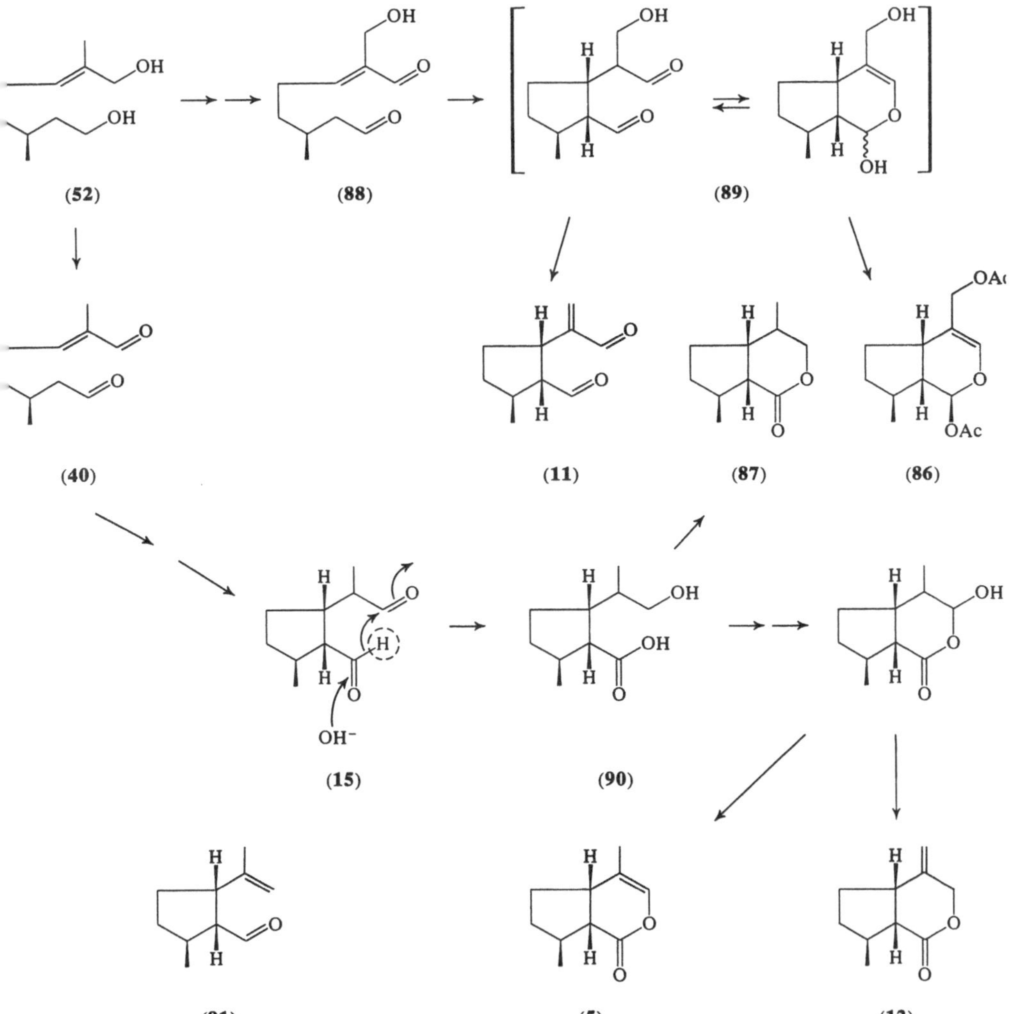

Fig. 14

(12) in *T. marum* as well as nepetalactone (5) and dihydroncpctalactonc (87) in *N. cataria* has also been raised (55). However, a feeding experiment using *N. cataria* demonstrated that [10-^{3}H]-10-hydroxygeraniol (49) was incorporated more efficiently than $(R,S,)$-($\pm$)-[10-^{3}H]-10-hydroxycitronellol (52) into 1,5,9-epideoxyloganic acid (13), which coexists with nepetalactone (5) in this plant (93). Thus, further studies seem

to be necessary to permit a full understanding of the cyclization mechanism in the biosynthesis of the aforementioned non-glycosidic iridoids.

Thus, contrary to earlier expectations, the routes leading to the iridane skeleton in the biosynthesis of iridoids and secoiridoids are not identical and sometimes differ depending on the kinds of compounds as well as on the plants containing them. Thus, new cyclization mechanisms may still await discovery.

IV. Biosynthetic Processes After Cyclopentane Ring Formation

1. Intermediacy of Deoxyloganic Acid, 8-Epideoxyloganic Acid and Loganic Acid in Biosynthesis of Iridoids

As mentioned in Chapter I, intervention of the iridoid glucoside loganin (10) in indole alkaloid biosynthesis was first demonstrated by BATTERSBY et al. (26, 27). On the other hand, INOUYE et al. found that deoxyloganic acid (92) was a precursor for asperuloside (1), loganin (10), aucubin (74) and other related glycosides by demonstrating incorporation of [10-³H]deoxyloganic acid (92) into verbenalin (18) in Verbena officinalis, loganin (10) in Lonicera japonica, scandoside (93) in Paederia scandens var. mairei, geniposide (94) and gardenoside (22) in Gardenia jasminoides f. grandiflora, asperuloside (1) in Daphniphyllum macropodum, aucubin (74) in Aucuba japonica and into a secoiridoid glucoside, jasminin (95) in Jasminum primulinum (94, 95). Afterwards, RIMPLER et al. isolated deoxyloganic acid (92) for the first time from the natural source, Physostegia virginiana (96). Subsequently, BATTERSBY and coworkers found incorporation of [carbo-³H-methoxy]-deoxyloganin (57) into loganin (10) and the indole alkaloids of C. roseus and actually isolated (57) from this plant (97). In this manner it was established that 57 and 92 are intermediates in the biosynthesis of secoiridoids and indole alkaloids.

COSCIA and coworkers (66, 67) found that loganic acid (43) and secologanic acid (44) occurred together in C. roseus and demonstrated that [¹⁴C]loganic acid (43), biosynthetically prepared from [2-¹⁴C]MVA was incorporated into loganin (10), secologanic acid (44) and secologanin (23) and that [¹⁴C]loganin (10) was incorporated into secologanic acid (44) in C. roseus. This led them to assume that methylation of the carboxy group proceeds reversibly sometime after loganic acid (43). Later, COSCIA's group reported that a methyl transferase fraction prepared from *C. roseus* catalyzed methylation of loganic acid (**43**) and secologanic acid (**44**) to loganin (**10**) and secologanin (**23**), but did not catalyze the conversion of deoxyloganic acid (**92**) into deoxyloganin (**57**) (*98, 99*). However, a report by BATTERSBY *et al.* (*100*) reporting incorporation of [carbo-³H-methoxy, 7-³H]loganin (**10**) into indole alkaloids without loss of ³H label seems to contradict these results. In any event, it was established that secologanin (**23**) and indole alkaloids are biosynthesized successively through deoxyloganic acid (**92**) and/or deoxyloganin (**57**), loganic acid (**43**) and loganin (**10**). Additional experiments which prove the intermediary role of loganin (**10**) (or loganic acid (**43**)) in the biosynthesis of secoiridoids are the following: COSCIA and coworkers (*64, 65*) verified incorporation of loganic acid (**43**) biosynthetically prepared from [2-¹⁴C, 4R, 4-³H]MVA into gentiopicroside (**2**) in *Swertia caroliniensis*, while INOUYE *et al.* (*94, 101*) and GRÖGER and SIMCHEN (*102*) demonstrated incorporation of [10-³H]loganin (**10**) into gentiopicroside (**2**) in *Gentiana triflora* var. *japonica* and [9-¹⁴C]-(**10**) into (**2**) in *S. petiolata*, respectively. INOUYE and coworkers (*101, 103*) further demonstrated incorporation of [7-³H]loganin (**10**) into morroniside (**26**) in *G. thunbergii* and jasminin (**95**) in *Jasminum primulinum*, whereas Battersby's group verified retention of the label from [7-³H]loganin (**10**) on C-7 of secologanin (**23**) in *C. roseus* (*104*).

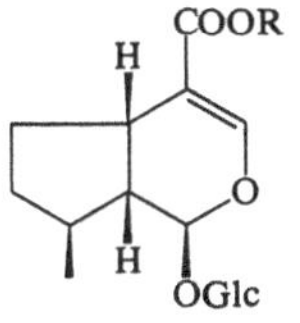

(**57**) R = CH₃
(**92**) R = H

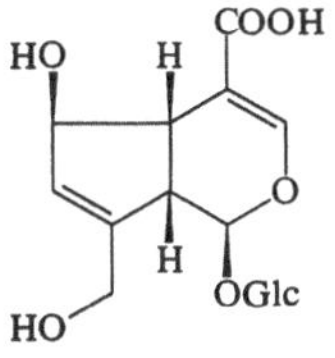

(**93**)

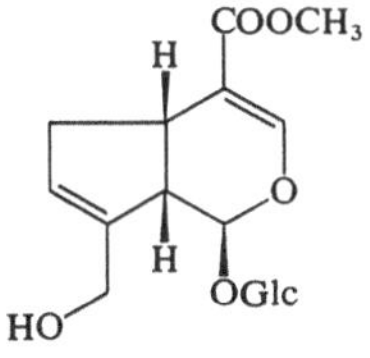

(**94**)

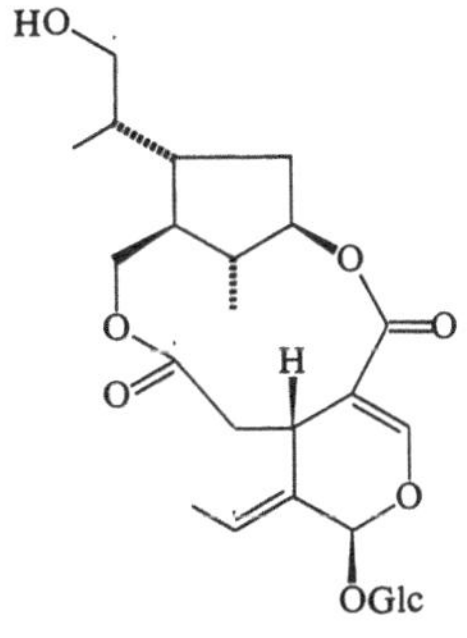

(**95**)

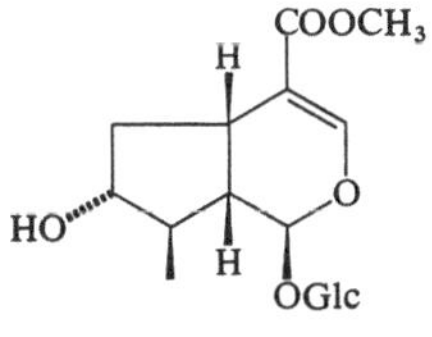

(**96**)

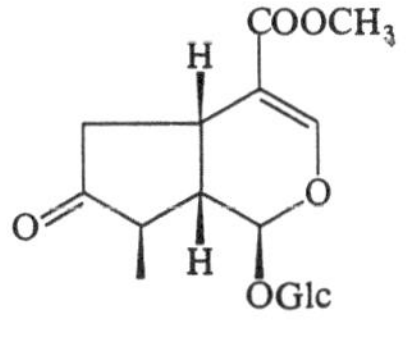

(**97**)

On the assumption that loganin (10) serves as the precursor not only of the secoiridioids and indole alkaloids, but also of iridoid glucosides possessing a highly oxidized cyclopentane ring, INOUYE et al. administered [10-³H]loganin (10), [10-³H]-7-epiloganin (96), [7-³H]loganin (10) and [7-³H]-7-epiloganin (96) to Daphniphyllum macropodum, separately. All these labelled compounds except [7-³H]-7-epiloganin (96) were incorporated into asperuloside (1), though with fluctuation in their incorporation ratios (95, 105), this leading to the conclusion that loganic acid (43) (or loganin (10)) formed by stereospecific hydroxylation of deoxyloganic acid (92) (or deoxyloganin (57)) is a normal precursor of asperuloside (1). The apparent incorporation of [10-³H]-7-epiloganin (96) into (1) is most likely due to successive conversion of (96) into the normal precursor 10 via the 7-dehydro derivative (97).

Postulating that deoxygeniposidic acid (98) and scandoside (93) (or aucubin (74)) were intermediates following loganin (10) (or loganic acid (43)) in the biosynthesis of iridoide glucosides possessing a highly oxidized cyclopentane ring such as asperuloside (1), theviridoside (99), gardenoside (22), aucubin (74) and catalpopside (100), INOUYE and coworkers (38) fed [10-³H]-10-deoxygeniposidic acid (98), chemically derived from (1), and [10-³H]geniposide (94), biosynthetically prepared from (98), to Daphniphyllum macropodum and substantiated the incorporation of (98) and (94) into (1). They also demonstrated incorporation of [10-³H]-geniposide (94) into theviridoside (99) of Cerebra manghas as well as of [10-³H]scandoside (93), biosynthetically prepared from [10-³H]deoxyloganic acid (92), into gardenoside (22) of Gardenia jasminoides f. grandiflora and aucubin (74) of Aucuba japonica. Biosynthetically obtained [10-³H]aucubin (74) was also incorporated into catalposide (100) of Catalpa ovata.

These results led to the proposal shown in Fig. 15 which shows loganic acid (43) (or loganin (10)) as a key intermediate in the biosynthesis of the iridoid glycosides mentioned above. However, because of evidence to be described below, part of this pathway needs to be revised. For example, aucubin (74) is biosynthesized via 8-epideoxyloganic acid (73), and geniposide (94) and gardenoside (22) is formed in Gardenia jasmonioides via precursors possessing an 8α-methyl group.

Taking into consideration the occurrence of griselinoside (101) in Verbena plants, JENSEN et al. presumed that deoxygeniposide (102) which contains potentially oxidizable allylic carbon atoms at C-6 and C-10 constitutes a reasonable biosynthetic precursor of verbenalin (= cornin) (18) and griselinoside (101). Administration of [7,8,10,carbo-²H₃-methoxy]deoxyloganin (57) to V. officinalis demonstrated that deoxyloganin (57) was incorporated into verbenalin (18) in high yield (5%) with retention of the label at C-8 and that methyl ester exchange

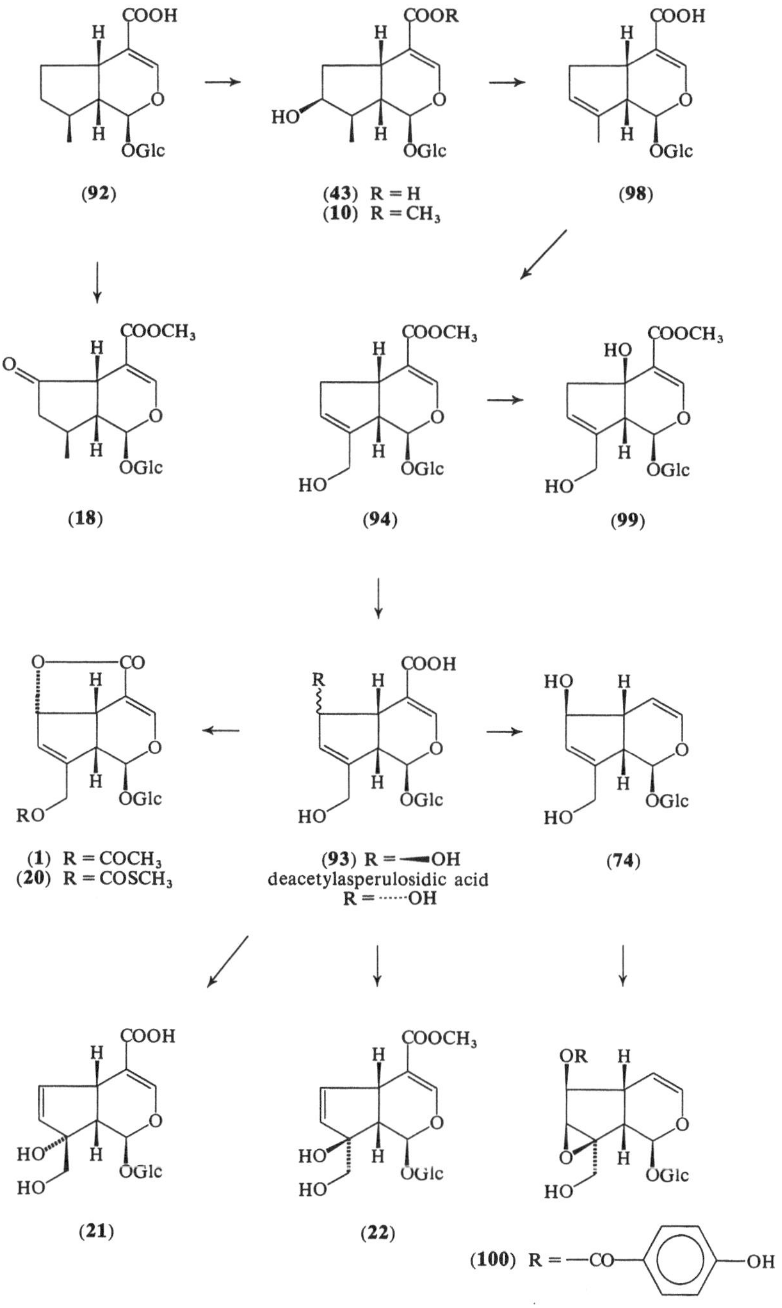

Fig. 15

did not occur during the conversion of (57) into (18). Administration of ^{2}H-labelled deoxygeniposide (102), loganin (10) and mussaenoside (103) to the same plant, however, resulted in non-incorporation of the fed substances into (18) (106, 107). Therefore, the intermediacy of deoxygeniposide (102) was excluded. In the same way, incorporation of deoxyloganin (57) and dihydrocornin (104) into verbenalin (18) and hastatoside (105) was demonstrated in V. hastata, while deoxygeniposide (102) again gave no observable incorporation. Forsythide dimethyl ester (106) was shown to be an efficient precursor for griselinoside (101) in V. hispida, while deoxygeniposide (102), geniposide (94), deoxyloganin (57) and dihydrocornin (104) were not incorporated (107). That iridodial glucoside (62) and dihydrocornin (104) were precursors of verbenalin (18) was also demonstrated in V. officinalis (108).

DAMTOFT further found that [6,7,8,10-^{2}H$_4$, carbo-^{2}H$_3$-methoxy]-8-epideoxyloganin (72) was incorporated in rather high yield into ipolamiide (59) (16%) and lamiide (19) (1.0%) of Hebenstreitia dentata (43); similarly incorporation of [6,7,8,10-^{2}H$_4$]-8-epideoxyloganic acid (73) into aucubin (74) (10% and 7%) of Plantago major and Scrophularia racemosa and antirrhinoside (75) (13%) of Antirrhinum majus (44) was good, whereas incorporation of the corresponding ^{3}H-labelled 8β-methyl compounds (57) and (92) into these glucosides was very poor. ^{2}H-Labelled 8-epideoxyloganin (72) was not incorporated into (74) in Melampyrum cristatum because the biosynthesis was impeded by the

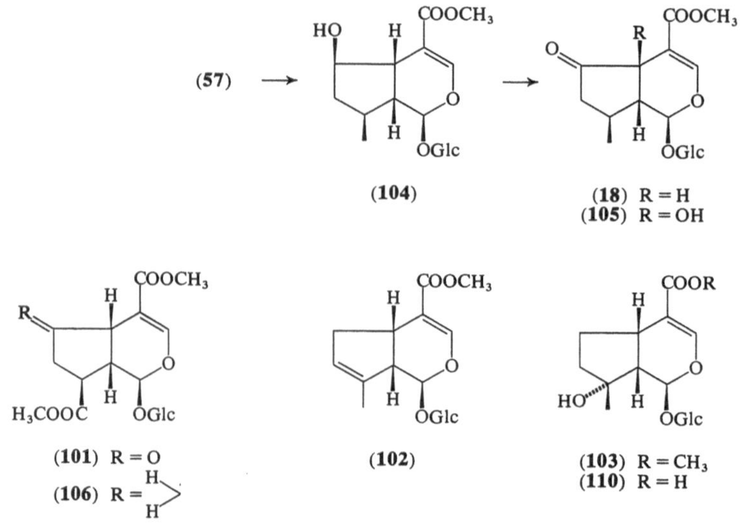

Fig. 16

methyl ester group of (72). These results suggested the intermediacy of the 8α-methyl compounds (72) for (59) and (19) and (73) for (74) and (75) (see Fig. 8). Previously, INOUYE *et al.* (*39, 40, 41, 42*) had suggested, through feeding experiments using the ³H-labelled putative precursors iridodial glucoside (62), 11-hydroxyiridodial glucoside (107) and deoxyloganic acid (92), a pathway involving elaboration of the cyclopentane ring of (62), but not accompanied by oxidation of the C-11 methyl group, leading to lamioside (58) in *Lamium amplexicaule* and deutzioside (17) in *Deutzia crenata* as well as a route involving the successive oxidation of (62) to (107) and (92) leading to ipolamiide (59) and lamiide (19) in the former plant. Later, however, the ³H-labelled (62), (107) and (92) used in these experiments, which had been prepared from geniposide (94) or asperuloside (1) by Pd/C-catalyzed reduction, were found to contain approximately 10% of the corresponding 8α-methyl compounds 8-epiiridodial glucoside (108), 8-epi-11-hydroxyiridodial glucoside (109) and 8-epideoxyloganic acid (73) (*49*). In the light of the low incorporation yields of (107) (0.0037%) and (92) (0.025%) into (19), the possibility can not be excluded that the ³H label in (19) did not originate in labelled (107) and (92), but in the contaminants (109) and (73). Even if (107) and (92) were actually responsible for the label found in (19), it would be desirable to reexamine the intermediacy of both glucosides, since the incorporation ratios were too low. In general, incorporation of fed compounds is affected by such factors as the growth stage of the plants, the season, period of administration and the stabilities, concentrations and cell membrane permeabilities of the administered compounds. As will be shown later,

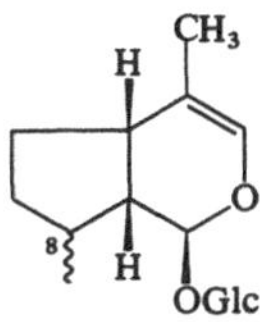

(62) 8—CH₃
(108) 8‒‒‒CH₃

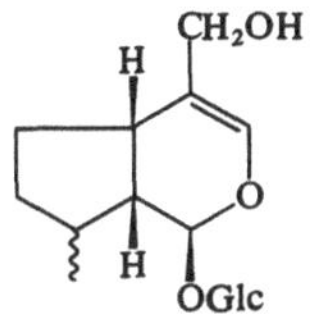

(107) 8—CH₃
(109) 8‒‒‒CH₃

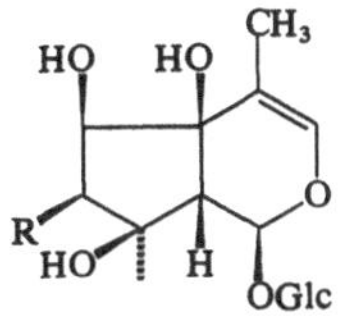

(59) R = H
(19) R = OH

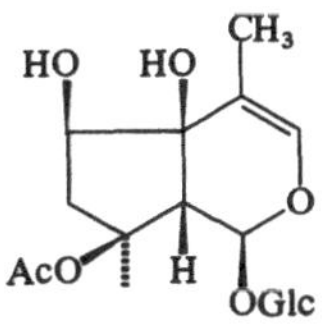

(58)

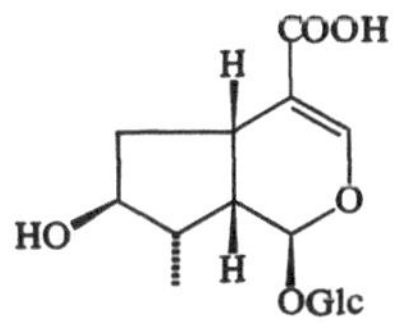

(111)

the compounds fed are usually incorporated into iridoid glucosides in yields higher than 2–3%, provided that they are normal intermediates close to the glucosides in question and administered under proper conditions.

JENSEN et al. studied the incorporation of putative 8-epideoxyloganic acid (73) derived precursors labelled with H_2, such as mussaenosidic acid (110), 8-epiloganic acid (111) and 10-deoxygeniposidic acid (98) into aucubin (74) and found incorporation of (98), but not (110) and (111) (109). Another biosynthetic pathway involving 8α-methyl glucosides is the one for tarennoside (63) and gardenoside (22) in Gardenia jasminoides cell suspension cultures. As mentioned in Chapter III, INOUYE and co-workers (46, 48) demonstrated that 8-epiiridodial (71) and boschnaloside (70) were precursors of (63) and (22) and concluded that dehydrogenation of (70) and successive hydroxylation of the resulting dehydroiridotrial glucoside (69) leads to tarennoside (63).

As in the case of ^{3}H-labelled (62), (92) and (107), ^{3}H-labelled iridotrial glucoside (68) and iridodial (15) used in the tracer studies of INOUYE et al., both of which possess an 8β-methyl group, also each contained ca. 10% of the corresponding 8α-methyl isomer, arising as a result of catalytic hydrogenation. Thus, to confirm that 7-deoxyloganic acid (92) and loganic acid (43) are precursors of iridoid glucosides possessing a highly oxidized cyclopentane ring the following experiments were carried out (49). [7,8,10-^{2}H$_5$]Deoxyloganic acid (92), [7,8,10-^{2}H$_5$]-8-epideoxyloganic acid (73), [7,8,10-^{2}H$_5$]iridotrial glucoside (68), [7-^{2}H]loganic acid (43) and [7-^{2}H]-7-epiloganic acid (112) were administered separately to Galium mollugo. Deoxyloganic acid (92) and loganic acid (43) each were incorporated into asperuloside (1), secogalioside (28) and geniposidic acid (113) in yields higher than 10%, whereas the other three compounds fed were not incorporated. Similarly, administration of [7,8,10-^{2}H$_5$]deoxyloganic acid (92) and [10-^{2}H$_2$]-8-epideoxyloganic acid (73) to Galium spurium var. echinospermon, gave a high yield (14.9%) of incorporation of the former and a low yield (2.7%) of incorporation of the latter into asperuloside (1). This confirmed that the iridoid and secoiridoid glucosides of both Galium species are biosynthesized via deoxyloganic acid (92) as shown in Fig. 17. Furthermore, non-incorporation of iridotrial glucoside (68) into the glucosides of G. mollugo led to the deduction that in this plant glucosylation takes place at the stage of deoxyloganic acid aglucone (114).

Incidentally, JENSEN et al. have also reported recently that deoxyloganic acid (92) is incorporated into asperuloside (1) (2% incorporation yield) in Theligonum cynocrambe (Rubiaceae), whereas 8-epideoxyloganic acid (73) is not (110).

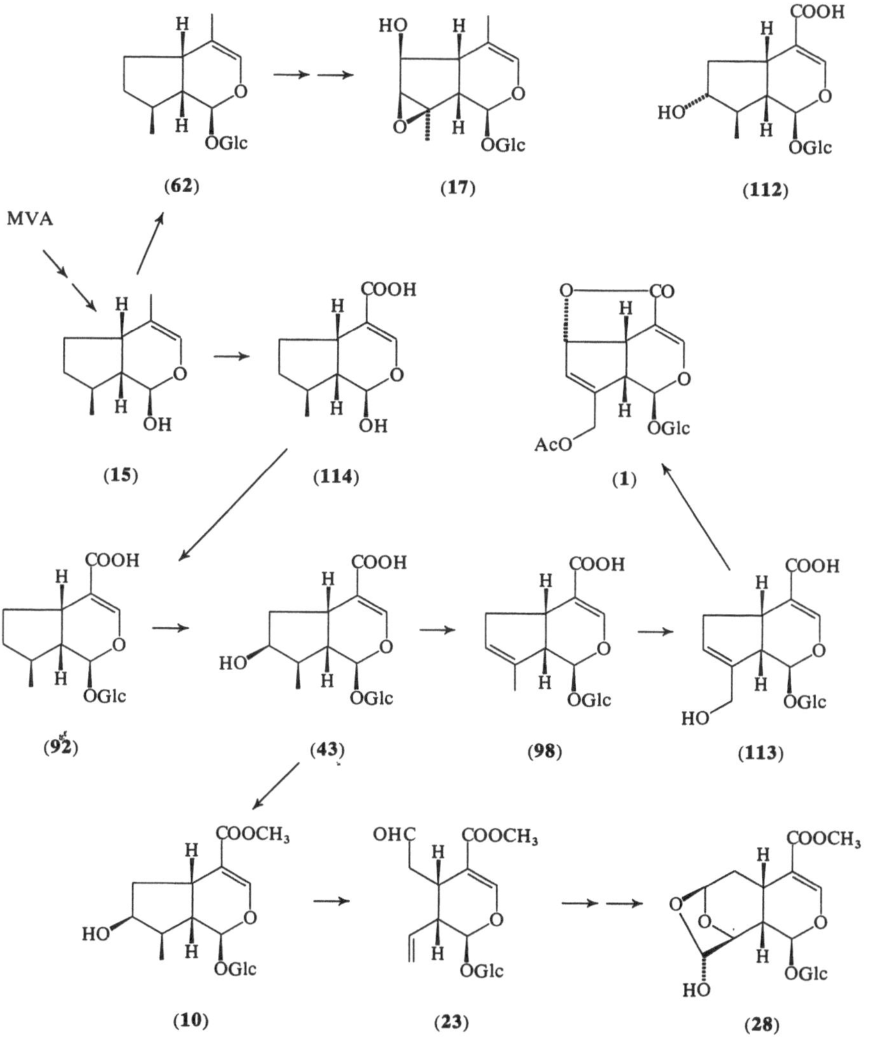

Fig. 17

These results seem to differ from the evidence (46, 48) obtained in the feeding experiments with *Gardenia jasminoides* f. *grandiflora* cell suspension cultures cited in Chapter III, where tarennoside (63) and its further transformation product geniposidic acid (113) are biosynthesized *via* the 8α-methyl glucoside, boschnaloside (70). In the *Galium* plants (113) is formed *via* an 8β-methyl glucoside such as deoxyloganic

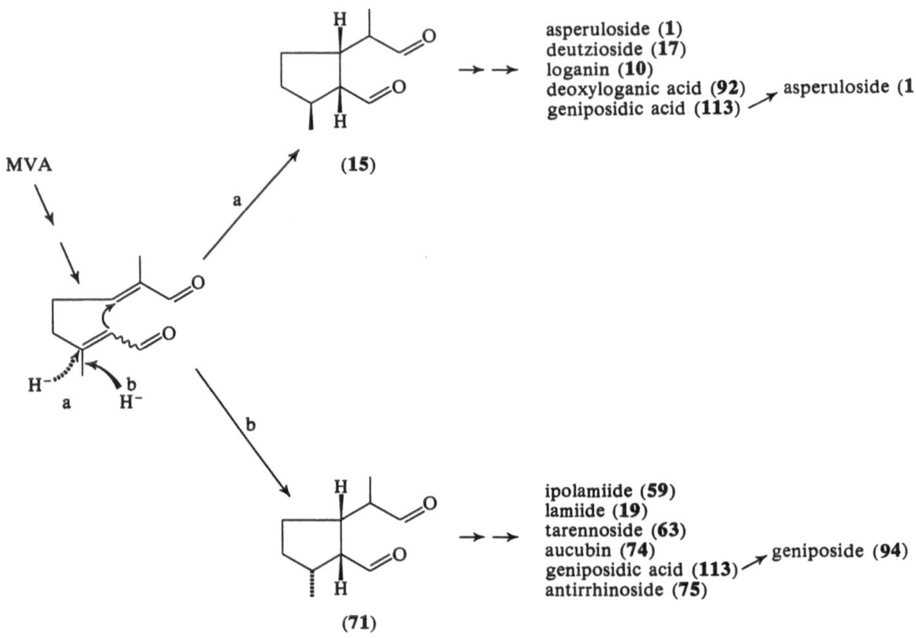

Fig. 18

acid (92), an example of the fact that the same iridoid glucoside may be biosynthesized by different cyclization processes in different plant species.

Finally, administration of [11-^{2}H]iridodial glucoside (62) and [11-^{2}H]-8-epiiridodial glucoside (108) to *Deutzia crenata* resulted in much higher incorporation (16.5%) of (62) into deutzioside (17) than of (108) (4.1%), in accordance with the proposal that (17) is formed *via* iridodial glucoside (62) (49).

From the evidence mentioned so far, the two pathways shown in Fig. 18 can be deduced for iridoid glucoside biosynthesis: one involves iridodial (15) as a pivotal precursor (route a) and the other, 8-epiiridodial (71) as a key intermediate (route b).

Whether aucubin (74) is biosynthesized by route a or b in *Aucuba japonica* could not be determined by feeding ^{2}H-labelled deoxyloganic acid (92) and 8-epideoxyloganic acid (73) (110, 111), probably because of the toxic effect of the precursor on the plant. However, it has been demonstrated recently through radioimmunoassay experiments that in *Aucuba japonica* (74) is biosynthesized through 8-epideoxyloganic acid (73) (route b) (*vide infra*).

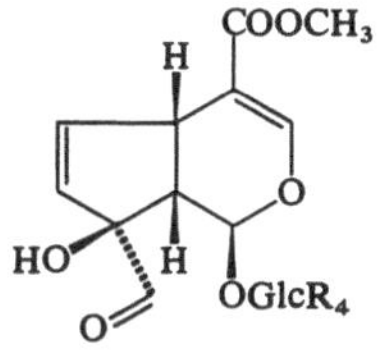

COOCH₃

(115) R = H
(116) R = Ac

Plumieride (**6**) is thought to be formed through condensation of the 10-dehydro derivative (**115**) of gardenoside (**22**) with an acetoacetyl unit, the former having been isolated from *Randia canthioides* (*112*). INOUYE *et al.* actually succeeded in carrying out a biogenetic-type synthesis of plumieride (**6**) from 10-dehydrogardenoside tetraacetate (**116**) and ethyl acetoacetate (*113*).

A number of compounds formed through the same route as **6** are illustrated below, such as allamandin (**117**) in *Allamanda catharitica* (Apocynaceae) (*114*), oruwacin (**118**) in *Morinda lucida* (Rubiaceae) (*115*), penstemide (**119**) in *Penstemon deutas* (Scrophulariaceae) (*116*). Besides allamandin (**117**) and oruwacin (**118**), there exist diverse nonglycosidic iridoids and secoiridoids which seem to be formed by hydrolysis (and successive elaboration) of the corresponding glycosides, such as genipin (**120**) in *Genipa americana* (Rubiaceae) (*117*), sarracenin

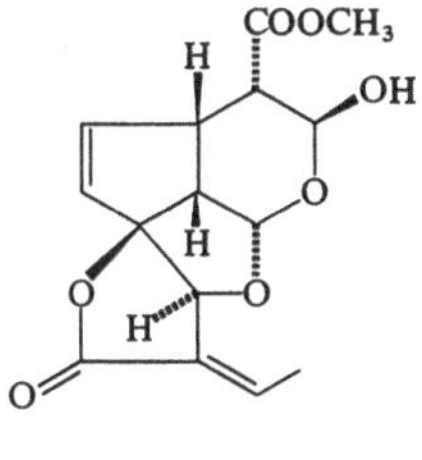

(117) 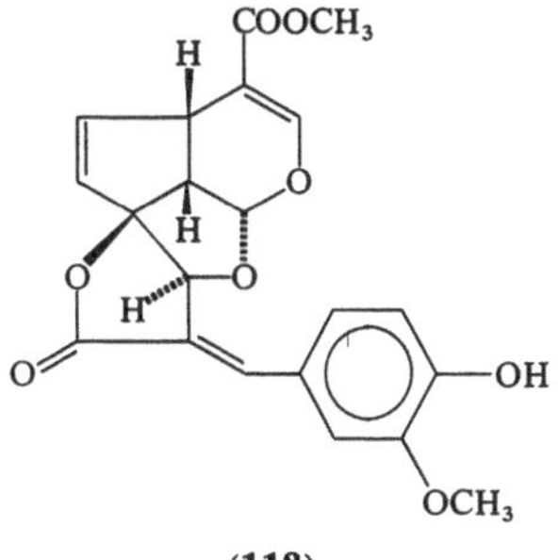   (118) 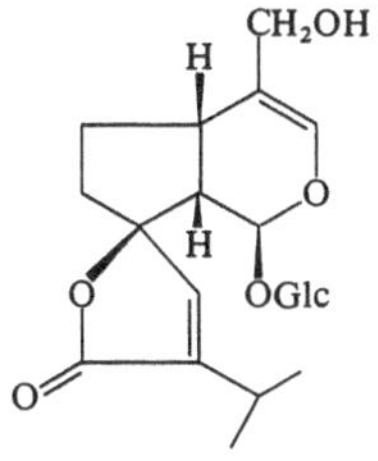   (119)

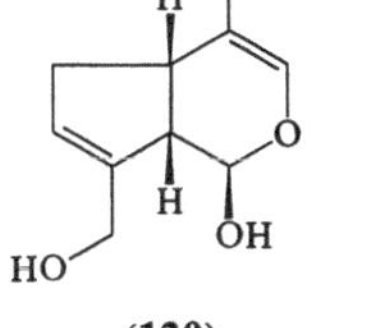

(120) 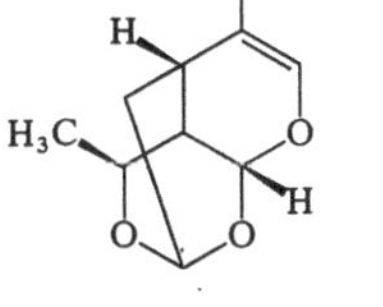   (121) 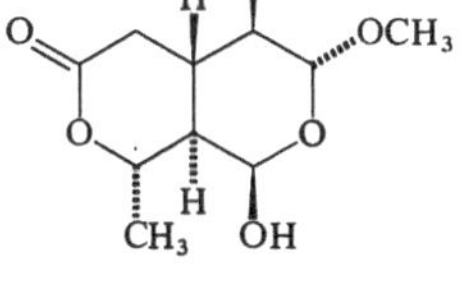   (122)

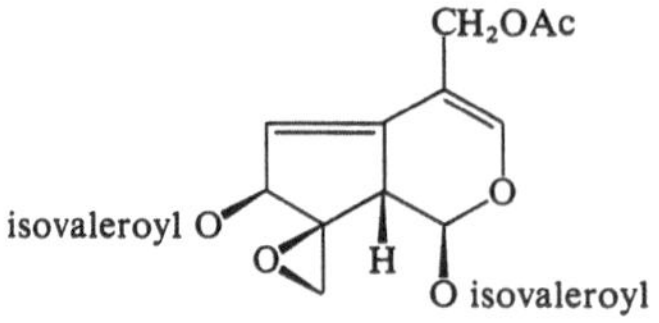

CH$_2$OAc

isovaleroyl O

O isovaleroyl

(123)

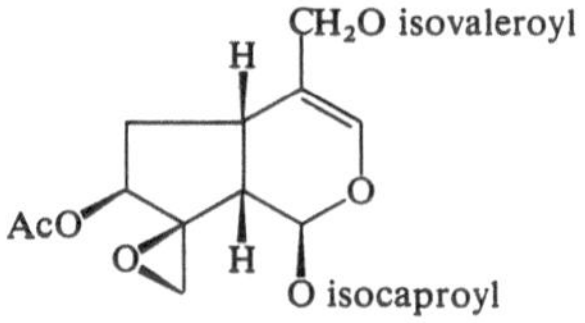

CH$_2$O isovaleroyl

AcO

O isocaproyl

(124)

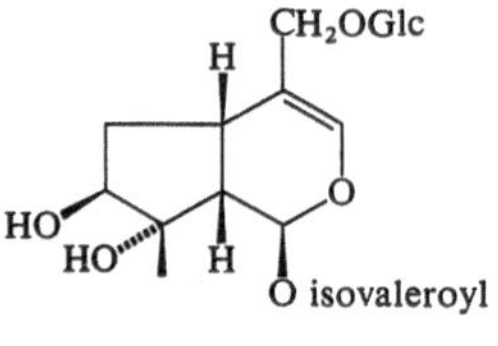

CH$_2$OGlc

HO
HO

O isovaleroyl

(125)

CH$_2$OGlc

HO

HO

O isovaleroyl

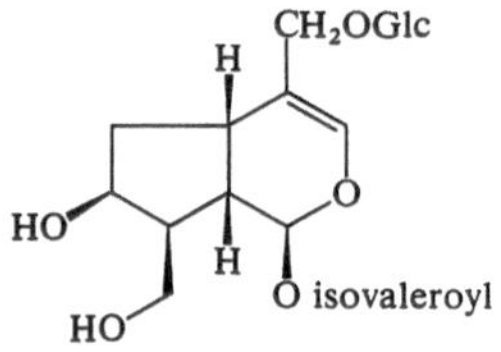

(126)

(121) in *Sarracenia flava* (Sarraceniaceae) (*118*) and xylomollin (122) in *Xylocarpus molluscensis* (Meliaceae) (*119*).

Furthermore, iridoids of known structure with a CH_2-OR function at C-11 and isovaleroyloxy or isocaproyloxy group(s) at various-position(s) are only found in Valerianaceae except for some compounds (*120, 121*) which occur in *Viburnum* plants (Caprifoliaceae). The typical iridoids of this group, valtrate (123), homodihydrovaltrate (124) and valerosidate (125) are illustrated. Although no biosynthetic experiments have been carried out, iridodial (15) seems to be a likely precursor for these compounds also, because loganin (10) and morroniside (26) were isolated from *Patrinia villosa* (*122*) which is congeneric with the patrinoside (126) producing *P. scabiosaefolia* (*123*).

2. Approach to Elucidation of the Mechanism of Cyclopentane Ring Cleavage of Loganin to Secologanin

In the course of his work on monoterpenoid indole and isoquinoline alkaloids, BATTERSBY suggested that secologanin (23) might serve as an important biosynthetic precursor for both alkaloid groups and actually demonstrated the correctness of this suggestion through isolation of (23) from *C. roseus* and incorporation of (23) into the indole and isoquinoline alkaloids of *C. roseus* and *Cephaelis ipecacuanha* plants (*74, 104, 124, 125, 126, 127, 127a*).

Subsequently, (23) was also isolated from various other species including *Lonicera morrowii* (*128*) in fairly good yield. In spite of several

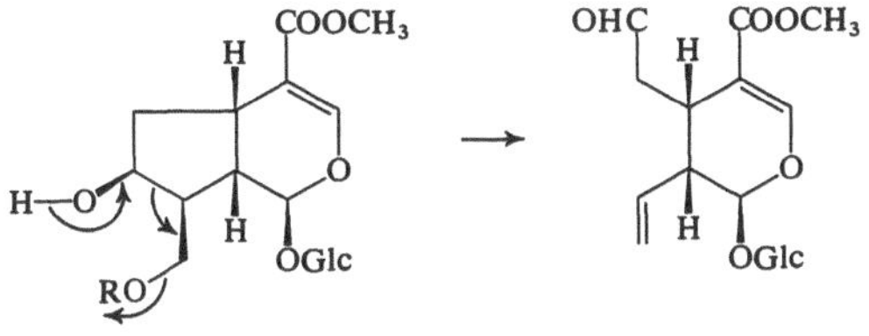

Fig. 19

efforts, however, the mechanism of conversion of loganin (10) into secologanin (23) has so far remained unsolved. BATTERSBY proposed in 1967 that cleavage of the cyclopentane ring of loganin (10) proceeded through 10-hydroxyloganin (127) and its phosphate (128) to secologanion (23) as shown in Fig. 19 (124). This mechanism was believed to be true until feeding experiments with 10-hydroxyloganin (127) and related compounds were performed.

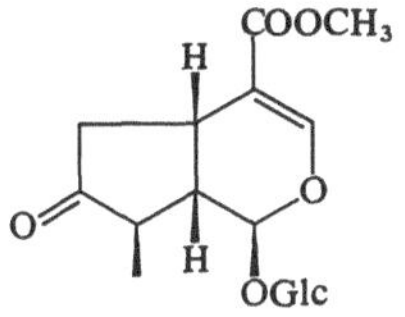

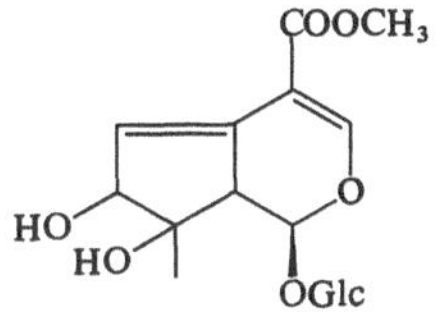

Possible candidates for the precursor directly undergoing cleavage to secologanin (23) are 7-dehydrologanin (97), 7,8-dihydroxy compounds such as gentioside (129) (129) and loganin (10) itself besides 10-hydroxyloganin (127). However, intermediacy of 7-dehydrologanin (97) and 7,8-dihydroxy compounds was excluded for the following reasons: [7-^{3}H]loganin (10) was incorporated into secologanin (23) and indole alkaloids of C. roseus with retention of the label from (10) on C-7 of (23) (104) and on the corresponding carbons of indole alkaloids (130). Moreover, [7-^{3}H]loganin (10) was incorporated into morroniside (26) of Gentiana thunbergii with retention of the label on C-7 of (26) (101, 103). These results ruled out the possibility of (97) serving as an intermediate. Later, TAKEDA and INOUYE demonstrated, by studying the incorporation of [7,8-^{3}H$_2$]deoxyloganic acid (92) into secologanin (23) of L. morrowii, loganin (10) and morroniside (26) of Cornus officinalis and (26) of G. thunbergii, that the label of (92) was retained on both C-7 and C-8 of these glucosides (131). HUTCHINSON et al. (132)

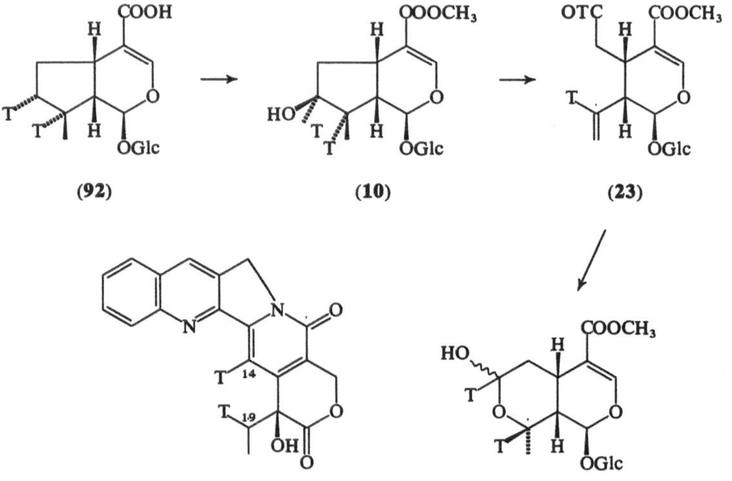

Fig. 20

also demonstrated that [6,8-^{3}H$_3$]loganin (10) is incorporated into camptothecin (130) with retention of 8-^{3}H from (10) on C-19 of (130) in *Camptotheca acuminata*.

These findings are favourable to the 10-hdroxyloganin (127) hypothesis, but not to the 7-dehydrologanin (97) or 7,8-dihydroxy compound hypothesis, since intervention of the latter two should be accompanied by elimination of the C-7 and/or C-8 protons of (10) (Fig. 20).

In 1973, TIETZE succeeded in the synthesis of 10-hydroxyloganin (127) (*133, 133a*). He attempted to cleave the cyclopentane ring of 10-hydroxyloganin aglucone-1-O-methyl ether (131) or its 7-epimer (132) by base treatment of their tosylates (133) and (134). Whereas (133) was converted into the oxetane-type compound (135), (134) was transformed into substances (136) and (137) of secologanin type. Thus, the stereochemistry at C-7 and C-8 of (132) was shown to be favourable for chemical cleavage of the cyclopentane ring (*134, 134a*) (Fig. 21).

In 1981, INOUYE et al. succeeded in preparing 10-hydroxyloganin (127) and 7-epi-10-hydroxyloganin (138) from geniposide (94) in fairly high yield (*135*) and attempted incorporation of [7-^{3}H]-(127) and (138) into secologanin (23) of *Lonicera morrowii* and *Adina pilulifera* (*136*). They also tried to carry out dilution analyses of 10-hydroxyloganin (127) and 7-epi-10-hydroxyloganin (138), after administering [10-^{3}H]deoxyloganic acid (92) and [7-^{3}H]loganin (10) to *L. morrowii* as well as [7-^{3}H]loganin (10) to *Adina pilulifera* (*136*). The very low incor-

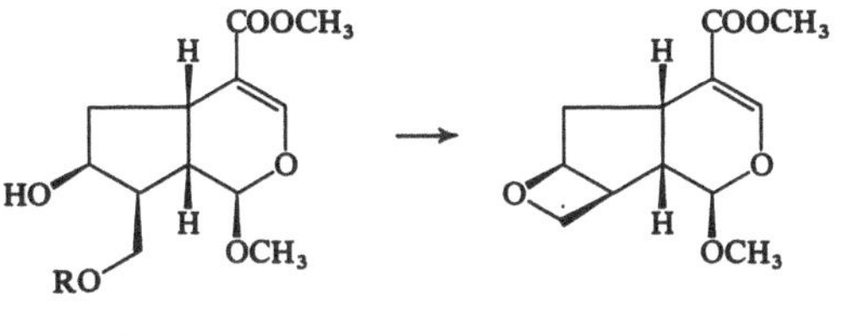

(131) R = H
(133) R = Tos

(135)

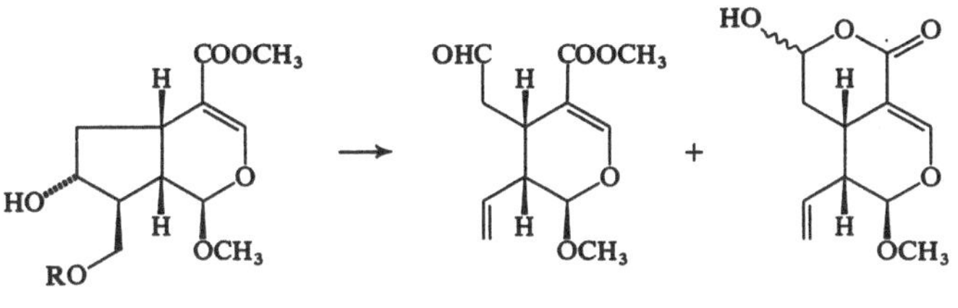

(132) R = H
(134) R = Tos

(136)

(137)

Fig. 21

porations of both (127) and (138) into secologanin (23) as well as those of (92) and (10) into (127) and (138) as compared with those of (92) and (10) into (23), indicated that neither (127) nor (138) acts as a precursor of (23). The group of TIETZE and BATTERSBY also came to the same conclusion from the results of the incorporation of [carbo-^{14}C-methoxy,7-^{3}H]-(127) and [carbo-^{14}C-methoxy]-(138) into the indole alkaloids of *C. roseus* (*137*). Based on the evidence outlined so far, intervention of 10-hydroxyloganin (127) in the cleavage process of the cyclopentane ring of loganin (10) to secologanin (23) seems to be ruled out.

Recently, INOUYE *et al.* postulated that 6β-hydroxyloganin (139) or its phosphate (140) might be cleaved to secologanin (23) in line with the mechanism shown in Fig. 22 (*138*). Of the four possible C-6 and C-7 stereoisomers of 6-hydroxyloganin, 6β-hydroxyloganin (139) seems to satisfy best the stereochemical demands necessary for the proposed cyclopentane ring cleavage. They synthesized [7-^{2}H]-labelled 6β-hydroxyloganin (139), 6α-hydroxy-7-epiloganin (141) and 6β-hydroxy-7-epiloganin (142), and examined incorporation of the former two glucosides into secologanin (23) of *Adina pilulifera* and swertiamarin (7) of *Swertia japonica* as well as of the latter glucoside into eustomoside (143) of *Eustoma rusellianum* and into swertiamarin (7) of *S. japonica*. However, none of the fed compounds were incorporated into the glucosides in question. The proposed cleavage mechanism involving (139) should be accompanied by the loss of one of the C-6 methylene protons of loganin (10). Thus, in order to confirm the above result, [6,6,7,8-^{2}H$_4$]-(10) was administered to *E. rusellianum* and *S. japonica* and [6,6,7,8,carbomethoxy-^{2}H$_7$]-(10) to *Lonicera morrowii* and *A. pilulifera*. The ^{2}H NMR spectra of the isolated eustomoside (143), swertiamarin (7), secologanin (23) and morroniside (26) and their acetates showed that two deuterons were retained at C-6 and one deuteron at both C-7 and C-8 of the respective glucosides. Therefore, a cleavage mechanism involving 6-hydroxyloganin was ruled out (*138*). In the above-mentioned experiments, the stereochemistry involved in the reduction process of the aldehyde group of (23) which eventually leads to (7) and (143) was also elucidated, since the C-7 deuteron of labelled (10) was incorporated into the C-7 pro-*R* position of both (7) and (143). This finding leads to the conclusion that the C-7 aldehyde group of secologanin (23) undergoes attack by a hydride ion from the *si* face yielding secologanol (144), which further gives (7) and (143) *via* lactonization to 24.

On the whole, then, it seems most probable that the cyclopentane ring of loganin (10) is directly cleaved through a radical or ionic process. TIETZE *et al.* also suggested the possibility that ring cleavage might occur *via* the thio analogue of 127 (*137*).

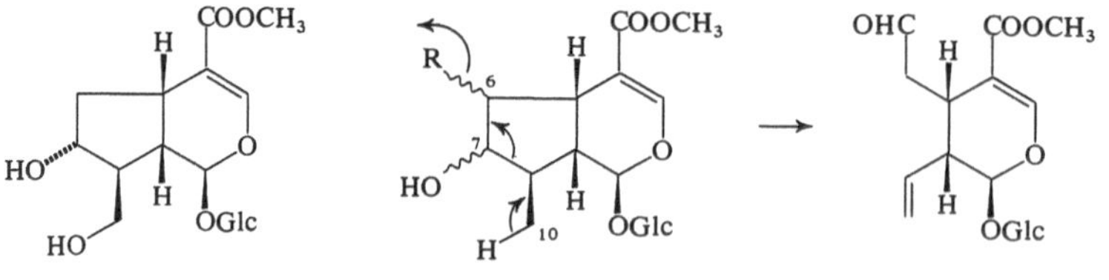

(138)

(139) R = —OH, 7—OH
(140) R = —OPi, 7—OH
(141) R = ⁗OH, 7⁗OH
(142) R = —OH, 7⁗OH

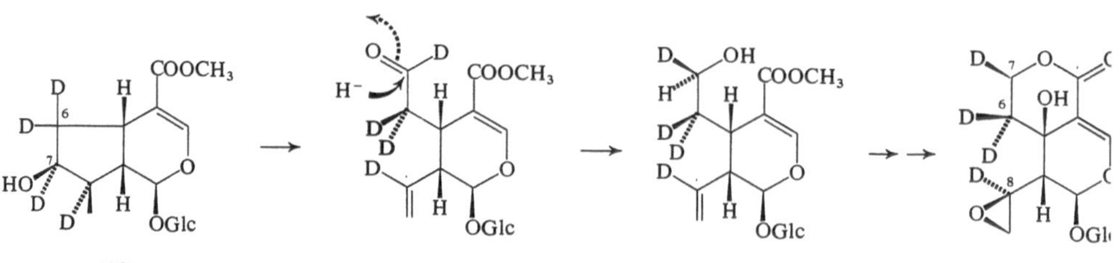

(23)

(10)

(23)

(144)

(143)

Fig. 22

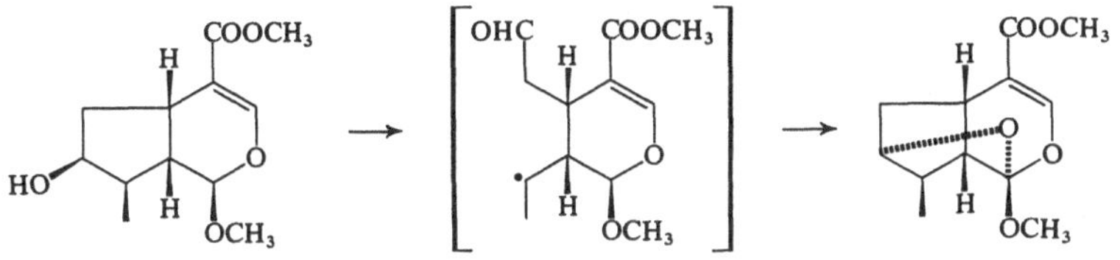

Fig. 23

Attempted conversion of loganin aglucone-1-O-methyl ether (145) into seco-type compounds through lead tetraacetate oxidation by PARTRIDGE *et al.* (*139*) resulted in the production of a tricyclic compound (146), suggestive of ring cleavage having occurred *via* the oxidative radical process shown in Fig. 23. This could be a reaction, mimicking a biosynthesis step.

V. Biosynthetic Relationships Between Groups of Secoiridoid Glycosides

1. Sweroside-Morroniside Type Glycosides

The glycosides of this type are widespread in plants of the families Gentianaceae, Caprifoliaceae, Dipsacaceae, Loganiaceae, Menyanthaceae, Apocynaceae, Icacinaceae, etc. Secologanin (23), sweroside (24), swertiamarin (7), gentiopicroside (2), morroniside (26), eustomoside (143) (*140*), bakankosin (147) (*141*) and abelioside A (148) (*142*) are representatives of this group.

As mentioned in the preceding sections, the gross outline of the biosynthetic pathways leading to these glycosides has already been established by the groups of INOUYE and COSCIA. INOUYE *et al.* (*143, 144*) confirmed through high yield incorporation (40%) of [10-^{14}C]sweroside (24) into gentiopicroside (2) in *Gentiana scabra* the biosynthetic sequence: sweroside (24) → swertiamarin (7) → gentiopicroside (2). In addition to this experiment, they also demonstrated incorporation of [10-^{14}C]-(24) into vindoline (9) of *C. roseus,* reserpinine (149) of *Vinca major* and quinine (150) of *Cinchona succirubra* (*144, 145*). Although no feeding experiments with labelled secologanin (23) have been carried out, (24) is considered to be incorporated into the above compounds after transformation to (23).

Bakankosin (**147**), a unique lactam glucoside of *Strychnos vacacoua*, has been chemically synthesized by reductive amination of secologanin (**23**) with NaBH$_3$CN (*146, 147*).

Eustomoside (**143**), an epoxide of swertiamarin (**7**), coexists with the corresponding chlorohydrin eustoside (**151**) and the diol, eustomorusside (**152**), in *Eustoma russellianum* (*140*). Thus, the latter two glycosides (**151**) and (**152**) are most likely to be formed by attack of a chloride anion and a hydroxy anion on the epoxide ring of (**143**), respectively. Eustoside (**151**) is the second example of an iridoid glycosides which possesses a chlorohydrin moiety; linarioside (**153**) was the first compound of this type (*148*).

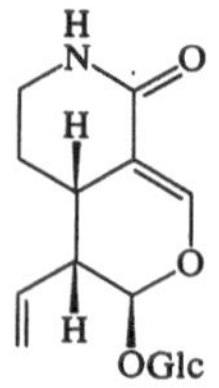

(**147**)

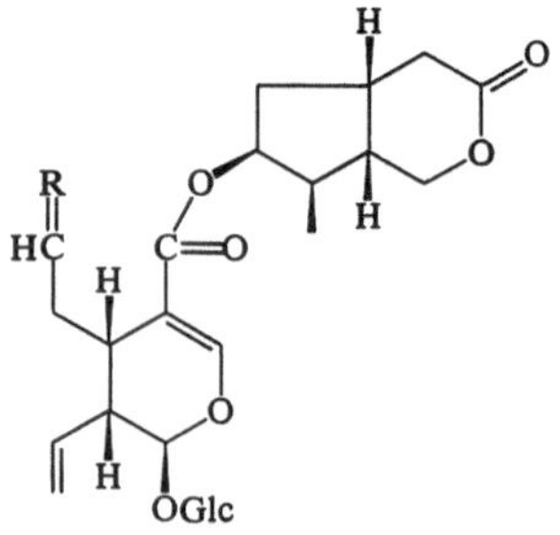

(**148**) R = O
(**158**) R = <OH, H

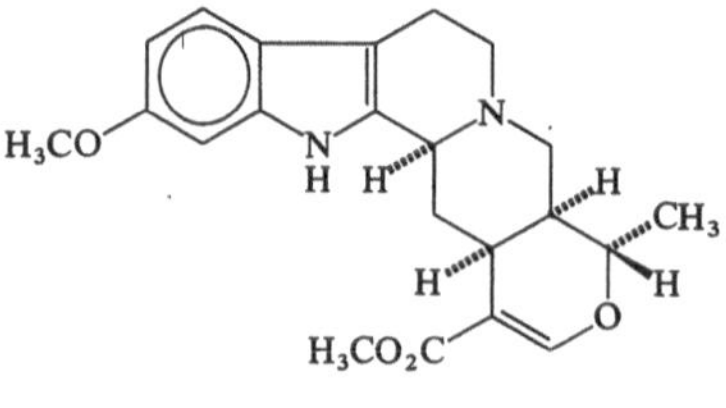

(**149**)

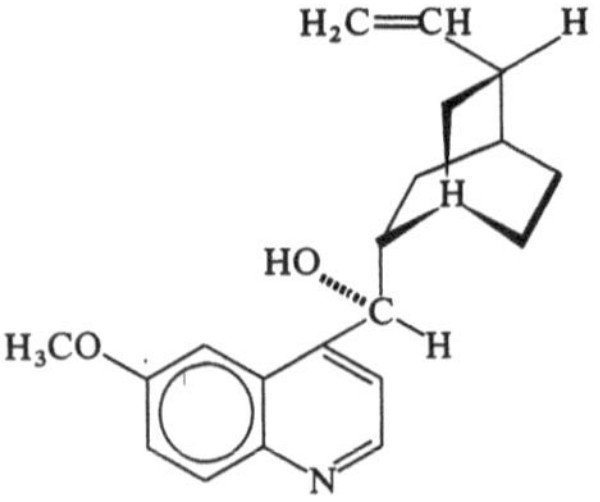

(**150**)

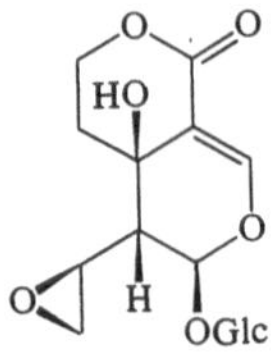

(**143**) 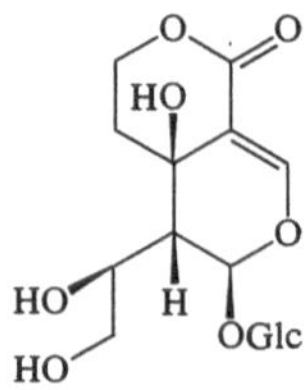   (**151**) 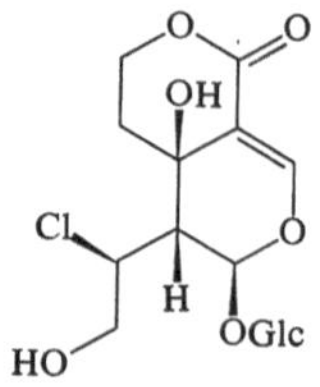   (**152**) 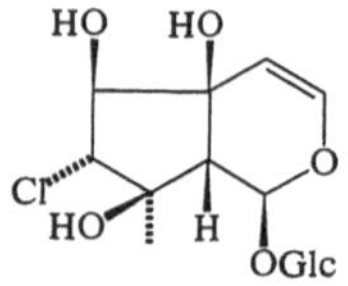   (**153**)

Amarogentin (**154**), amaroswerin (**155**) (*149, 149a, 150*) and amaropanin (desoxyamarogentin) (**156**) (*151*) are very bitter secoiridoid glucosides occurring in some gentianaceous plants. The biphenylcarboxylic acid moiety of these glucosides seems to be biosynthesized *via* the pathway shown in Fig. 24 (*149a*). That *m*-hydroxybenzoic acid is the precursor of this portion is suggested by the occurrence of *m*-hydroxybenzoic acid conjugated secoiridoid glucosides, centapicrin (**157**) (*152*) and congeneric glucosides (*153*), in several Gentianaceous plants.

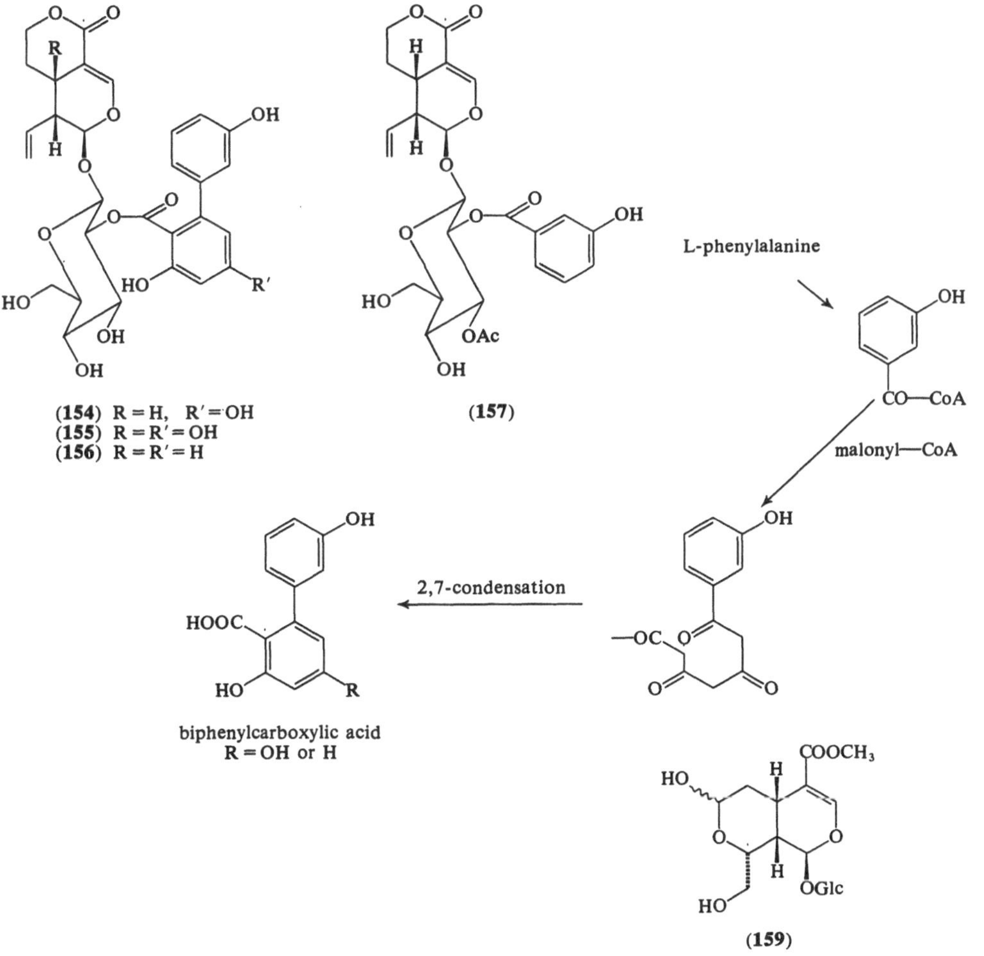

(**154**) R = H, R′ = OH
(**155**) R = R′ = OH
(**156**) R = R′ = H

(**157**)

L-phenylalanine

malonyl—CoA

2,7-condensation

biphenylcarboxylic acid
R = OH or H

(**159**)

Fig. 24

Abelioside A (148) and B (158) (*142*), secoiridoid glucosides of *Abelia grandiflora*, have bisiridoid structures which are composed of iridoid lactone and secoiridoid moieties. By comparing the structures of loganin (10) and the lactone moiety, the latter is easily seen to be derived from (10).

Morroniside (26) was first isolated from *Lonicera morrowii* (Caprifoliaceae) (*128*) and then from plants of genera *Gentiana* (*150*), *Cornus* (*154*), etc. Kingiside (27) occurs together with (26) in *Lonicera morrowii* (*128*). Structural comparison of (26) and (27) as well as the incorporation of [carbo-^{14}C-methoxy]secologanin (23) into (26) of *Cornus officinalis*, and of [7-^{3}H]loganin (10) into (26) of *Gentiana thunbergii* mentioned in the preceding Chapter allows delineation of the biosynthetic sequence: loganin(10) → secologanin(23) → morroniside(26) → kingiside(27).

The morroniside analogues secogalioside (28) and 10-hydroxymorroniside (159) were isolated from *Galium mollugo* (= *G. album*) (*155, 156*). This plant is an unusual one in that it contains not only both secoiridoid glucosides, but also highly oxidized iridoid glucosides such as asperuloside (1) and geniposidic acid (113). The isolation of 10-hydroxyloganin (127) from *G. mollugo* further suggests the possibility that the secoiridoid glucosides (28) and (159) of this plant could be formed by cyclopentane ring cleavage of (127). This has, however, not yet been substantiated. The proposed biosynthetic relationship between morroniside (26) and secogalioside (28) also remains to be verified.

2. Oeloside-10-Hydroxyoleoside Type Glycosides

The glycosides of this type occur only in oleaceous plants, and comprise an oleoside (160) or 10-hydroxyoleoside (161) moiety as a framework. Oleuropein (29), nüzhenide (33), jasminin (95), ligustroside (162), 10-acetoxyligustroside (163) and 10-acetoxyoleuropein (164), etc. belong to this group (*157*). INOUYE et al. at first assumed that the biosynthesis (*101, 158*) of oleoside (160) involves a Baeyer-Villiger type oxidation of dehydrologanin (97) to 8-epikingiside (165) and subsequent dehydration to 160, on the following basis: 1) products which seem to be formed by a Baeyer-Villiger-type oxidation occur frequently in nature (*159, 160*); 2) The arrangement of the C-8 and C-9 substituents of (165) satisfies the stereochemical demands necessary for dehydration to form oleoside (160). However, equal incorporation of [8-^{3}H]-8-epikingiside (165) and [8-^{3}H]kingiside (27) into jasminin (95) as well as incorporation of a mixture of [carbo-^{14}C-methoxy]-(165) and [carbo-^{3}H-methoxy]-(27) into (29) retaining intact ^{3}H/^{14}C ratio are

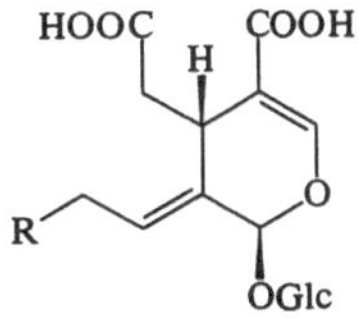

(160) R = H
(161) R = OH

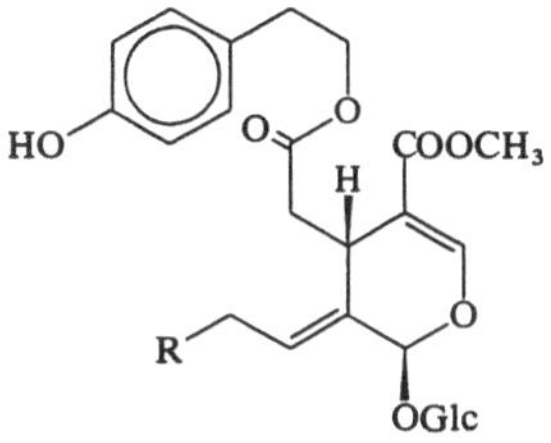

(162) R = H
(163) R = OAc

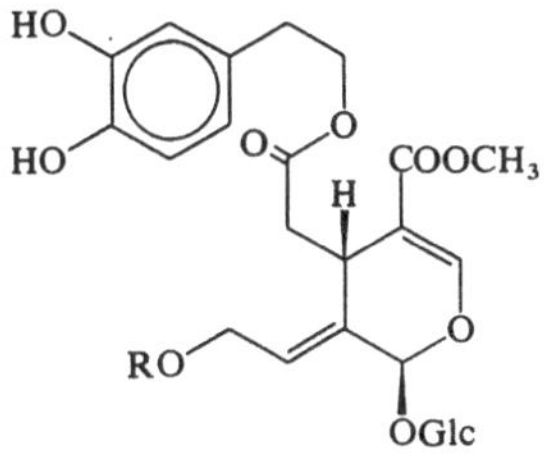

(164) R = Ac
(166) R = H

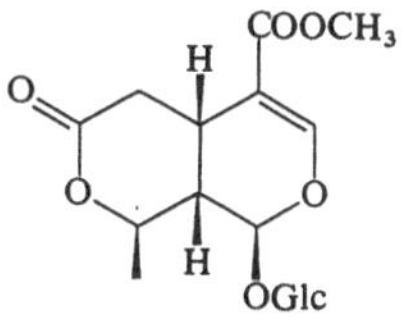

(165)

not associated with the general concept of a stereoselective enzymatic reaction. The possibility that 8-epikingiside (165) is an intermediate in the biosynthesis of oleoside-type glucosides was therefore ruled out. Since secologanin (23) was incorporated into (29) of *O. europaea* more efficiently than (27) and (165), it is most probable that the oleuropein-type glycosides are biosynthesized *via* (23) rather than 8-epikingiside (165).

Although details of the route from secologanin (23) to the glycosides of oleoside (160) and 10-hydroxyoleoside (161) types are still somewhat obscure, INOUYE *et al.* have proposed the biogenetic pathway depicted in Fig. 25 ((*161*).

Important clues for this proposal were given by the isolation of the 10-aldehyde glucosides ligustaloside A (31) and ligustaloside B (32) (both of the ligustaloside type), together with 10-hydroxyoleuropein (166) (10-hydroxyoleoside type), from *Ligustrum japonicum* (Oleaceae). By postulating epoxide (167) as the pivotal intermediate, the biogenesis of the glucosides of these series could reasonably be explained as follows: i) cleavage of the epoxide ring of the intermediate (167) with synchronous hydride shift leads to ligustaloside A (31) and B (32). ii) Cleavage of the epoxide ring of (167) to the 8-ol (168) followed by dehydration leads to oleoside (160) type glucosides. iii) Cleavage

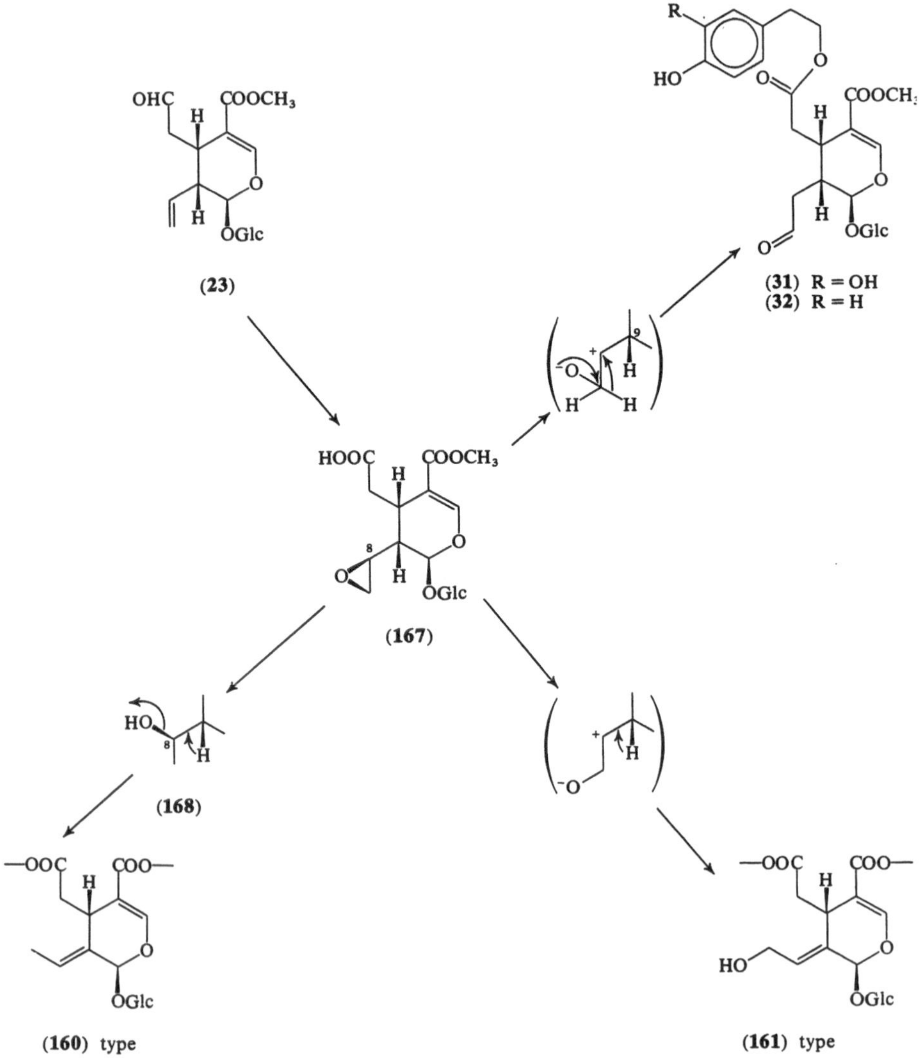

Fig. 25

of the epoxy ring of (167) with simultaneous deprotonation on C-9 leads to 10-hydroxyoleoside (161) type glucosides. The C-8 configuration of the pivotal intermediate (167) should be S, in view of the stereochemistry involved in the formation of the ethylidene group of the glucosides of oleoside type.

References, pp. 226–236

VI. Biosynthetic Pathways of Alkaloidal Glycosides and Hydrangenosides

We discuss in this Chapter alkaloidal glycosides and hydrangenosides, in which C-7 of secologanin (**23**) is bonded to a unit of different biosynthetic origin through a C-C bond.

1. Biosynthesis of Alkaloidal Glycosides and the Role of Strictosidine and Deacetylisoipecoside in the Biosynthesis of Alkaloids

About forty alkaloidal glycosides of the iridoid series are indole alkaloids and isoquinoline alkaloids. The glycosides belonging to the former group are further subdivided into members of the tryptamine and tryptophan series. Tryptamine series glycosides include strictosidine (**34**), vincoside (**35**), vincoside lactam (**169**), strictosamide (**170**), isodihydrocadambine (**171**) etc., while the tryptophan series includes cordifoline (**172**), rubenine (**173**), etc. The first series occurs in Apocynaceae and Rubiaceae and the latter mostly in the genus *Adina* (Rubiaceae). Isoquinoline series glycosides include ipecoside (**36**) of *Cephaelis ipecacuanha,* alangiside (**174**) of *Alangium lamarckii,* etc. Among glycosides of these groups, derivatives of strictosidine (**34**) and ipecoside (**36**) have been the main objects of studies related to the biosynthesis of indole and isoquinoline alkaloids.

Based on the assumption that the condensation product of tryptamine and secologanin (**23**) is the first nitrogenous intermediate of indole alkaloids (*124*), BATTERSBY et al. synthesized in 1969 vincoside (**35**) and isovincoside (**34**) and isolated both glucosides from *C. roseus.* They also demonstrated the incorporation of loganin (**10**) into both compounds and that of vincoside (**35**) into indole alkaloids (*127, 127a*). On the other hand, Smith in 1968 isolated strictosidine from *Rhazya stricta* and assumed that it was the pivotal intermediate in the biogenesis of these alkaloids (*162*). Later, his group also identified strictosidine with isovincoside (**34**) (*163*).

Although the C-3 stereochemistry of vincoside (**35**) and strictosidine (**34**) was at first an object of intensive controversy (*33*), it was later established in several ways that vincoside (**35**) has a 3β-H, while strictosidine (**34**) has a 3α-H (*164, 165, 166, 167*).

In 1974, HUTCHINSON et al. (*168*) reported that strictosidine (**34**) was incorporated into camptothecin (**130**) of *Camptotheca acuminata.* Later in 1977, BROWN et al. succeeded in an one-pot synthessis of ak-

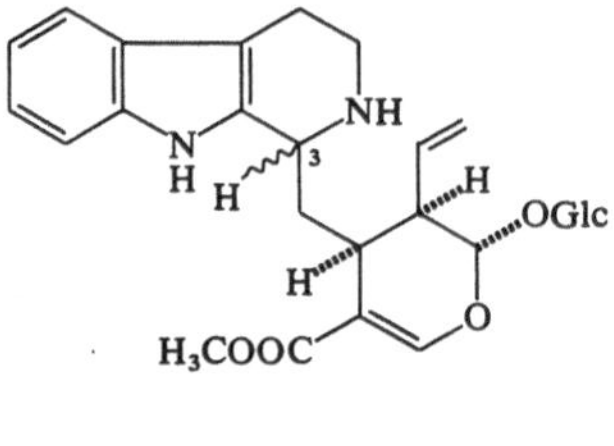

(34) 3····H
(35) 3—H

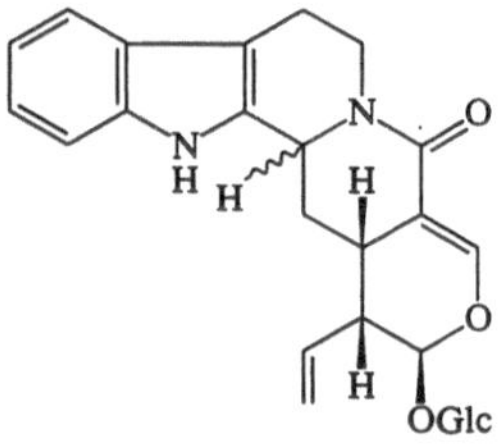

(169) 3—H
(170) 3····H

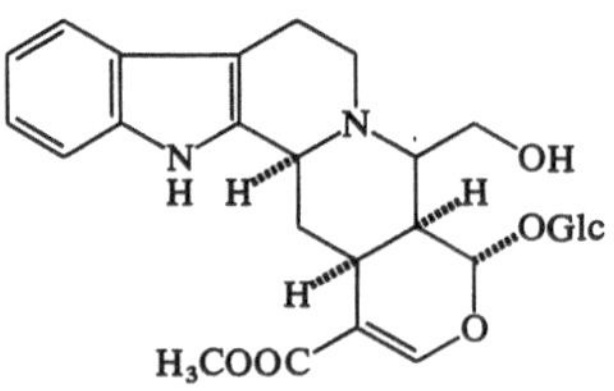

(171)

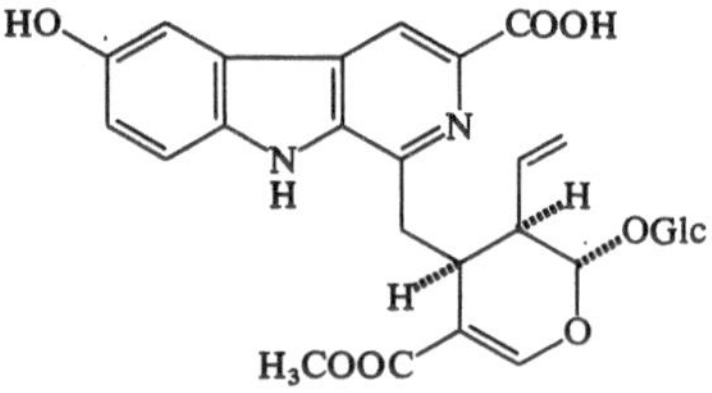

(172)

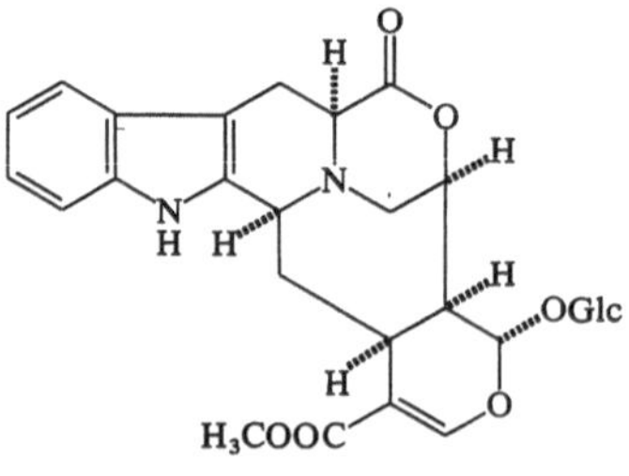

(173)

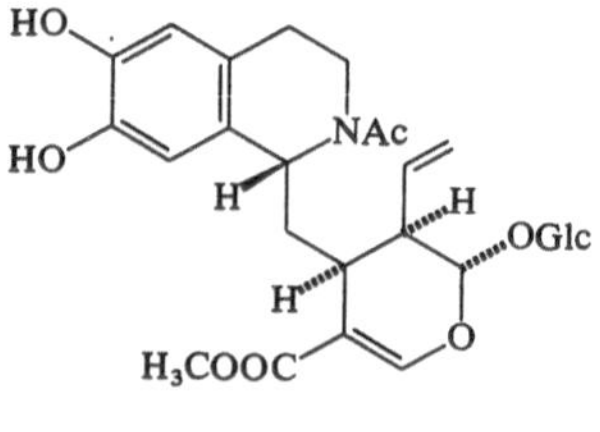

(36)

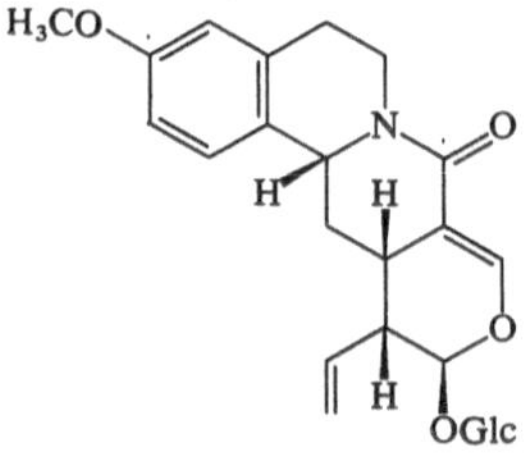

(174)

uammigine (175) (with 3β-H), tetrahydroalstonine (176) and ajmalicine (50) (with 3α-H) from tryptamine and secologanin (23) through treatment with β-glucosidase and NaBH₃CN and considered that this synthesis duplicated the *in vivo* process; strictosidine (34) is converted into

3α-indole alkaloids, and vincoside (**35**) into 3β-alkaloids, each without inversion of the C-3 proton (*169*).

On the other hand, ZENK *et al.* succeeded in enzymatic conversion of [2-^{14}C]tryptamine and secologanin (**23**) into radioactive ajmalicine (**50**), 19-epiajmalicine (**177**) and tetrahydroalstonine (**176**) in the presence of NADH or NADPH using cell-free extracts prepared from *C. roseus* cell suspension cultures (*170*). Moreover, they obtained strictosidine (**34**) (*171*) and 20,21-didehydroajmalicine (cathenamine) (**178**) (*172*) in the absence of the reduced pyridine nucleotides. Strictosidine (**34**) was also derived from incubating the mixture with a crude enzyme preparation originating from *Rhazya stricta* cell suspension cultures (*172*). Reincubation of strictosidine (**34**) thus obtained with the cell-free system from the *C. roseus* cell cultures led to the accumulation of (**178**) in the absence of the reduced pyridine nucleotides, whereas incubation in the presence of NADH or NADPH yielded (**50**), (**177**) and (**176**). Therefore, it was concluded that the aforementioned *Corynanthe* type of indole alkaloids are biosynthesized through condensation of tryptamine and secologanin (**23**) to strictosidine (**34**), which is further metabolized *via* cathenamine (**178**) to the *Corynabthe* type of alkaloids with retention of the C-3 configuration in the presence of the reduced pyridine nucleotides.

Since this conclusion was inconsistent with that of BATTERSBY *et al.*, ZENK's group performed various tracer experiments in order to confirm the position of strictosidine (**34**) as an intermediate and obtained the following results: Enzymatically prepared [6-^{14}C]strictosidine (**34**) was incubated with the cell free preparation from the *C. roseus* cell suspension cultures in the presence of NADPH, leading to the formation of radioactive (**50**), (**177**) and (**176**). Furthermore, feeding of [6-^{14}C]strictosidine (**34**), [O-methyl-^{3}H]strictosidine (**34**) and [O-methyl-^{3}H]vincoside (**35**) to intact plants of *C. roseus* resulted in incorporation of (**34**) but not of (**35**) into *Iboga* and *Aspidosperma* type alkaloids possessing 3β-H, as well as *Corynanthe* type alkaloids possessing 3α-H.

	19-H	20-H
(**50**)	β	β
(**176**)	β	α
(**177**)	α	β

(**175**)   (**178**)

Subsequently, in order to examine the feasibility of extending the intermediacy of strictosidine (34) to other plants containing monoterpenoid
alkaloids, incubation of [2-^{14}C]tryptamine and secologanin (23) was
carried out with cell free extracts prepared from *C. roseus, Rhazya
orientalis, R. stricta* and *Vinca minor* cell suspension cultures (*173*).
In all instances, radioactive strictosidine (34) accumulated in high yield.
The intermediacy of strictosidine (34) was thus established at least
in 3α-indole alkaloid biosynthesis.

Previous failures to incorporate isovincoside (strictosidine) (34) into
the indole alkaloids of *C. roseus* were explained by ZENK *et al.* by
assuming that the small amount of (34) administered to the plant
was decomposed through non-specific hydrolysis with β-glucosidase
before reaching the site where indole alkaloids are biosynthesized.

Another question related to the biosynthesis of indole alkaloids
is whether vincoside (35) or strictosidine (34) serves as a precursor
of 3β-indole alkaloids. ZENK *et al.* showed through administration of
[^{3}H/^{14}C]strictosidine (34) and vincoside (35) to *Rauwolfia canescens*
and *Mitragyna speciosa* that (34) was incorporated not only into the
3α-indole alkaloids (α-yohimbine (179) and mitragynine (180)), but
also into the 3β-alkaloids (reserpiline (181) and speciociliatine (182)),
whereas incorporation of (35) into these alkaloids was very low. Additionally, the 3α-^{3}H of (34) was retained on C-3 of the 3α-alkaloids,
but not on C-3 of the 3β-alkaloids (*174*). The same group further
showed that (34) can act as a precursor of *Cinchona, Corynanthe, Iboga*
and *Aspidosperma* type of alkaloids in *Cinchona, Stemmadenia, Uncaria*
and *Catharanthus* plants, as well as of the more complicated indole
alkaloids in *Gelsemium* and *Strychnos* plants (*175*).

These results were in accord with reports on the biosynthesis of
camptothecin (130) by HUTCHINSON *et al.* (*168, 176*) who concluded
on the basis of feeding of [14-^{3}H$_2$,5-^{14}C]strictosamide (170), [14-^{3}H$_2$]-
isovincoside (34), [14-^{3}H$_2$]vincoside (35) and [14-^{3}H$_2$]isovincoside aglucone (183) to *Camptotheca acuminata* that (34) rather than (35) is
a precursor of (130). SCOTT *et al.* (*177*) also corroborated the intermediacy of (34) in the biosynthesis of the indole alkaloids of *C. roseus*.

On the other hand, BROWN *et al.* (*178*) found that [ar-^{3}H]-strictosi-
dine (34) was incorporated into 3α-indole alkaloids such as tetrahydroalstonine (176), ajmalicine (50), catharanthine (51) and vindoline
(9), but not into the 3β-indole alkaloid, akuammigine (175).

Ipecoside (36), an isoquinoline series glucoside, was the first substance shown to be a nitrogenous secoiridoid glucoside formed by condensation of dopamine and secologanin (23) (*104, 126, 166, 179*). Based
on feeding of [2-^{3}H,2-^{14}C]geraniol, [carbo-^{3}H-methoxy]loganin (10),
[carbo-^{3}H-methoxy,6-^{3}H$_2$]secologanin (23), [3′-^{14}C]deacetylipecoside

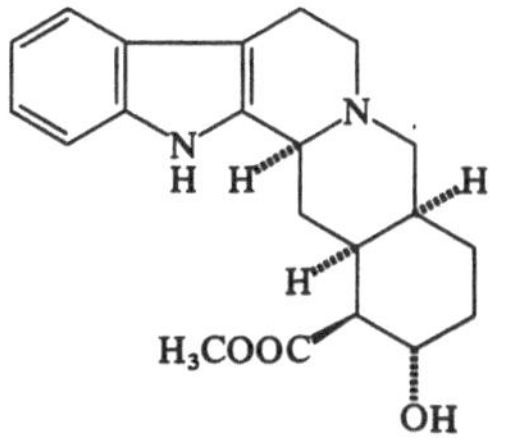

(179)

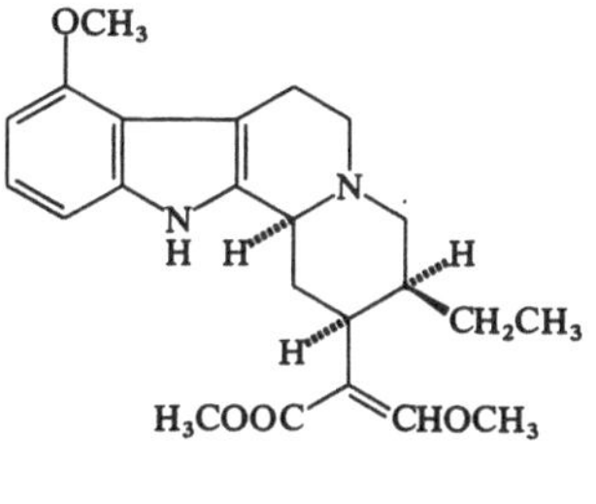

(180)

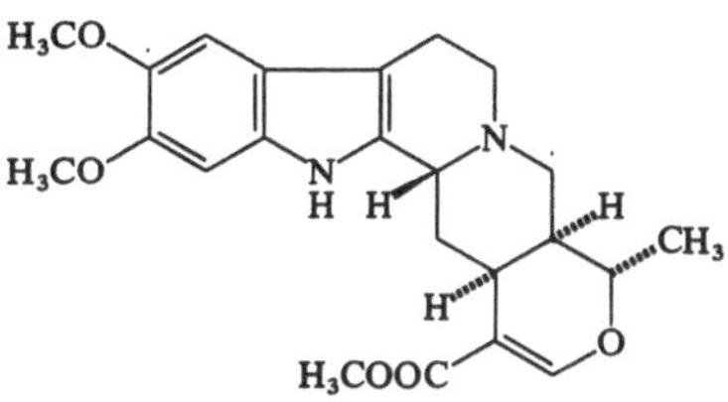

(181)

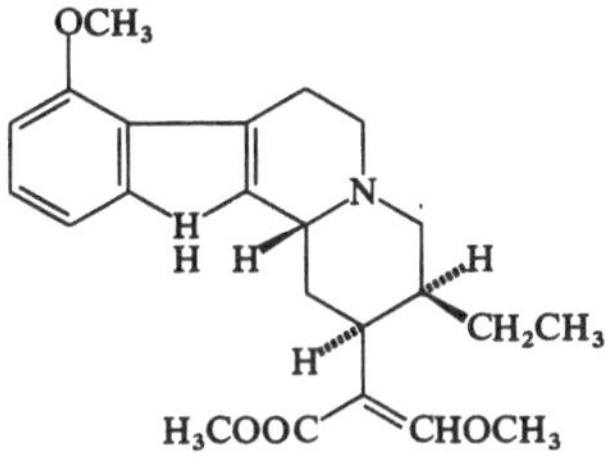

(182)

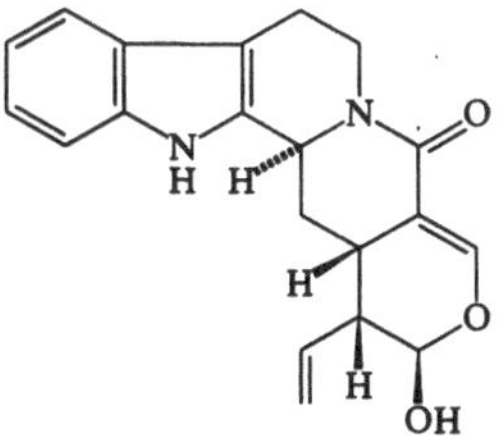

(183)

(184), [3'-^{14}C]deacetylisoipecoside (185) and [3'-^{14}C]dihydrodeacetyl-ipecoside (186) to *Cephaelis ipecacuanha*, BATTERSBY *et al.* inferred that condensation of dopamine and (23) affords deacetylipecoside (184); N-acetylation of this compound terminates the biosynthesis of ipeco-side (36), whereas elaboration of (184) involving inversion of the C-3 configuration leads to cephaeline (187) and emetine (188) (*180, 75*).

On the other hand, ZENK *et al.* showed that [1-^{3}H, 3-^{14}C]deacetyl-ipecoside (184) (with a 1 β-proton) was incorporated into ipecoside (36) and alangiside (174) (both with a 1 β-proton), but not into cephaeline (187) and emetine (188) (both with an 11 α-proton). Contrasting with these results, [1-^{3}H, 3-^{14}C]deacetylisoipecoside (185) (with a 1 α-proton) was incorporated into (187) and (188), but not into (36) and (174).

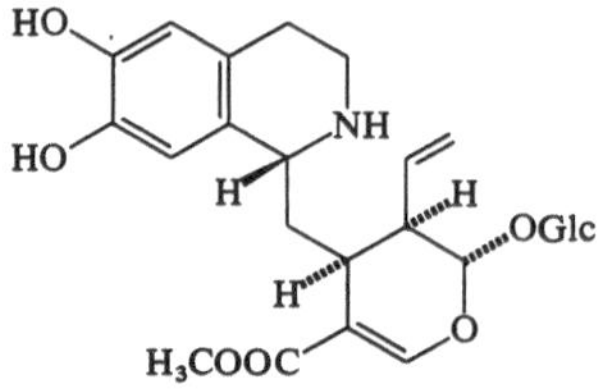

(184)
(186) vinyl reduced

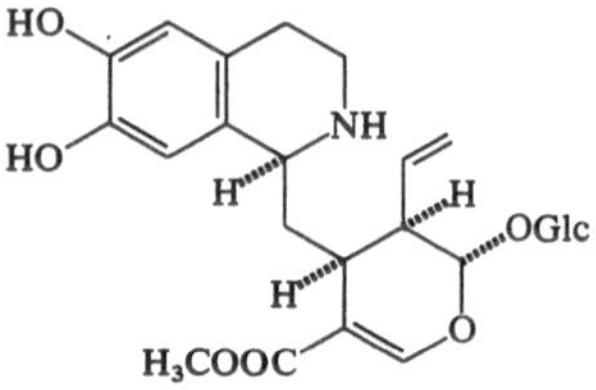

(185)

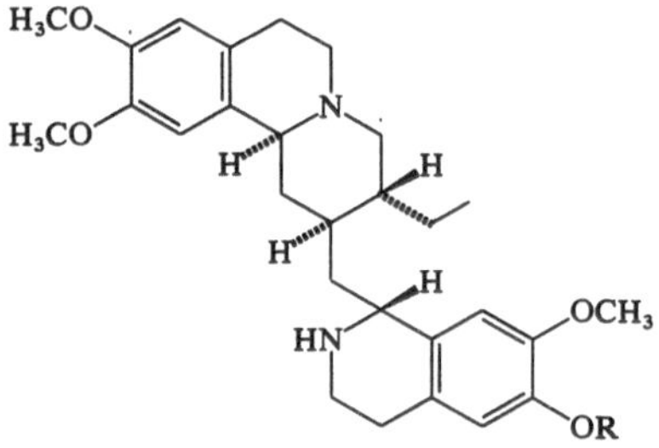

(187) R = H
(188) R = CH₃

Based on these findings, they concluded that after biosynthesis of deace-tylipecoside **(184)** and deacetylisoipecoside **(185)** by condensation of dopamine and secologanin **(23)**, **(184)** serves as a precursor of ipecoside **(36)** and alangiside **(174)**, both of which possess a 1 β-proton, whereas **(185)** is a precursor of cephaeline **(187)** and emetine **(188)** with 11 α-ste-reochemistry (*181, 182*).

2. Biogenesis of Hydrangenosides

Hydrangenosides have hitherto been isolated only from *Hydrangea* species. Hydrangenosides A **(37)**, B **(189)**, C **(190)** and D **(191)** are contained in *Hydrangea macrophylla* var. *macrophylla* and *H. macro-phylla* f. *normalis* (*183, 184, 185*). Hydrangenosides C**(190)**, D**(191)**, E**(38)**, F**(192)** and G**(39)** are found in *H. scandens* (*185, 186*). These compounds are a new type of secoiridoid glucosides consisting of seco-loganin **(23)** and a C_{13} or C_{15} unit originating from the shikimate-malo-nate route, which are joined through a C−C bond.

Although no biosynthetic experiments have been carried out on this type of glucosides, cooccurrence of them and secologanin **(23)** with hydrangenol **(193)** and hydrangenol glucoside **(194)** leads to the assumption that intramolecular condensation of a C_{15} unit ($C_6 - C_3 +$ 3 × malonyl) affords (193) and (194), whereas intramolecular aldol-type condensation of the C_{15} unit with (23) followed by decarboxylation and ether formation furnishes hydrangenoside A(37) and B(189). On the other hand, since *H. scandens* does not contain (193) and (194), it is assumed that condensation of a one-malonate-fewer C_{13} unit ($C_6 - C_3 + 2 \times$ malonyl) with (23) eventually affords hydrangeno-sides G(39), C(190), D(191), E(38) and F(192) (*186*) (Fig. 26). It is noteworthy that hydrangenoside G(39) has the same *S*-chirality at C-15 as the other glucosides with a tetrahydropyran ring. This strongly sug-gests that (39) is the biosynthetic intermediate of the other compounds.

VII. Epilogue

Our understanding of the processes involved in biosynthesis of the iridoids and secoiridoids has been deepened considerably. However, there still remain several unsolved questions. For example, the mecha-nisms of cyclopentane ring formation from an acyclic monoterpene have not yet been wholly elucidated and the mechanism of the cyclopen-tane ring cleavage of loganin(10) to secologanin(23) still remains to be clarified. As for studies with cell-free systems or enzymes, only the oxidation of geraniol (or nerol), the cyclization to iridodial (15) and the methylation of loganic acid (43) and secologanic acid (44) have so far been studied. In addition, biosynthetic pathways leading to indi-vidual compounds are sometimes very complicated as was shown in Chapter IV in the case of 10-deoxygeniposidic acid (98) which is formed *via* deoxyloganic acid (92) or 8-epideoxyloganic acid (73) depending on the plant. Thus, solution of such intricate problems awaits further detailed studies on individual plant species.

Since iridoids and related compounds occur abundantly in the dico-tyledons, especially in certain orders of sympetalous plants, they have been adopted as useful markers for chemotaxonomic studies (*58*, *187–197*). However, most of the studies related to this topic were made in the days when the knowledge of iridoids was still rather poor. How-ever, now, with an increase in our understanding of the biosynthesis of iridoids reexamination of earlier chemotaxonomic results may be in order.

Let us provide an instance from Rubiaceae. Iridoid glucosides, se-coiridoid glucosides and indole alkaloids are widespread in this family. KOOIMAN *et al.* classified this family in 1969 from the chemotaxonomi-

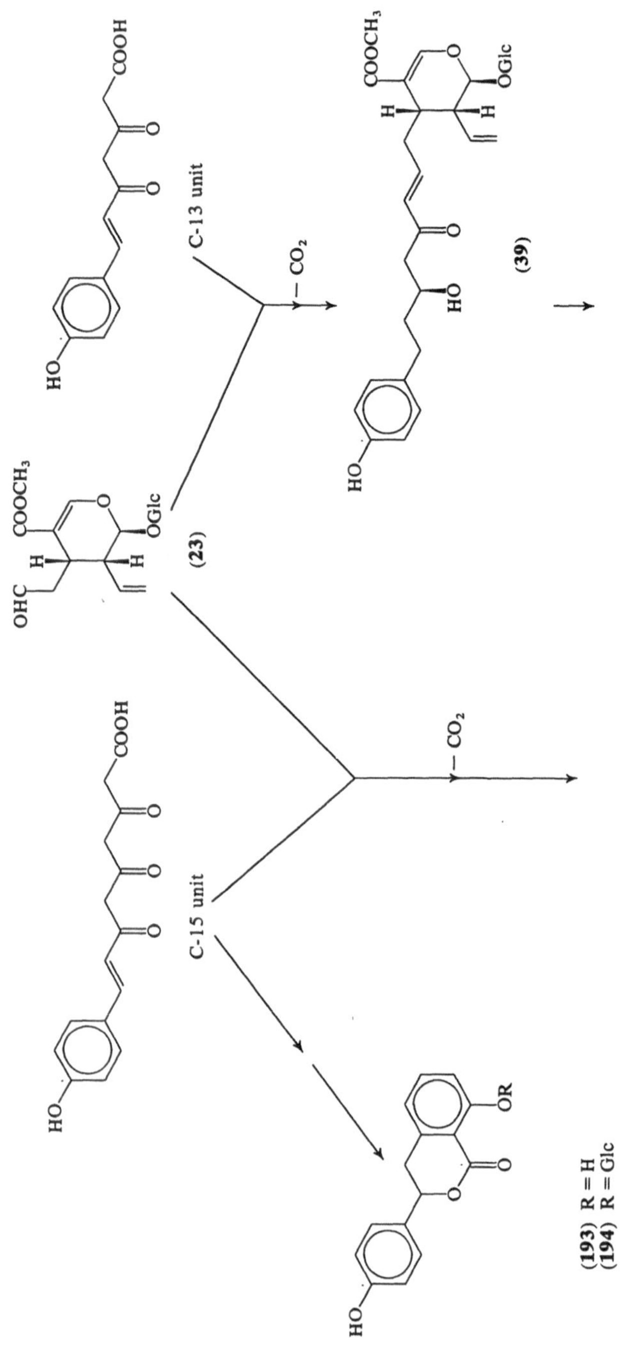
COOH
C-13 unit
$- CO_2$
COOCH$_3$
OGlc
(39)
OHC
COOCH$_3$
OGlc
(23)
COOH
$- CO_2$
C-15 unit
OR
(193) R = H
(194) R = Glc

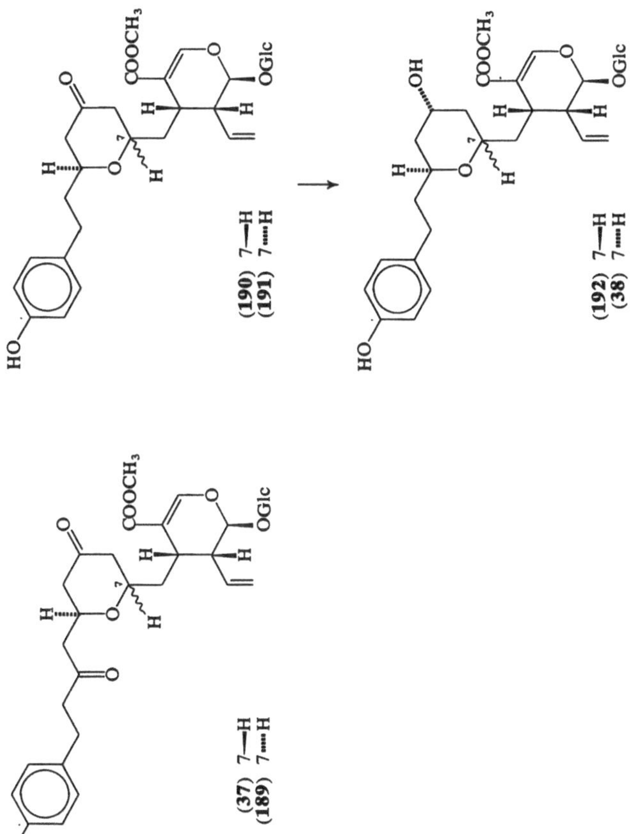

Fig. 26

cal point of view using iridoid glucosides as markers (191). Complementary studies by Inouye et al. (198) using GC and GC-mass spectroscopy confirmed that the family can be divided into three subfamilies: 1) subfamily Ixoroideae, which contains various type of iridoid glucosides such as gardenoside (22), 2) subfamily Rubioideae, which contains asperuloside (1) and/or deacetylasperulosidic acid and 3) subfamilies Cinchonoideae, Urophylloideae, Pomazotoideae and Guettaroideae etc., which contain loganin (10), secologanin (23) and/or indole alkaloids. Based on the cyclization patterns described in Chapter III, it is presumed that glucosides of group 1 such as gardenoside (22) are biosynthesized via 8-epiridodial (71) and 8-epideoxyloganic acid (73), whereas glucosides of groups 2 and 3 are formed via iridodial (15) and deoxyloganic acid (92). The validity of this hypothesis should be subject to verification by performing immunoassays for iridoid constituents of these groups with 8-epideoxyloganic acid (73) and deoxyloganic acid (92) antibodies.

 Zenk et al. have already succeeded in the quantitative determination of loganin (10) and secologanin (23) in plants and cultivated plant cells using the radioimmunoassay technique (199). By means of this technique, the pivotal biosynthetic intermediates of iridoids such as deoxyloganic acid (92), 8-epideoxyloganic acid (73) and 10-deoxygeniposidic acid (98) might be determined rapidly and accurately. Such efforts are now under way and evidence for the intermediacy of 8-epideoxyloganic acid (73) in aucubin (74) biosynthesis in *Aucuba japonica* has also been obtained in this manner (see p. 202). The immunoassay method thus promises to be a useful tool not only for chemotaxonomic, but also for biosynthetic studies.

List of Compounds

1 asperuloside
2 gentiopicroside
3 iridomyrmecin
4 isoiridomyrmecin
5 nepetalactone
6 plumieride
7 swertiamarin
8 gentianine
9 vindoline
10 loganin
11 dolichodial
12 dolicholactone
13 1,5,9-epideoxyloganic acid

14 iridodialogentiobioside
15 iridodial
16 β-skytanthine
17 deutzioside
18 verbenalin
19 lamiide
20 paederoside
21 monotropein
22 gardenoside
23 secologanin
24 sweroside
25 foliamenthin
26 morroniside

27 kinigiside
28 secogalioside
29 oleuropein
30 10-hydroxyligustroside
31 ligustaloside A
32 ligustaloside B
33 nüzhenide
34 strictosidine
35 vincoside
36 ipecoside
37 hydrangenoside A
38 hydrangenoside E
39 hydrangenoside G
40 10-oxocitronellal
41 iridotrial
42 (S)-(−)-citronellal ethylene acetal
43 loganic acid
44 secologanic acid
45 secologanoside
46 actinidine
47 dihydrofoliamenthin
48 10-hydroxynerol
49 10-hydroxygeraniol
50 ajmalicine
51 catharanthine
52 10-hydroxycitronellol
53 10-hydroxycitronellal
54 10-oxocitronellol
55 10-hydroxylinalool
56 9,10-dioxoneral
57 deoxyloganin
58 lamioside
59 ipolamiide
60 10-oxogeranial
61 10-oxoneral
62 iridodial glucoside
63 tarennoside
64 9,10-dihydroxygeraniol
65 9,10-dihydroxynerol
66 9,10-dihydroxycitronellol
67 9,10-dioxogeranial
68 iridotrial glucoside
69 dehydroiridotrial glucoside
70 boschnaloside
71 8-epiiridodial
72 8-epideoxyloganin
73 8-epideoxyloganic acid
74 aucubin
75 antirrhinoside
76 10-oxogeraniol
77 10-oxonerol
78 10-hydroxygeranial

79 10-hydroxyneral
80 9-oxo-10-hydroxygeraniol
81 9,9,10,10-tetraethoxygeranial
82 9-hydroxy-10-oxogeranial
83 9-oxo-10-hydroxygeranial
84 vomilenine
85 ajmaline
86 teucrein
87 dihydronepetalactone
88 9-hydroxy-10-oxocitronellal
89 11-hydroxyiridodial
90 hydroxy-acid
91 photocitral A
92 deoxyloganic acid
93 scandoside
94 geniposide
95 jasminin
96 7-epiloganin
97 7-dehydrologanin
98 deoxygeniposidic acid
99 theviridoside
100 catalposide
101 griselinoside
102 10-deoxygeniposide
103 mussaenoside
104 dihydrocornin
105 hastatoside
106 forsythide dimethyl ester
107 11-hydroxyiridodial glucoside
108 8-epiiridodial glucoside
109 8-epi-11-hydroxyiridodial glucoside
110 mussaenosidic acid
111 8-epiloganic acid
112 7-epiloganic acid
113 geniposidic acid
114 aglucone of deoxyloganic acid
115 10-dehydro derivative of gardenoside
116 10-dehydrogardenoside tetraacetate
117 allamandin
118 oruwacin
119 penstemide
120 genipin
121 sarracenin
122 xylomollin
123 valtrate
124 homodihydrovaltrate
125 valerosidate
126 patrinoside
127 10-hydroxyloganin
128 10-hydroxyloganin phosphate
129 gentioside
130 camptothecin

131 10-hydroxyloganin aglucone-
　　1-O-methyl ether
132 7-epi-10-hydroxyloganin aglucone-
　　1-O-methyl ether
133 tosylate of 131
134 tosylate of 132
135 oxetane-type compound
136 secologanin type substance
137 secologanin type substance
138 7-epi-10-hydroxyloganin
139 6β-hydroxyloganin
140 6β-hydroxyloganin phosphate
141 6α-hydroxy-7-epiloganin
142 6β-hydroxy-7-epiloganin
143 eustomoside
144 secologanol
145 loganin aglucone methyl ether
146 tricyclic compound
147 bakankosin
148 abelioside A
149 reserpinine
150 quinine
151 eustoside
152 eustomorrusside
153 linarioside
154 amarogentin
155 amaroswerin
156 amaropanin (desoxyamarogentin)
157 centapicrin
158 abelioside B
159 10-hydroxymorroniside
160 oleoside
161 10-hydroxyoleoside

162 ligustroside
163 10-acetoxyligustroside
164 10-acetoxyoleuropein
165 8-epikingiside
166 10-hydroxyoleuropein
167 epoxide
168 8-ol
169 vincoside lactam
170 strictosamide
171 isodihydrocadambine
172 cordifoline
173 rubenine
174 alangiside
175 akuammigine
176 tetrahydroalstonine
177 19-epiajmalicine
178 cathenamine
179 α-yohimbine
180 mitragynine
181 reserpiline
182 speciociliatine
183 isovincosamide aglucone
184 deacetylipecoside
185 deacetylisoipecoside
186 dihydrodeacetylipecoside
187 cephaeline
188 emetine
189 hydrangenoside B
190 hydrangenoside C
191 hydrangenoside D
192 hydrangenoside F
193 hydrangenol
194 hydrangenol glucoside

References

1. PAVAN, M.: The Extraction and Crystallization of Iridomyrmecin. Chim. et Ind. **37**, 625 (1955).
2. FUSCO, R., R. TRAVE, and A. VERCELLONE: Constitution of Iridomyrmecin, a Natural Insecticide. Chim. et Ind. **37**, 251 (1955).
2a. – – – Structure of Iridomyrmexin. ibid. **37**, 958 (1955).
3. CAVILL, G.W.K., D.L. FORD, and H.D. LOCKSLEY: Iridodial and Iridolactone. Chem. & Ind. **1956**, 465.
3a. – – – – The Chemistry of Ants I. Terpenoid Constituents of Some Australian *Iridomyrmex* Species. Austral. J. Chem. **9**, 288 (1956).
4. CAVILL, G.W.K., and H.D. LOCKSLEY: The Chemistry of Ants II. Structure and Configuration of Iridolactone (Isoiridomyrmecin). Austral. J. Chem. **10**, 352 (1957).
5. BATES, R.B., E.J. EISENBRAUN, and S.M. McELVAIN: The Configurations of the Nepetalactones and Related Compounds. J. Amer. Chem. Soc. **80**, 3420 (1958).

6. HALPERN, O., and H. SCHMID: Zur Kenntnis des Plumierids. Helv. Chim. Acta 41, 1109 (1958).

7. EL-NAGGAR, L.J., and J.L. BEAL: Iridoids, a Review. J. Natl. Prod. 43, 649 (1980).

8. BARGER, G., and C. SCHOLZ: Über Yohimbin. Helv. Chim. Acta 16, 1343 (1933).

9. HAHN, G., and H. WERNER: Synthese von Tetrahydroharman(4-Carbolin)-Systemen unter physiologischen Bedingungen. III. Mitteilung. Synthese des Yohimbin-Gerüstes. Liebig's Ann. Chem. 520, 123 (1935).

10. WENKERT, E., and N.V. BRINGI: A Stereochemical Interpretation of the Biosynthesis of Indole Alkaloids. J. Amer. Chem. Soc. 81, 1474 (1959).

11. WENKERT, E.: Alkaloid Biosynthesis. Experientia 15, 165 (1959).

12. THOMAS, R.: A Possible Biosynthetic Relationship between the Cyclopentanoid Monoterpenes and the Indole Alkaloids. Tetrahedron Letters 1961, 544.

13. WENKERT, E.: Biosynthesis of Indole Alkaloids. The Aspidosperma and Iboga Bases. J. Amer. Chem. Soc. 84, 98 (1962).

14. LEETE, E., and S. GHOSAL: Further Studies on the Biosynthesis of the Non-Tryptophan Derived Portion of Ajmaline and Related Alkaloids. Tetrahedron Letters 1962, 1179.

15. LEETE, E., S. GHOSAL, and P.N. EDWARDS: Biosynthesis of the Non-Tryptophan Derived Portion of Ajmaline. J. Amer. Chem. Soc. 84, 1068 (1962).

16. MONEY, T., I.G. WRIGHT, F. MCCAPRA, and A.I. SCOTT: Biosynthesis of the Indole Alkaloids. Proc. Natl. Acad. Sci. U.S. 53, 901 (1965).

17. MCCAPRA, F., T. MONEY, A.I. SCOTT, and I.G. WRIGHT: Biosynthesis of the Indole Alkaloids: Vindoline. Chem. Commun. 1965, 537.

18. MONEY, T., I.G. WRIGHT, F. MCCAPRA, E.S. HALL, and A.I. SCOTT: Biosynthesis of Indole Alkaloids. Vindoline. J. Amer. Chem. Soc. 90, 4144 (1968).

19. GOEGGEL, H., and D. ARIGONI: The Mevalonoid Nature of Vindoline and Reserpine. Chem. Commun. 1965, 538.

20. BATTERSBY, A.R., R.T. BROWN, R.S. KAPIL, A.D. PLUNKETT, and J.B. TAYLOR: Biosynthesis of the Indole Alkaloids. Chem. Commun. 1966, 46.

21. BATTERSBY, A.R., R.T. BROWN, J.A. KNIGHT, J.A. MARTIN, and A.O. PLUNKETT: Biosynthesis of the Indole Alkaloids from a Monoterpene. Chem. Comun. 1966, 346.

22. BATTERSBY, A.R., R.T. BROWN, R.S. KAPIL, J.A. KNIGHT, J.A. MARTIN, and A.O. PLUNKETT: Further Evidence Concerning the Biosynthesis of Indole Alkaloids and Quinine. Chem. Commun. 1966, 810.

23. LOEW, P., H. GOEGGEL, and D. ARIGONI: A Monoterpene Precursor in the Biosynthesis of Indole Alkaloids. Chem. Commun. 1966, 347.

24. HALL, E.S., F. MACCAPRA, T. MONEY, K. FUKUMOTO, J.R. HANSON, B.S. MOOTOO, G.T. PHILLIPS, and A.I. SCOTT: Concerning the Terpenoid Origin of Indole Alkaloids: Biosynthetic Mapping by Direct Mass Spectrometry. Chem. Commun. 1966, 348.

25. LEETE, E., and S. UEDA: Biosynthesis of the Vinca Alkaloids. The Incorporation of Geraniol-3-^{14}C into Catharanthine and Vindoline. Tetrahedron Letters 1966, 4915.

26. BATTERSBY, A.R., R.T. BROWN, R.S. KAPIL, J.A. MARTIN, and A.O. PLUNKETT: Role of Loganin in the Biosynthesis of Indole Alkaloids. Chem. Commun. 1966, 812.

26a. – – – – – Role of Loganin in the Biosynthesis of Indole Alkaloids. ibid. 1966, 890.

27. BATTERSBY, A.R., R.S. KAPIL, J.A. MARTIN, and L. MO: Loganin as Precursor of the Indole Alkaloids. Chem. Commun. 1968, 133.

28. LOEW, P., and D. ARIGONI: The Biological Conversion of Loganin into Indole Alkaloids. Chem. Commun. 1968, 137.

29. BOBBITT, J.M., and K.P. SEGEBARTH: Iridoid Glycosides and Similar Substances. In: Cyclopentanoid Terpene Derivatives. p. 1. Edited by TAYLOR, W.I., and A.R. BATTERSBY. New York: Marcel Dekker, Inc. 1969.

30. INOUYE, H.: Biosynthesis of Iridoid- and Secoiridoid Glucosides. In: Pharmacognosy and Phytochemistry 1970, p. 290. Edited by Wagner, H., and L. Hörhammer. Springer, Berlin-Heidelberg-New York: 1970.

31. GROSS, D.: Die Biosynthese Iridoider Naturstoffe. Fortschritte der Botanik 32, 93 (1970).

32. PLOUVIER, V., and J. FAVRE-BONVIN: Phytochemistry 10, 1697 (1971).

33. CORDELL, G.A.: The Biosynthesis of Indole Alkaloids. Lloydia 37, 219 (1974).

34. INOUYE, H., S. UEDA, and Y. TAKEDA: Biosynthesis of Secoiridoid Glucosides. Heterocycles 4, 527 (1976).

35. INOUYE, H.: Neuere Ergebnisse über die Biosynthese der Glucoside der Iridoidreihe. Planta Med. 33, 193 (1978).

36. TIETZE, L.-F.: Secologanin, a Biogenetic Key Compound – Synthesis and Biogenesis of the Iridoid and Secoiridoid Glycosides. Angew. Chem. Int. Ed. 22, 828 (1983).

37. INOUYE, H., S. UEDA, Y. AOKI, and Y. TAKEDA: Studies on Monoterpene Glucosides and Related Natural Products. XVII. The Intermediacy of 7-Desoxyloganic Acid and Loganin in the Biosynthesis of Several Iridoid Glucosides. Chem. Pharm. Bull. (Japan) 20, 1287 (1972).

38. INOUYE, H., S. UEDA, and Y. TAKEDA: Studies on Monoterpene Glucosides and Related Natural Products. XVIII. Formation Sequences of Iridoid Glucosides in Highly Oxidized Levels. Chem. Pharm. Bull. (Japan) 20, 1305 (1972).

39. INOUYE, H., S. UEDA, and S. UESATO: Zum Mechanismus der Methylcyclopentan-Gerüstbildung bei der Biosynthese einiger Iridoidglucoside. Tetrahedron Letters 1977, 709.

40. – – – Über die Biosynthese des Deutziosids. Tetrahedron Letters 1977, 713.

41. – – – Intermediacy of Iridodial in the Biosynthesis of Some Iridoid Glucosides. Phytochem. 16, 1669 (1977).

42. INOUYE, H., S. UEDA, S. UESATO, and K. KOBAYASHI: Studies on Monoterpene Glucosides and Related Natural Products. XXXVII. Biosynthesis of the Iridoid Glucosides in Lamium amplexicaule, Deutzia crenata and Galium spurium var. echinospermon. Chem. Pharm. Bull. (Japan) 26, 3384 (1978)

43. DAMTOFT, S.: Biosynthesis of Lamiide and Ipolamiide from 8-epi-Deoxyloganin Studied by ^{2}H N.M.R. Spectroscopy. J.C.S. Chem. Comm. 1981, 228.

44. – Biosynthesis of the Iridoids Aucubin and Antirrinoside from 8-Epi-Deoxyloganic Acid. Phytochem. 22, 1929 (1983).

45. UESATO, S., S. UEDA, K. KOBAYASHI, and H. INOUYE: Mechanism of Iridane Skeleton Formation in the Biosynthesis of Iridoid Glucosides in Gardenia jasminoides Cell Cultures. Chem. Pharm. Bull. (Japan) 31, 4185 (1983).

46. UESATO, S., S. UEDA, K. KOBAYASHI, M. MIYAUCHI, and H. INOUYE: Biosynthetic Pathway of Iridoid Glucosides in Gardenia jasminoides f. grandiflora Cell Suspension Cultures after Iridodial Cation Formation. Tetrahedron Letters 25, 573 (1984).

47. KOBAYASHI, K., S. UESATO, S. UEDA, and H. INOUYE: Studies on Monoterpene Glucosides and Related Natural Products. LV. Iridane Skeleton Formation from Acyclic Monoterpenes in the Biosynthesis of Iridoid Glucosides in Gardenia jasminoides f. grandiflora Cell Suspension Cultures. Chem. Pharm. Bull. (Japan) 33, 4228 (1985).

48. UESATO, S., S. UEDA, K. KOBAYASHI, M. MIYAUCHI, H. ITOH, and H. INOUYE: Intermediacy of 8-Epiiridodial in the Biosynthesis of Iridoid Glucosides of Gardenia jasminoides Cell Cultures. Phytochem. 25, 2309 (1986).

49. UESATO, S., M. MIYAUCHI, H. ITOH, and H. INOUYE: Biosynthesis of Iridoid Glucosides in Galium mollugo, Galium spurium var. echinospermon and Deutzia crenata.

Intermediacy of Deoxyloganic Acid, Loganin and iridodial Glucoside. Phytochem. 25, in press (1986).

50. BELLESIA, F., U.M. PAGNONI, A. PINETTI, and R. TRAVE: The Biosynthesis of Dolichodial in *Teucrium marum*. Phytochem. 22, 2197 (1983).

51. GRANDI, R., U.M. PAGNONI, A. PINETTI, and R. TRAVE: Biosynthesis of Dolicholactone in *Teucrium marum*. Phytochem. 22, 2723 (1983).

52. BELLESIA, F., U.M. PAGNONI, A. PINETTI, and R. TRAVE: Teucrein, a New Iridolactol from *Teucrium marum*, and its Biosynthetic Relationship with Dolichodial. J. Chem. Research(s) 1983, 328.

53. – – – – The Intermediacy of Iridodial in the Biosynthesis of Dolicholactone in *Teucrium marum*. J. Chem. Research(s) 1984, 192.

54. BELLESIA, F., R. GRANDI, U.M. PAGNONI, A. PINETTI, and R. TRAVE: Biosynthesis of Nepetalactone in *Nepeta cataria*. Phytochem. 23, 83 (1984).

55. GRANDI, R., U.M. PAGNONI, A. PINETTI, and R. TRAVE: The Possible Role of Photocitral-A in the Biosynthesis of Cyclopentane Monoterpenes. J. Chem. Research(s) 1984, 194.

56. MURAI, F., M. TAGAWA, S. DAMTOFT, S.R. JENSEN, and B.J. NIELSEN: (1R,5R,8S,9S)-Deoxyloganic Acid from *Nepeta cataria*. Chem. Pharm. Bull. (Japan) 32, 2809 (1984).

57. TAGAWA, M., and F. MURAI: A New Iridoid Glucoside from *Actinidia polygama*. Abstract Papers (II) of the 50th Annual Meeting of the Chemical Society of Japan, p. 918 (1985).

58. JENSEN, S.R., B.J. NIELSEN, and R. DAHLGREN: Iridoid Compounds, Their Occurrence and Systematic Importance in Angiosperms. Bot. Notiser. 128, 148 (1975).

59. YEOWELL, D.A., and H. SCHMID: Zur Biosynthese des Plumierids. Experientia 20, 250 (1964).

60. CLARK, K.J., G.I. FRAY, R.H. JAEGER, and R. ROBINSON: Synthesis of D- and L-Isoiridomyrmecin and Related Compounds. Tetrahedron 6, 217 (1959).

61. COSCIA, C.J., and R. GUARNACCIA: Biosynthesis of Gentiopicroside, a Novel Monoterpene. J. Amer. Chem. Soc. 89, 1280 (1967).

62. COSCIA, C.J., R. GUARNACCIA, and L. BOTTA: Monoterpene Biosynthesis. I. Occurrence and Mevalonoid Origin of Gentiopicroside and Loganic Acid in *Swertia caroliniensis*. Biochemistry 8, 5036 (1969).

63. INOUYE, H., S. UEDA, and Y. NAKAMURA: Zur Biosynthese der Bitteren Glucoside der Genzianaceen, des Gentiopicrosids, des Swertiamarins und des Swerosids. Tetrahedron Letters 1967, 3221.

63a. – – – Studies on Monoterpene Glucosides XII. Biosynthesis of Gentianaceous Secoiridoid glucosides. Chem. Pharm. Bull. (Japan) 18, 2043 (1970).

64. GUARNACCIA, R., L. BOTTA, and C.J. COSCIA: Mechanism of Secoiridoid Monoterpene Biosynthesis. J. Amer. Chem. Soc. 91, 204 (1969).

65. COSCIA, C.J., L. BOTTA, and R. GUARNACCIA: On the Mechanism of Iridoid and Secoiridoid Monoterpene Biosynthesis. Arch. Biochem. Biophys. 136, 498 (1970).

66. GUARNACCIA, R., and C.J. COSCIA: Occurrence and Biosnthesis of Secologanic Acid in *Vinca rosea*. J. Amer. Chem. Soc. 93, 6320 (1971).

67. GUARNACCIA, R., L. BOTTA, and C.J. COSCIA: Biosynthesis of Acidic Iridoid Monoterpene Glucosides in *Vinca rosea*. J. Amer. Chem. Soc. 96, 7079 (1974).

68. AUDA, H., H.R. JUNEJA, E.J. EISENBRAUN, G.R. WALLER, W.R. KAYS, and H.H. APPEL: Biosynthesis of Methylcyclopentane Monoterpenoids. I. Skytanthus Alkaloids. J. Amer. Chem. Soc. 89, 2476 (1967).

69. Auda, H., G.R. Waller, and E.J. EISENBRAUN: Biosynthesis of Methylcyclopentane Monoterpenoids. III. Actinidine. J. Biol. Chem. 242, 4157 (1967).

70. HORODYSKY, A.G., G.R. WALLER, and E.J. EISENBRAUN: Biosynthesis of Methylcyclopentane Monoterpenoids. IV. Verbenalin. J. Biol. Chem. 244, 3110 (1969).

71. Hüni, J.E.S., H. Hiltebrand, H. Schmid, D. Gröger, S. Johne, and K. Mothes: Zur Biosynthese des Verbenalins und Aucubins. Experientia 22, 656 (1966).
72. Regnier, F.E., G.R. Waller, E.J. Eisenbraun, and H. Auda: The Biosynthesis of Methylcyclopentane Monoterpenoids – II. Nepetalactone. Phytochem. 7, 221 (1968).
73. Loew, P., Ch. von Szczepanski, C.J. Coscia, and D. Arigoni: The Structure and Biosynthesis of Foliamenthin. Chem. Commun. 1968, 1276.
74. Battersby, A.R., A.R. Burnett, G.D. Knowles, and P.G. Parsons: Seco-cyclopentane Glucosides from Menyanthes trifoliata: Foliamenthin, Dihydrofoliamenthin, and Menthiafolin. Chem. Commun. 1968, 1277.
75. Battersby, A.R., and R.J. Parry: Biosynthesis of the Ipecac Alkaloids and of Ipecoside. Chem. Commun. 1971, 901.
76. Coscia, C.J., and R. Guarnaccia: Natural Occurrence and Biosynthesis of a Cyclopentanoid Monoterpene Carboxylic Acid. Chem. Commun. 1968, 138.
77. Escher, S., P. Loew, and D. Arigoni: The Role of Hydroxygeraniol and Hydroxynerol in the Biosynthesis of Loganin and Indole Alkaloids. Chem. Commun. 1970, 823.
78. Battersby, A.R., S.H. Brown, and T.G. Payne: Biosynthesis of Loganin and the Indole Alkaloids from Hydroxygeraniol-Hydroxynerol. Chem. Commun. 1970, 827.
79. Meehan, T.D., and C.J. Coscia: Hydroxylation of Geraniol and Nerol by a Monooxygenase from Vinca rosea. Biochem. Biophys. Res. Comm. 53, 1043 (1973).
80. Bowman, R.M., and E. Leete: Observations on the Administration of Iridodial-7-^{14}C to Vinca rosea. Phytochem. 8, 1003 (1969).
81. Ueda, S., K. Kobayashi, T. Muramatsu, and H. Inouye: Studies on Monoterpene Glucosides and related Natural Products. Part XL. Iridoid Glucosides of Cultured Cells of Gardenia jasminoides f. grandiflora. Planta Med. 41, 186 (1981).
82. Uesato, S., K. Kobayashi, and H. Inouye: Studies on Monoterpene Glucosides and Related Natural Products. XLV. Synthesis of ^{13}C-Labeled Acyclic Monoterpenes for Studies on the Mechanism of the Iridane Skeleton Formation in the Biosynthesis of Iridoid Glucosides. Chem. Pharm. Bull. (Japan) 30, 927 (1982).
83. Battersby, A.R., M. Thompson, K.-H. Glüsenkamp, and L.-F. Tietze: Untersuchungen zur Biogenese der Indolalkaloide. Synthese und Verfütterung radioaktiv markierter Monoterpenaldehyde. Chem. Ber. 114, 3430 (1981).
84. Balsevich, J., and W.G.W. Kurz: The Role of 9- and/or 10-Oxygenated Derivatives of Geraniol, Geranial, Nerol, and Neral in the Biosynthesis of Loganin and Ajmalicine. Planta Med. 49, 79 (1983).
85. Uesato, S., S. Matsuda, and H. Inouye: Mechanism for Iridane Skeleton Formation from Acyclic Monoterpenes in the Biosynthesis of Secologanin and Vindoline in Catharanthus roseus and Lonicera morrowii. Chem. Pharm. Bull. (Japan) 32, 1671 (1984).
86. – – – Studies on Monoterpene Glucosides and Related Natural Products. LII. Mechanism for Iridane Skeleton Formation from Acyclic Monoterpenes in the Biosynthesis of Secoiridoid Glucosides and Indole Alkaloids. Yakugaku Zasshi 104, 1232 (1984).
87. Stöckigt, J., A. Pfitzner, and J. Firl: Indole Alkaloids from Cell Suspension Cultures of Rauwolfia serpentina Benth. Plant Cell Reports 1, 36 (1981).
88. Uesato, S., S. Matsuda, A. Iida, H. Inouye, and M.H. Zenk: Intermediacy of 10-Hydroxygeraniol, 10-Hydroxynerol and Iridodial in the Biosynthesis of Ajmaline and Vomilenine in Rauwolfia serpentina Suspension Cultures. Chem. Pharm. Bull. (Japan) 32, 3764 (1984).
89. Uesato, S., S. Kanomi, A. Iida, H. Inouye, and M.H. Zenk: Mechanism for Iridane Skeleton Formation in the Biosynthesis of Secologanin and Indole Alkaloids in

Plants of *Lonicera tatarica* and *Catharanthus roseus* and Suspension Cultures of *Rauwolfia serpentina*. Phytochem. **25**, 839 (1986).

90. UESATO, S., Y. OGAWA, H. INOUYE, K. SAIKI, and M.H. ZENK: Synthesis of Iridodial by Cell Free Extracts from *Rauwolfia serpentina* Cell Suspension Cultures. Tetrahedron Letters **27**, 2893 (1986).

91. COOKSON, R.C., J. HUDEC, S.A. KNIGHT, and B.R.D. WHITEAR: The Photochemistry of Citral. Tetrahedron **19**, 1995 (1963).

92. COOKSON, R.C.: The Photochemistry of Some Allylic Compounds. Quart. Rev. **22**, 423 (1968).

93. UESATO, S., H. ITOH, S. XIE, and H. INOUYE: unpublished results.

94. INOUYE, H., S. UEDA, Y. AOKI, and Y. TAKEDA: Zur Biosynthese der Iridoidglucoside. Tetrahedron Letter **1969**, 2351.

95. – – – – Studies on Monoterpene Glucosides and Related Natural Products. XVII. The Intermediacy of 7-Desoxyloganic Acid and Loganin in the Biosynthesis of Several Iridoid Glucosides. Chem. Pharm. Bull. (Japan) **20**, 1287 (1972).

96. RIMPLER, H., and B. VON LEHMANN: Bisdesoxydihydromonotropein aus *Physostegia virginiana*. Phytochem. **9**, 641 (1970).

97. BATTERSBY, A.R., A.R. BURNETT, and P.G. PARSONS: Preparation and Isolation of Deoxyloganin: Its Role as Precursor of Loganin and the Indole Alkaloids. Chem. Commun. **1970**, 826.

98. MADYASTHA, K.M., R. GUARNACCIA, and C.J. COSCIA: Enzymic Synthesis of Loganin by Carboxyl Group Methylation of Loganic Acid. FEBS Letters **14**, 175 (1971).

99. MADYASTHA, K.M., R. GUARNACCIA, C. BAXTER, and C.J. COSCIA: S-Adenosyl-L-methionine: Loganic Acid Methyltransferase. J. Biol. Chem. **248**, 2497 (1973).

100. BATTERSBY, A.R., A.R. BURNETT, E.S. HALL, and P.G. PARSONS: The Rearrangement Process in Indole Alkaloid Biosynthesis. Chem. Commun. **1968**, 1582.

101. INOUYE, H., S. UEDA, K. INOUE, and Y. TAKEDA: Studies on Monoterpene Glucosides and Related Natural Products. XXIII. Biosynthesis of the Secoiridoid Glucosides, Gentiopicroside, Morroniside, Oleuropein, and Jasminin. Chem. Pharm. Bull. (Japan) **22**, 676 (1974).

102. GRÖGER, D., and P. SIMCHEN: Über den Einbau von Loganin in Gentiopicrosid. Naturforsch. **24b**, 356 (1969).

103. INOUYE, H., S. UEDA, and Y. TAKEDA: Zur Biosynthese des Morronisids. Tetrahedron Letters **1971**, 4069.

104. BATTERSBY, A.R., A.R. BURNETT, and P.G. PARSONS: Alkaloid Biosynthesis. Part XIV. Secologanin: Its Conversion into Ipecoside and Its Role as Biological Precursor of the Indole Alkaloids. J. Chem. Soc. (C) **1969**, 1187.

105. INOUYE, H., S. UEDA, and Y. TAKEDA: Loganin als precursor in der Biosynthese des Asperulosids. Naturforsch. **24b**, 1666 (1969).

106. DAMTOFT, S., S.R. JENSEN, and B.J. NIELSEN: Application of ^{2}H N.M.R. Spectroscopy to a Study of the Biosynthesis of the Iridoid Glucosides Cornin in *Verbena officinalis*. J. C. S. Chem. Commun. **1980**, 42.

107. – – – Biosynthesis of the Iridoid Glucosides Cornin, Hastatoside, and Gliselinoside in *Verbena* Species. J. Chem. Soc., Perkin I **1983** 1943.

108. DAMTOFT, S., M.U. JARS, S.R. JENSEN, O. KIRK, and B.J. NIELSEN: The Effect of Metabolic Period, Dose and Application Method on the Incorporation of Deoxyloganin into Cornin in *Verbena officinalis*. Phytochemistry **22**, 695 (1983).

109. JENSEN, S.R., and B.J. NIELSEN: private communication.

110. DAMTOFT, S., S.R. JENSEN, and B.J. NIELSEN: The Biosynthesis of Iridoid Glucosides from 8-*epi*-Deoxyloganic Acid. Biochem. Soc. Transactions **11**, 593 (1983).

111. UESATO, S., H. ITOH, and H. INOUYE: unpublished results.

112. UESATO, S., E. ALI, H. NISHIMURA, I. KAWAMURA, and H. INOUYE: Four Iridoids from *Randia canthioides*. Phytochem. **21**, 353 (1982).

113. INOUE, K., Y. TAKEDA, H. NISHIMURA, and H. INOUYE: Studies on Monoterpene Glucosides and Related Natural Products. XXXIX. Biogenetic-type Transformation of Geniposide into Plumieride. Chem. Pharm. Bull. (Japan) **27**, 3115 (1979).

114. KUPCHAN, S.M., A.L. DESSERTINE, B.T. BLAYLOCK, and R.F. BRYAN: Isolation and Structural Elucidation of Allamandin, an Antileukemic Iridoid Lactone from *Allamanda cathartica*. J. Org. Chem. **39**, 2477 (1974).

115. ADESOGAN, E.K.: Oruwacin, A New Iridoid Ferulate from *Morinda lucida*. Phytochem. **18**, 175 (1979).

116. JENSEN, S.R., B.J. NIELSEN, C.B. MIKKELSEN, J.J. HOFFMANN, S.D. JOLAD, and J.R. COLE: The Revised Structure of Penstemide. Tetrahedron Letters **1979**, 3261.

117. DJERASSI, C., T. NAKANO, A.N. JAMES, L.H. ZALKOW, E.J. EISENBRAUN, and J.N. SHOOLERY: Terpenoids. XLVII. The Structure of Genipin. J. Org. Chem. **26**, 1192 (1961).

118. MILES, D.H., U. KOKPOL, J. BHATTACHARYYA, J.L. ATWOOD, K.E. STONE, T.A. BRYSON, and C. WILSON: Structure of Sarracenin. An unusual Enol Diacetal Monoterpene from the Insectivorous Plant *Sarracenia flava*. J. Amer. Chem. Soc. **98**, 1569 (1976).

119. KUBO, I., I. MIURA, and K. NAKANISHI: The Structure of Xylomollin, a Secoiridoid Hemiacetal Acetal. J. Amer. Chem. Soc. **98**, 6704 (1976).

120. BOCK, K., S.R. JENSEN, B.J. NIELSEN, and V. NORN: Iridoid Allosides from *Viburunum opulus*. Phytochem. **17**, 753 (1978).

121. S.R. Jensen, B.J. Nielsen, and V. Norn: Iridoids from *Viburnum betulifolium*. Phytochem. **24**, 487 (1985).

122. TAGUCHI, H., Y. YOKOKAWA, and T. ENDO: Studies on the Constituents of *Patrinia villosa*. Yakugaku Zasshi **93**, 607 (1973).

123. TAGUCHI, H., and T. ENDO: Patrinoside, a New Iridoid Glycoside from *Patrinia scabiosaefolia*. Chem. Pharm. Bull. (Japan) **22**, 1935 (1974).

124. BATTERSBY, A.R.: Biosynthesis of the Indole and Colchicum Alkaloids. Pure and Appl. Chem. **14**, 117 (1967).

125. BATTERSBY, A.R., B. GREGORY, H. SPENCER, J.C. TURNER, M.-M. JANOT, P. POTIER, P. FRANCOIS, and J. LEVISALLES: Constitution of Ipecoside: A Monoterpenoid Isoquinoline. Chem. Commun. **1967**, 219.

126. BATTERSBY, A.R., A.R. BURNETT, and P.G. PARSONS: Preparation of Secologanin: its Conversion into Ipecoside and its Role in Indole Alkaloid Biosynthesis. Chem. Commun. **1968**, 1280.

127. – – – Partial Synthesis and Isolation of Vincoside and Isovincoside: Biosynthesis of the Three Major Classes of Indole Alkaloids from the β-Carboline Systen. Chem. Commun. **1968**, 1282.

127a. – – – Alkaloid Biosynthesis. Part XV. Partial Synthesis and Isolation of Vincoside and Isovincoside: Biosynthesis of the Three Major Classes of Indole Alkaloids from Vincoside. J. Chem. Soc. (*C*) **1969**, 1193.

128. SOUZU, I., and H. MITSUHASHI: Structures of Iridoids from *Lonicera morrowii* A. Gray. II. Tetrahedron Letters **1970**, 191.

129. POPOV, S., and N. MAAREKOV: A New Iridoid Precursor of Gentiopicroside. Phytochem. **10**, 3077 (1971).

130. BATTERSBY, A.R., and K.H. GIBSON: Further Studies on Rearrangements during Biosynthesis of Indole Alkaloids. Chem. Commun. **1971**, 902.

131. TAKEDA, Y., and H. INOUYE: Studies on Monoterpene Glucosides and Related Natural Products. XXX. The Fate of the C-8 Proton of 7-Deoxyloganic Acid in the Biosynthesis of Secoiridoid Glucosides. Chem. Pharm. Bull. (Japan) **24**, 79 (1976).

132. HUTCHINSON, C.R., A.H. HECKENDORF, and P.E. DADDONA: Biosynthesis of Camptothecin. I. Definition of the Overall Pathway Assisted by Carbon-13 Nuclear Magnetic Resonance Analysis. J. Amer. Chem. Soc. **96**, 5609 (1974).

133. TIETZE, L.-F.: Totalsynthese von Hydroxyloganin und Hydroxyloganinsäure. Angew. Chem. **85**, 763 (1973).

133a. – Iridoide, IV Totalsynthese von Hydroxyloganin und Hydroxyloganinsäure. Chem. Ber. **107**, 2499 (1974).

134. – Fragmentation of Hydroxyloganin Derivatives. An Assay to Secologanin Type Compounds. J. Amer. Chem. Soc. **96**, 946 (1974).

134a. – Iridoide, V Biogenetische Synthese von Secologanin- und Swerosid-aglyconmethyläther. Chem. Ber. **109**, 3626 (1976).

135. INOUE, K., Y. TAKEDA, T. TANAHASHI, and H. INOUYE: Studies on Monoterpene Glucosides and Related Natural Products. XLI. Chemical Conversion of Geniposide into 10-Hydroxyloganin. Chem. Pharm. Bull. (Japan) **29**, 970 (1981).

136. – – – – Studies on Monoterpene Glucosides and Related Natural Products. XLII. On the Possibility of the Intermediacy of 10-Hydroxyloganin in the Biosynthesis of Secologanin. Chem. Pharm. Bull. (Japan) **29**, 98 (1981).

137. BATTERSBY, A.R., N.D. WESTCOTT, K.-H. GLÜSENKAMP, and L.-F. TIETZE: Untersuchungen zur Biogenese der Indolalkaloide. Synthese und Verfütterung radioaktiv markierter Hydroxyloganin-Derivate. Chem. Ber. **114**, 3439 (1981).

138. INOUE, K., H. KUWAJIMA, K. TAKAISHI, and H. INOUYE: unpublished results.

139. PARTRIDGE, J.J., N.K. CHADHA, S. FABER, and M.R. USKOKOVIC: Lead Tetraacetate Fragmentation of Loganin Aglucone O-Methyl Ether and Its stereoisomers. Synth. Commun. **1**, 233 (1971).

140. UESATO, S., T. HASHIMOTO, and H. INOUYE: Three New Secoiridoid Glucosides from *Eustoma russellianum*. Phytochem. **18**, 1981 (1979).

141. BALENOVIC, K., H.U. DÄNIKER, R. GOUTAREL, M.M. JANOT, and V. PRELOG: Über Bakankosin. Helv. Chim. Acta **35**, 2519 (1952).

142. MURAI, F., M. TAGAWA, S. MATSUDA, T. KIKUCHI, S. UESATO, and H. INOUYE: Abeliosides A and B, Secoiridoid Glucosides from *Abelia grandiflora*. Phytochem. **24**, 2329 (1985).

143. INOUYE, H., S. UEDA, and Y. TAKEDA: The Biological Conversion of Sweroside into Gentiopicroside and Vindoline. Tetrahedron Letters **1968**, 3453.

144. – – – Studies on Monoterpene Glucosides and Related Natural Products. XIII. Incorporation of [10-14C]Sweroside into Gentiopicroside and the Alkaloids in *Vinca* and *Cinchona* Plants. Chem. Pharm. Bull. (Japan) **19**, 587 (1971).

145. INOUYE, H., S. UEDA, and Y. TAKEDA: Zur Biosynthese der Vinca- sowie der Cinchonaalkaloide. Inkorporation des Swerosids in Reserpinin und Chinin. Tetrahedron Letters **1969**, 407.

146. INOUYE, H., S. TOBITA, and M. MORIGUCHI: Studies on Monoterpene Glucosides and Related Natural Products. XXXIII. Structure of Bakankosin. Chem. Pharm. Bull. (Japan) **24**, 1406 (1976).

147. TIETZE, L.-F.: Synthese und Strukturbeweis von Bakankosin. Tetrahedron Letters **1976**, 2535.

148. KITAGAWA, I., T. TANI, K. AKITA, and I. YOSIOKA: On the Constituents of *Linaria japonica* Miq. I. The Structure of Linarioside, a New Chlorinated Iridoid Glucoside and Identification of Two Related Glucosides. Chem. Pharm. Bull. (Japan) **21**, 1978 (1973).

149. INOUYE, H., and Y. NAKAMURA: Zwei stark bittere Glucoside aus *Swertia japonica* Makino: Amarogentin und Amaroswerin. Tetrahedron Letters **1968**, 4919.

149a. – – Über die Monoterpenglucoside und verwandte Naturstoffe. XIV. Die Struktur der beiden stark bitter schmeckenden Glucoside Amarogentin und Amaroswerin aus *Swertia japonica*. Tetrahedron **27**, 1951 (1971).

150. INOUYE, H., and Y. NAKAMURA: Studies on Monoterpene Glucosides and Related

Natural Products. XVI. Occurrence of Secoiridoid Glucosides in Gentianaceous Plants especially in the Genera *Gentiana* and *Swertia*. Yakugaku Zasshi **91**, 755 (1971).

151. WAGNER, H., and K. VASIRIAN: Desoxyamarogentin, ein neuer Bitterstoff aus *Gentiana pannonica* Scop. Phytochem. **13**, 615 (1974).

152. SAKINA, K., and K. AOTA: Studies on the Constituents of *Erythraea centaurium* (Linne) Persoon. I. The Structure of Centapicrin, a New Bitter Secoiridoid Glucoside. Yakugaku Zasshi **96**, 683 (1976).

153. VAN DER SLUIS, W.G., and R.P. LABADIE: Secoiridoids and Xanthones in the Genus *Centaurium* Part III: Decentapicrins A, B and C, New *m*-Hydroxybenzoyl Esters of Sweroside from *Centaurium littorale*. Planta Med. **41**, 150 (1981).

154. ENDO, T., and H. TAGUCHI: Study on the Constituents of *Cornus officinalis* Sieb. et Zucc. Yakugaku Zasshi **93**, 30 (1973).

155. BOCK, K., S.R. JENSEN, and B.J. NIELSEN: Secogalioside, an Iridoid Glucoside from *Galium album* Mill. and ^{13}C NMR Spectra of some Seco-iridoid Glucosides. Acta Chem. Scand. B **30**, 743 (1976).

156. UESATO, S., M. UEDA, H. INOUYE, H. KUWAJIMA, M. YATSUZUKA, and K. TAKAISHI. Iridoids from *Galium mollugo*. Phytochem. **23**, 2535 (1984).

157. INOUYE, H., T. YOSHIDA, S. TOBITA, K. TANAKA, and T. NISHIOKA: Über die Monoterpenglucoside und verwandte Naturstoffe. XXII. Absolutstruktur des Oleuropeins, Kingisids und Morronisids. Tetrahedron **30**, 201 (1974).

158. INOUYE, H., S. UEDA, K. INOUE, and Y. TAKEDA: Über die Biosynthese der Oleuropein-Typ-Secoiridoidglucoside der Oleaceae. Tetrahedron Letters **1971**, 4073.

159. LAVIE, D., and E.C. LEVY: Oxidative Reactions of Biogenetic Interest. Tetrahedron Letters **1970**, 1315.

160. CONRAD, H.E., R. DUBUS, M.J. NAMTVEDT, and I.C. GUNSALUS: Mixed Function Oxidation. II. Separation and Properties of the Enzymes Catalyzing Camphor Lactonization. J. Biol. Chem. **240**, 495 (1965).

161. INOUE, K., T. NISHIOKA, T. TANAHASHI, and H. INOUYE: Three Secoiridoid Glucosides from *Ligustrum japonicum*. Phytochem. **21**, 2305 (1982).

162. SMITH, G.N.: Strictosidine: A Key Intermediate in the Biosynthesis of Indole Alkaloids. Chem. Commun. **1968**, 912.

163. DE SILVA, K.T.D., G.N. SMITH, and K.E.H. WARREN: Biochemistry of Strictosidine. Chem. Commun. **1971**, 905.

164. – – – Stereochemistry of Strictosidine. Chem. Commun. **1971**, 905.

165. BLACKSTOCK, W.P., R.T. BROWN, and G.K. LEE: Configuration at C-3 in Vincoside. Chem. Commun. **1971**, 910.

166. KENNARD, O., P.J. ROBERTS, N.W. ISAACS, F.H. ALLEN, W.D.S. MOTHERWELL, K.H. GIBSON, and A.R. BATTERSBY: X-Ray Determination of the Structure of O,O-Dimethylipecoside. Chem. Commun. **1971**, 899.

167. MATTES, K.C.,C.R. HUTCHINSON, J.P. SPRINGER, and J. CLARDY: Absolute Configuration of Vincoside. J. Amer. Chem. Soc. **97**, 6270 (1975).

168. HUTCHINSON, C.R., A.H. HECKENDORF, and P.E. DADDONA: Biosynthesis of Camptothecin. I. Definition of the Overall Pathway Assisted by Carbon-13 Nuclear Magnetic Resonance Analysis. J. Amer. Chem. Soc. **96**, 5609 (1974).

169. BROWN, R.T., J. LEONARD, and S.K. SLEIGH: One-pot Biomimetic Synthesis of 19β-Heterojohimbine Alkaloids. J.C.S. Chem. Comm. **1977**, 636.

170. STÖCKIGT, J., J. TREIMER, and M.H. ZENK: Synthesis of Ajmalicine and Related Indole Alkaloids by Cell Free Extracts of *Catharanthus roseus* Cell Suspension Cultures. FEBS Letters **70**, 267 (1976).

171. STÖCKIGT, J., and M.H. ZENK: Isovincoside (Strictosidine), the Key Intermediate in the Enzymatic Formation of Indole Alkaloids. FEBS Letters **79**, 233 (1977).

172. STÖCKIGT, J., H.P. HUSSON, C. KAN-FAN, and M.H. ZENK: Cathenamine, a Central Intermediate in the Cell Free Biosynthesis of Ajmalicine and Related Indole Alkaloids. J.C.S. Chem. Commun. 1977, 164.

173. STÖCKIGT, J., and M.H. ZENK: Strictosidine (Isovincoside): the Key Intermediate in the Biosynthesis of Monoterpenoid Indole Alkaloids. J.C.S. Chem. Commun. 1977, 646.

174. RÜFFER, M., N. NAGAKURA, and M.H. ZENK: Strictosidine, the Common Precursor for Monoterpenoid Indole Alkaloids with 3α and 3β Configuration. Tetrahedron Letters 1978, 1593.

175. NAGAKURA, N., M. RÜFFER, and M.H. ZENK: The Biosynthesis of Monoterpenoid Indole Alkaloids from Strictosidine. J. Chem. Soc., Perkin I 1979, 2308.

176. HECKENDORF, A.H., and C.R. HUTCHINSON: Biosynthesis of Camptothecin. II. Confirmation that Isovincoside, not Vincoside, is the Penultimate Biosynthetic Precursor of Indole Alkaloids. Tetrahedron Letters 1977, 4153.

177. SCOTT, A.I., S.L. LEE, P. DE CAPILE, M.G. CULVER, and C.R. HUTCHINSON: The Role of Isovincoside (Strictosidine) in the Biosynthesis of the Indole Alkaloids. Heterocycles 7, 979 (1977).

178. BROWN, R.T., J. LEONARD, and S.K. SLEIGH: The Role of Strictosidine in Monoterpenoid Indole Alkaloid Biosynthesis. Phytochem. 17, 899 (1978).

179. ROBERTS, P.J., N.W. ISAACS, F.H. ALLEN, W.D.S. MOTHERWELL, and O. KENNARD: The Crystal Structure and Absolute Configuration of O,O-Dimethylipecoside. Acta Crystallogr. B 30, 133 (1974).

180. BATTERSBY, A.R., and B. GREGORY: Biosynthesis of the Ipecac Alkaloids and of Ipecoside, a Cleaved Cyclopentane Monoterpene. Chem. Commun. 1968, 134.

181. NAGAKURA, N., G. HÖFLE, and M.H. ZENK: Deacetylipecoside: the Key Intermediate in the Biosynthesis of the Alkaloids Cephaeline and Emetine. J.C.S. Chem. Commun. 1978, 896.

182. NAGAKURA, N., G. HÖFLE, D. COGGIOLA, and M.H. Zenk: The Biosynthesis of the Ipecac Alkaloids and of Ipecoside and Alangiside. Planta Med. 34, 381 (1978).

183. INOUYE, H., Y. TAKEDA, S. UESATO, K. UOBE, and T. HASHIMOTO: A Novel Secoiridoid Glucoside, Hydrangenoside A from Hydrangea macrophylla. Tetrahedron Letters 21, 1059 (1980).

184. UESATO, S., T. HASHIMOTO, K. UOBE, and H. INOUYE: Novel Type Secoiridoid Glucosides, Hydrangenoside B, C and D from Hydrangea macrophylla. Chem. Pharm. Bull. (Japan) 29, 3421 (1981).

185. UESATO, S., Y. TAKEDA, T. HASHIMOTO, K. UOBE, H. INOUYE, H. TAGUCHI, and T. ENDO: Studies on Monoterpene Glucosides and Related Natural Products. Part 49 Absolute Structures of Hydrangenosides A, B, C, D, E, F and G. Novel Type Secoiridoid Glucosides from Two Hydrangea plants. Helv. Chim. Acta 67, 2111 (1984).

186. UESATO, S., T. HASHIMOTO, Y. TAKEDA, and H. INOUYE: New Secoiridoid Glucosides, Hydrangenosides E, F and G from Hydrangea scandens. Chem. Pharm. Bull. (Japan) 30, 4222 (1982).

187. BATE-SMITH, E.C., and T. SWAIN: The Asperulosides and the Aucubins, p. 159 in Comparative Phytochemistry. (Ed. by T. SWAIN.) New York: Academic Press. 1966.

188. HEGNAUER, R.: Aucubinartige Glucoside. Über ihre Verbreitung und Bedeutung als systematisches Merkmal. Pharmaceutica Acta Helvetiae 41, 577 (1966).

189. WIFFERING, J.H.: Aucubinartige Glucoside (Pseudoindikane) und verwandte Heteroside als systematische Merkmale. Phytochem. 5, 1053 (1966).

190. FIKENSCHER, L.H., R. HEGNAUER, and H.W. RUIJGROK: Iridoide Pflanzenstoffe (Pseudoindikane) als systematische Merkmale. Pharm. Weekblad 104, 561 (1969).

191. KOOIMAN, P.: The Occurrence of Asperulosidic Glycosides in the Rubiaceae. Acta Bot. Neerl. **18**, 124 (1969).
192. KOOIMAN, P.: The Occurrence of Iridoid Glycosides in the Scrophulariaceae. Acta Bot. Neerl. **19**, 329 (1970).
193. HEGNAUER, R.: Pflanzenstoffe und Pflanzensystematik. Naturwissenschaften **58**, 585 (1971).
194. BATE-SMITH, E.C.: Chemistry and Phylogeny of the Angiosperms. Nature **236**, 353 (1972).
195. GRAYER-BARKMEIJER, R.J.: A Chemosystematic Study of Veronica: Iridoid Glucosides. Biochemical Systematics and Ecology **1**, 101 (1973).
196. BATE-SMITH, E.C., I.K. FERGUSON, K.H. HUTSON, S.R. JENSEN, B.J. NIELSEN, and T. SWAIN: Phytochemical Interrelationship in the Cornaceae. Biochemical Systematics and Ecology **3**, 79 (1975).
197. HEGNAUER, R., and P. KOOIMAAN: Die Systematische Bedeutung von Iridoiden Inhaltsstoffen im Rahmen von Wettstein's Tubiflorae. Planta Med. **33**, 1 (1978).
198. INOUYE, H., Y. TAKEDA, S. KANOMI, and T. OKUDA: unpublished results.
199. TANAHASHI, T., N. NAGAKURA, H. INOUYE, and M.H. ZENK: Radioimmunoassay for the Determination of Loganin and the Biotransformation of Loganin to Secologanin by Plant Cell Cultures. Phytochem. **23**, 1917 (1984).
200. INOUYE, H., K. INOUE, and M. ONO: unpublished results.

(*Received 10 July, 1986*)

Author Index

Page numbers printed in *italics* refer to References

Subject Index

Fortschritte der Chemie organischer Naturstoffe

Progress in the Chemistry of Organic Natural Products

Volume 49:

1986. VIII, 400 pages. Cloth DM 290,–, öS 2030,–. ISBN 3-211-81910-X

Contents: R. A. Hill: Naturally Occurring Isocoumarins. – R. Wijnsma and R. Verpoorte: Anthraquinones in the Rubiaceae. – H. Chr. Krebs: Recent Developments in the Field of Marine Natural Products with Emphasis on Biologically Active Compounds.

Volume 48:

1985. 33 figures. IX, 285 pages. Cloth DM 220,–, öS 1540,–.
ISBN 3-211-81886-3

Contents: P. S. Steyn and R. Vleggaar: Tremorgenic Mycotoxins. – R. E. Moore: Structure of Palytoxin. – P. Crews and S. Naylor: Sesterterpenes: An Emerging Group of Metabolites from Marine and Terrestrial Organisms.

Volume 47:

1985. 16 figures. VIII, 290 pages. Cloth DM 198,–, öS 1390,–.
ISBN 3-211-81864-2

Contents: R. Southgate and S. Elson: Naturally Occurring β-Lactams. – I. Howe and M. Jarman: New Techniques for the Mass Spectrometry of Natural Products. – P. G. McDougal and N. R. Schmuff: Chemical Synthesis of the Trichothecenes. – J. Polonsky: Quassinoid Bitter Principles II.

Volume 46:

1984. 7 figures. IX, 253 pages. Cloth DM 178,–, öS 1250,–.
ISBN 3-211-81804-9

Contents: O. Tanaka and R. Kasai: Saponins of Ginseng and Related Plants. – E. Fujita, M. Node: Diterpenoids of *Rabdosia* Species. – S. Johne: The Quinazoline Alkaloids.

Volume 45:

1984. 2 figures. VIII, 288 pages. Cloth DM 194,—, öS 1360,—.
ISBN 3-211-81755-7

Contents: D. A. H. Taylor: The Chemistry of the Limonoids from Meliaceae. —
J. A. Elix, A. A. Whitton, and M. V. Sargent: Recent Progress in the Chemistry
of Lichen Substances. — Y. Shimizu: Paralytic Shellfish Poisons.

Volume 44:

1983. 72 partly coloured figures. IX, 326 pages.
Cloth DM 208,—, öS 1460,—. ISBN 3-211-81754-9

Contents: F. J. Evans and S. E. Taylor: Pro-Inflammatory, Tumour-Promoting
and Anti-Tumour Diterpenes of the Plant Families Euphorbiaceae and
Thymelaeaceae. — A. Mondon and B. Epe: Bitter Principles of Cneoraceae. —
S. Naylor, F. J. Hanke, L. V. Manes, and P. Crews: Chemical and Biological
Aspects of Marine Monoterpenes. — J. G. Buchanan: The C-Nucleoside Anti-
biotics.

Volume 43:

1983. VIII, 383 pages. Cloth DM 208,—, öS 1460,—. ISBN 3-211-81741-7

Contents: J. L. Ingham: Naturally Occurring Isoflavonoids (1855—1981). —
A. Koskinen and M. Lounasmaa: The Sarpagine-Ajmaline Group of Indole
Alkaloids.

All Volumes and Cumulative Index 1—20 available

Price reduction for subscribers: 10%

**Special reduced price (20% reduction) for the complete Series
Vols. 1—50 incl. the Cumulative Index to Vols. 1—20**

Springer-Verlag Wien New York

Mölkerbastei 5, A-1011 Wien
175 Fifth Avenue, New York, NY 10010, U.S.A.
Heidelberger Platz 3, D-1000 Berlin 33
37-3, Hongo 3-chome, Bunkyo-ku, Tokyo 113, Japan